Millets

Nutritional Value and Processing Technology

The Authors

Dr. U.D. Chavan obtained his B. Sc. and M. Sc. (Agri. in Biochemistry) degrees from Mahatma Phule Krishi Vidyapeeth Rahuri in 1985 and 1987 respectively. He received his Ph.D. degree in Food Science from Memorial University of Newfoundland, St. John's Canada in 1999. He has done International training on "Global Nutrition 2002" at Uppsala University Uppsala, Sweden in 2002. He also attended follow-up International workshop on "Global Nutrition 2002" at Hanoi, Vietnam in 2002. Dr. Chavan visited Denmark, Finland, Ireland, France, Switzerland, Poland, Spain, Vietnam, Thailand, England and U.S.A. under "Global Nutrition 2002" programme sponsored by Swedish International Development Agency (SIDA). During this programme he worked on human nutritional diet and disorders as well as on genetically modified organisms (GMO).

Dr. Chavan worked as a Senior Research Assistant in the Department of Biochemistry and Food Science and Technology at Mahatma Phule Krishi Vidyapeeth, Rahuri from 1988 to 2000. During his Ph. D., he worked as Technician/Research Associate at Atlantic Cool Climate Crop Research Center and Agriculture and Agri-Food Canada. He received D.Sc. degree in 2006 from USA. He has guided 22 students for M.Sc. (Agri.) in Biochemistry and Food Science and Technology. From 2000 to 2004 he worked as Assistant Professor of Biochemistry at Mahatma Phule Krishi Vidyapeeth Rahuri.

Dr. Chavan received Korgavkar Trust Fellowship, under-graduate and post-graduate merit scholarship as well as Senior Research Fellowship from ICAR, New Delhi and United States Department of Agriculture U. S. A. He received University Graduate Fellowship from Memorial University of Newfoundland, St. John's Canada for his Ph. D. programme. Dr. Chavan also received international Scholar Award and Excellentiam Award for his Ph.D. research work from Memorial University of Newfoundland St. John's Canada. He has received a "Certificate of Appreciation" from the U.S. Department of Agriculture in 1992 for his work on processing of groundnut under the guidance of Dr. S. S. Kadam (Principal Investigator). He has written 129 research papers and 135 popular articles. He has authored 19 books in Marathi, 10 books in English and six book chapters in English. He has been awarded "Literary Award" for his best book on "Growth Regular" on agriculture during 1997. Dr. Chavan was selected as a best group leader and best presentation for "Global Nutrition 2002" by SIDA. He has been awarded "Life Time Achievement Award 2004" for his outstanding contribution in post-harvest technology of fruits and vegetables and allied fields by United Writers' Association of India. He has been elected as a Fellow of United Writers' Association of India in 2004. Dr. Chavan received pride of nation and Maharashtra Gunigan Ratna Award for 2006. He also received Jewels of India Award for 2006 for his contribution in the field of Food Science and Technology. He has been awarded "Literary Awards" for his best book on "Sorghum Grain Processing" on agriculture during 2009-2010 and Pulses cultivation to processing and value addition for 2012. He is recipient of "Krishi Goorav Award-2011" from Bhartiya Krishak Samaj, Maharashtra State, Nasik. He also contributed in the development of crop varieties in wheat NIAW-917 and in Sorghum Phule Anuradha (RSV 423), Phule Chitra (RSV 1546), Phule Suchitra (RSV 1098), Phule Vasudha (RSV 423), Phule Revati (RSV 1006), SPV-1626, Phule Panchami (RSPOV-3) for Sorghum pops, Phule Godhan (SPV-2057) for forage single cut, CSH-50 (Sweet Sorghum), Phule Madhur (RSSGV – 46) for sorghum *hurda* and Phule Rohini (RPASV-3) for sorghum *papad*. Now, he is working as a Professor in the Department of Food Science and Technology, Senior Cereal Food Technologist in Sorghum Improvement Project and Foreign Student Advisor at Mahatma Phule Krishi Vidyapeeth, Rahuri, Dist, Ahmednagar, Maharashtra, India.

Dr. J.V. Patil obtained his B. Sc., M. Sc. and Ph. D. degrees (Agri. in Botany, Cytogenetics and plant breeding) from Mahatma Phule Krishi Vidyapeeth Rahuri in 1979, 1981 and 1992 respectively.

Dr. Patil worked on several positions since 1981 as Senior Research Assistant, Jr. Genetist-2, Associate Professor and Plant Breeder, Professor of Botany and now from 2007 he is working as a Head Department of Agricultural Botany and holding additional charge of Sorghum Breeder at Mahatma Phule Krishi Vidyapeeth, Rahuri. During his service career, he worked on several administrative posts in the University. He has guided 12 students for M. Sc. (Agri.) and 5 students for Ph. D. in the discipline of Agricultural Botany.

Dr. Patil received Merit Scholarships during his under and post-graduate study programme. He also received Baliraja: Late Annasaheb Shinde Smruti Krishi Sanshodhak Gaurao Puraskar (2004) and "Vasantrao Naik Krishi Puraskar" (2005) from Government of Maharashtra. He has written 73 research papers, 48 technical publications and 58 popular articles. He also organized 13 farmers' rallies and given 27 radio talks and 10 TV programmes. He has developed Girna (safflower), Virat, Vaibhav, Varun, Vipula, Digvijay, Rajas (chickpea), Phule Amruta, Uttara, CSV 19 SS, Chitra, CMS 1409 A/B, RSV 423, (sorghum) crop varieties during his service carrier. He also gave 5 recommendations for farming various crops on farmers' field. He also developed several innovative agro techniques such as planting methods and sprinkler method in irrigated chickpea, technique for *in-situ* rainwater conservation in *Rabi* sorghum etc. He worked as a Director of National Research Center for Sorghum, Rajendranagar, Hyderabad during August, 2010 to April, 2015. Now he is working as a Head, Department of Agricultural Botany at Mahatma Phule Krishi Vidyapeeth, Rahuri, Dist: Ahmednagar Maharashtra State.

Millets

Nutritional Value and Processing Technology

— Authors —

U.D. Chavan

Professor
Department of Food Science and Technology
Mahatma Phule Krishi Vidyapeeth,
Rahuri, Ahmednagar, Maharashtra

J.V. Patil

Head,
Department of Agricultural Botany,
Mahatma Phule Krishi Vidyapeeth,
Rahuri, Ahmednagar, Maharashtra

2016

Daya Publishing House®

A Division of

Astral International Pvt. Ltd.

New Delhi – 110 002

Publisher's Note:

Every possible effort has been made to ensure that the information contained in this book is accurate at the time of going to press and the publisher and author cannot accept responsibility for any errors or omissions, however caused. No responsibility for loss or damage occasioned to any person acting, or refraining from action, as a result of the material in this publication can be accepted by the editor, the publisher or the author. The Publisher is not associated with any product or vendor mentioned in the book. The contents of this work are intended to further general scientific research, understanding and discussion only. Readers should consult with a specialist where appropriate.

Every effort has been made to trace the owners of copyright material used in this book, if any. The author and the publisher will be grateful for any omission brought to their notice for acknowledgement in the future editions of the book.

Cataloging in Publication Data--DK
Courtesy: D.K. Agencies (P) Ltd. <docinfo@dkagencies.com>

Chavan, U. D., author.
Millets : nutritional value and processing technology / authors, U.D. Chavan, J.V. Patil.
pages cm
Includes index.

ISBN 978-93-5130-920-8 (International Edition)

1. Millets as feed. 2. Millets--Nutrition. 3. Millets--Processing. I. Patil, J. V., author. II. Title.

SB191.M5C43 2016 DDC 641.33171 23

Published by : **Daya Publishing House®**
A Division of
Astral International Pvt. Ltd.
– ISO 9001:2008 Certified Company –
4760-61/23, Ansari Road, Darya Ganj
New Delhi-110 002
Ph. 011-43549197, 23278134
E-mail: info@astralint.com
Website: www.astralint.com

Laser Typesetting : **Classic Computer Services,** Delhi - 110 035

Acknowledgement

Authors are thankful to the Hon. Dr. S. S. Kadam, Ex. Vice-Chancellor, Vasantrao Naik Marathwada Krishi Vidyapeeth, Parbhani for his critical and fruitful suggestions while the preparation of *"Millets: Nutritional Value and Processing Technology"* book. We are thankful to those scientists who worked on development of various processing technology, value-added products from millets and their contributions are acknowledged in this book as well their sources are cited in the reference section. Authors are also thankful to all scientists and friends who helped directly or indirectly while preparation of this manuscript. We would like to express over profound thanks to our family members for their constant support, patience and devotion, which inspired us to complete this manuscript.

U.D. Chavan

J.V. Patil

Preface

For millions of people in the semi-arid tropics of Asia and Africa sorghum and millets are the most important staple foods. These crops sustain the lives of the poorest rural people and will continue to do so in the foreseeable future. Sorghum and millets grow in harsh environments where other crops do not grow well. Improvements in production, availability, storage, utilization and consumption of these food crops will significantly contribute to the household food security and nutrition of the inhabitants of these areas.

The publication is broad in scope and coverage, starting with the history and nature of sorghum and millets and dealing with production, utilization and consumption. It provides extensive information on the nutritional value, chemical composition, processing and health benefits of these foods. In addition, the anti-nutritional factors present in these foods and ways of reducing their health hazards are discussed. The author has described formulations of various popular foods prepared from sorghum and millets and their nutritional composition and quality and they have compiled many recipes for the preparation of foods from regions where sorghum and millets are important dietary staples. An extensive bibliography is included as well.

Sorghum and millets in human nutrition is intended to provide up-to-date scientific and practical information to scientists, government officials, extension workers, university professors and others interested in these foods. *Millets: Nutritional Value and Processing Technology* book contains coarse as well as minor millets which have high value in human nutrition. It is hoped that this text will assist them in the development of new health food products as well as training programmes for their staff and students.

U.D. Chavan

J.V. Patil

Contents

Chapter 1

Introduction

Millets are a group of highly variable small-seeded grasses, widely grown around the world as cereal crops or grains for fodder and human food. They do not form a taxonomic group, but rather a functional or agronomic one. Millets are important crops in the semi-arid tropics of Asia and Africa (especially in India, Nigeria and Niger) with 97 percent of millet production in developing countries. The crop is favored due to its productivity and short growing season under dry, high-temperature conditions. The most widely grown millet is pearl millet, which is an important crop in India and parts of Africa. Finger millet, proso millet and foxtail millet are also important crop species. In the developed world, millets are less important. For example, in the United States only proso millet is significant and it is mostly grown for bird seed. While millets are indigenous to many parts of the world, it is believed that they had an evolutionary origin in tropical western Africa, as that is where the greatest number of both wild and cultivated forms exist. Millets have been important food staples in human history, particularly in Asia and Africa. They have been in cultivation in East Asia for the last 10,000 years.

Description

The height of the pearl millet plant may range from 0.5 to 4 metres. The pearl millet grain has great variation and can be nearly white, pale yellow, brown, grey, slate blue or purple. The kernel shape has five different classifications: obovate, hexagonal, lanceolate, globular and elliptical. Grains of pearl millet are about 3 to 4 mm long, much larger than those of other millets. The seeds usually weigh between 2.5 and 14 mg, with a typical mean of 8 mg. The size of the pearl millet kernel is about one-third that of sorghum. The relative proportion of germ to endosperm is higher in pearl millet than in sorghum. The height of finger millet plant ranges from 40 cm to 1 meter, with the spike length ranging from 3 to 13 cm. The colour of finger millet grains may vary from white through orange-red, deep brown, purple, to almost black. The grains are smaller than those of pearl millet. The typical mean weight of finger millet seed is about 2.6 mg.

Major Millets

Major millets (The most widely cultivated species)

Eragrostideae Tribe

- ✰ *Eleusine coracana*: Finger millet (also known as *ragi, nachani* or *mandwa* in India), fourth most cultivated millet.

Paniceae Tribe

- ✰ *Panicum miliaceum*: Proso millet (syn.: common millet, broom corn millet, hog millet or white millet), 3rd most cultivated millet.
- ✰ *Pennisetum glaucum*: Pearl millet (also known as *bajra* in India only in Hindi states), the most cultivated millet.
- ✰ *Setaria italica*: Foxtail millet, second most cultivated millet (also known as "kang or rala" in Maharashtra, India).

Andropogoneae Tribe

- ✰ Maize and sorghum are occasionally counted as major millets.

Minor Millets

Andropogoneae Tribe

- ✰ *Coix* spp. : Job's tears

Eragrostideae Tribe

- ✰ *Eragrostis tef* : Teff

Paniceae Tribe

- ✰ *Digitaria* spp. : white fonio, black fonio, raishan, Polish millet
- ✰ *Echinochloa* spp. : Japanese barnyard millet, Indian barnyard millet (syn. : sawa millet) (also known as "bhagar" or "varai" in Maharashtra, India), Burgu Millet
- ✰ *Panicum sumatrense* : Little Millet
- ✰ *Paspalum scrobiculatum* : Kodo millet
- ✰ *Urochloa* spp. (also known as Brachiaria) : Browntop millet, Guinea millet

History

Chinese legends attribute the domestication of millet to Shennong, the legendary Emperor of China. Specialized archaeologists called palaeoethnobotanists, relying on data such as the relative abundance of charred grains found in archaeological sites, hypothesize that the cultivation of millets was of greater prevalence in prehistory than rice, especially in northern China and Korea. Millets also formed important parts of the prehistoric diet in Indian, Chinese Neolithic and Korean

Barnyard Millet

Browntop Millet

Finger Millet

Foxtail Millet

Kodo Millet

Figure 1.1a: Seeds of Various Types of Millets.

Little Millet

Maize

Pearl Millet

Proso Millet

Rice

Figure 1.1b: Seeds of Various Types of Millets.

Sorghum

Wheat

Figure 1.1c: Seeds of Various Types of Millets.

Mumun societies. Broomcorn (*Panicum miliaceum*) and foxtail millet were important crops beginning in the Early Neolithic of China. For example, some of the earliest evidence of millet cultivation in China was found at Cishan (north). Cishan dates for common millet husk phytoliths and biomolecular components have been identified around 8300–6700 BC in storage pits along with remains of pit-houses, pottery and stone tools related to millet cultivation. Evidence at *Cishan* for foxtail millet dates back to around 6500 BC. A 4,000-year-old well preserved bowl containing well-preserved noodles made from foxtail millet and broomcorn millet was found at the Lajiaarchaeological site in China.

Palaeoethnobotanists have found evidence of the cultivation of millet in the Korean Peninsula dating to the Middle Jeulmun pottery period (c. 3500–2000 BC) (Crawford 1992; Crawford and Lee 2003). Millet continued to be an important element in the intensive, multicropping agriculture of the Mumun pottery period (c. 1500–300 BC) in Korea (Crawford and Lee 2003). Millets and their wild ancestors, such as barnyard grass and panic grass, were also cultivated in Japan during the Jômon period some time after 4000 BC (Crawford 1983, 1992). Millet made its way from China to the Black Sea region of Europe by 5000 BC. The cultivation of common millet as the earliest dry crop in East Asia has been attributed to its resistance to drought and this has been suggested to have aided its spread.

Research on millets is carried out by the International Crops Research Institute for the Semi-Arid Tropics in Telangana, India and by the USDA-ARS at Tifton, Georgia, USA.

Production

Pearl millet is one of the two major crops in the semiarid, impoverished, less fertile agriculture regions of Africa and Southeast Asia. Millets are not only adapted to poor, droughty and infertile soils, but they are also more reliable under these conditions than most other grain crops. This has, in part, made millet production popular, particularly in countries surrounding the Sahara Desert in western Africa.

Millets however, do respond to high fertility and moisture. On a per hectare basis, millet grain produced per hectare can be two to four times higher with use of irrigation and soil supplements. Improved breeds of millet improve their disease resistance and can significantly enhance farm yield productivity. There has been cooperation between poor countries to improve millet yields. For example, 'Okashana 1', a variety developed in India from a natural-growing millet variety in Burkina Faso, doubled yields. This breed was selected for trials in Zimbabwe. From there it was taken to Namibia, where it was released in 1990 and enthusiastically adopted by farmers. Okashana 1 became the most popular variety in Namibia, the only non-Sahelian country where pearl millet locally known as *mahangu* is the dominant food staple for consumers. 'Okashana 1' was then introduced to Chad. The breed has significantly enhanced yields in Mauritania and Benin. India is the world's largest producer of millet. In the 1970s, all of the millet crops harvested in India were used as a food staple. By the 2000s, the annual millet production had increased in India, yet per capita consumption of millet had dropped by between 50 per cent to 75 per cent in different regions of the country. As of 2005, most millet produced in India is being used for alternative applications such as livestock fodder and alcohol production. Indian organizations are discussing ways to increase millet use as food to encourage more production; however, they have found that some consumers now prefer the taste of other grains. In 2010, the average yield of millet crops worldwide was 0.83 tonnes per hectare. The most productive millet farms in the world were in France, with a nationwide average yield of 3.3 tonnes per hectare in 2010.

Table 1.1: Top Millet Producers of 2013

Country	*Production (Tonnes)*
India	10,910,000
Nigeria	5,000,000
Niger	2,955,000
China	1,620,000
Mali	1,152,331
Burkina Faso	1,109,000
Sudan	1,090,000
Ethiopia	807,056
Chad	582,000
Senegal	572,155
Russia	418,844
World	**29,870,058**

Utilization Pattern of Millets

The role of sorghum and millets as food staples at the aggregate level has declined since 1980. In Asia, sorghum for food use contributed 5.7 kg capita^{-1} yr^{-1} and millets 5.2 kg capita^{-1} yr^{-1} in 1980. In 2011, it declined to 1.9 kg capita^{-1} yr^{-1} for sorghum and 2.9 kg capita^{-1} yr^{-1} for millets. During the same period the per capita availability of all cereals in Asia also decreased marginally from 164 kg capita^{-1} yr^{-1} to 156.4 kg capita^{-1} yr^{-1}. Per capita availability of sorghum and millets declined in all countries including China, India and Pakistan. In India, the per capita availability of sorghum reduced by more than half and for millets by less than one-third (Table 1.2).

Table 1.2: Millets Consumption across Countries

Country/Region	*Total Cereals*			*Sorghum*			*Millets*		
	1980	*2000*	*2011*	*1980*	*2000*	*2011*	*1980*	*2000*	*2011*
Asia									
Bangladesh	168	179	190.4	0.1	0	0	0.4	0.4	0.1
China	186	183	151.5	4.5	0.9	0.7	4.4	0.9	0.5
India	140	159	152.1	13.4	6.6	5.1	12.2	9	8.9
Myanmar	192	215	137	0	0	1.6	1.9	2.9	0.4
Nepal	164	194	186.6	0	0	0	7.2	10.7	9.8
Pakistan	144	149	142.3	2.3	1.3	0.7	1.1	0.6	0.9
Saudi Arabia	145	154	158.3	12.4	9.7	3.9	0.7	0.5	0.6
Syrian Arab Republic	171	177	166.3	0	0	0	0.6	0	0.1
Thailand	153	126	133.4	0	0	0	0	0	
Yemen	148	164	162.8	64	18.3	16.6	7.8	3.3	3.4
Region									
Asia	164	171	156.4	5.7	2.3	1.9	5.2	3	2.9
South Asia	143	160	157	10.7	5.1	3.7	9.8	7.1	6.7
World	150	155	147.2	5.3	4	3.6	4.6	3.6	3.3

Source: FAO, http://faostat3.fao.org/download/FB/FBS/E.

Despite the low and declining per capita availability of sorghum and millets at the aggregate level their importance as food crops cannot be underestimated. For example, in the Indian state of Gujarat, Rajasthan, Maharashtra and Karnataka, coarse cereals accounted for 30-50 per cent of the total cereal production and consumption in 2012-13.

Consumption Trends of Nutri-Cereals in India

According to Parthasarathy Rao and Basavaraj the annual per capita consumption of sorghum at the all-India level has declined sharply by 74 per cent (8.5 to 2 kg) in urban areas and by 81 per cent (19.2 to 3.5 kg) in rural areas of India between 1972-73 and 2009-10. Similarly, the pearl millet consumption both in rural and urban areas has fallen very steeply from 11.5 kg to 3.06 kg (by 73 per cent) in rural areas and from 4 kg to 1.08 kg (by 73 per cent) in urban areas. This is due to increase in per capita income, growing urbanization, changing tastes and preferences rendering sorghum and millet as inferior goods with low to negative income elasticity of demand and positive price elasticity.

Apart from decline in consumption, these crops are gradually disappearing in traditional areas due to access to irrigation and markets, which have enabled farmers to shift their area under cultivation to high value crops. The food security strategy of the Government of India of supplying subsidized rice and wheat through the Public Distribution System (PDS) has been the major factor contributing to the decline in consumption of sorghum and other cereals like pearl millet and finger millet in rural areas and urban centers. Thus, the popularity of sorghum and pearl millet faded, resulting in negative growth in area, production practices and hence their cultivation became uneconomical. The PDS system in India is based on the wheat and rice model, which is less relevant in many areas and especially in the dryland farming areas, where millets, sorghum and pulses are traditionally the staple grains for household consumption. Despite the decline in per capita consumption, sorghum grain is an important staple for the low and middle income consumers in regions where they are grown. For example, in rural areas of central Maharashtra, per capita annual consumption of sorghum is around 60 kg, accounting for almost half (48 per cent) of the per capita consumption of all cereals. Similarly, among the major pearl millet producing regions, per capita consumption was highest (69 kg/ year) in rural Rajasthan and in the dry areas of Gujarat (59 kg/year). In these two regions, pearl millet accounts for more than 50 per cent of cereal consumption, contributing about 20 to 40 per cent of the total energy and protein intake.

The main reasons for this decline in demand for direct consumption (human consumption) is due to rising incomes, urbanization and subsequent changes in consumer tastes and food habits (Table 1.3). Apart from decreasing consumption demand other reasons that led to decrease in the area and production are due to lack of competitiveness or vice-versa other competing crops such as pulses, oilseeds and commercial crops, low profitability etc.

If the trend of sorghum and coarse cereals consumption continues in the current scenario, after the next two decades they will be completely vanished and replace by fine cereals such as rice and wheat. However, the prospects of sorghum and millets in direct food consumption tends to rise as interventions in its food processing and promotion has been successfully taken out by IIMR in collaboration with DAC and other millets research institutes.

Impending demand for sorghum stalks for meeting their domestic livestock fodder needs has become lifeline for continuation of sorghum farming in the dry-land regions despite low consumption demand for grain. Thus utilization of *Kharif* sorghum grain (by product) became a question in event of deterioration of sorghum grain quality (due to blackening (caught up in October rains)) making it unfit for human consumption. Further, the scale of sorghum farming is linked to the extent of domestic fodder requirements in *Kharif* season, thus resulting in meager marketed surpluses. Therefore, *Kharif* sorghum grain was evolved as a regular source of raw material for above mentioned alternate uses. However, due to smallness of marketed surplus volumes, supply chain management has become a major constraint and making it unsustainable. *Dayakar et al.*, reveals that decrease in consumption of sorghum was found to be proportional to increase in consumer expenditure. Increased income is accompanied by increased consumption of wheat and rice, as products made from these cereals are easy to prepare and have better keeping quality.

Table 1.3: Trend in Direct Consumption of Cereals and Food Grains (kg/person/year)

Commodity	*1973–74*	*1983–84*	*1993–94*	*2003–04*	*2013–14**
Rural					
Rice	84.0	80.7	85.4	82.7	83.80
Wheat	42.8	54.3	53.5	51.7	51.13
Sorghum	19.0	12.5	9.7	5.7	3.9
Coarse cereals	56.8	45.1	24.1	16.5	13.2
Total cereals	183.6	180.1	163.0	151.6	148.59
Food grains			172.8	162.2	159.9
Urban					
Rice	65.5	64.7	64.2	59.4	60.11
Wheat	52.6	58.6	57.4	56.8	53.42
Sorghum	11.0	6.0	4.9	3.3	2.5
Coarse cereals	19.7	14.1	7.7	6.0	5.4
Total cereals	137.7	137.5	129.3	122.4	119.37
Food grains			140.4	135.1	134.3

Source: Various NSSO reports, Gal and Dayakar Rao *et al.* (2014). *: Estimated utilization.

Note: Projection assumes 7.1 per cent growth rate in economy: 5 per cent for rural and 12 per cent for urban population.

Similarly, Pearl millet consumption also declined by 67 per cent in urban areas and 59 per cent in rural areas during 1972-73 to 2004-05 (Basavaraj *et al2010*). The food use of Pearl millet has declined sharply at the country level, but it is still an important crop in the major producing states. The increase in Pearl Millet production juxtaposed with its declining food use implies that its use in alternative uses has been increasing mainly from animal feed which includes mainly dairy and to some extent in poultry, alcohol industry, starch industry, processed food industry and export demand. The potential demand for food processing, though at early stage, presents encouraging prospects for value addition.

Another study conducted by Indian Institute of Millets Research (IIMR) on "Macro and Micro Level Changes in Consumption of Nutritious Cereals in India" (Dayakar *et al.*, 2004) also pointed that sorghum in all the States and country as a whole showed decline in sorghum consumption among urban and rural masses, whereas Pearl millet consumption in major producing States like Gujarat (129 per cent) and Rajasthan exhibited increased growth in urban areas, while in rural Gujarat, it has declined (-18.16 per cent) and the same has increased in rural Rajasthan

(39.1 per cent). The magnitude of reduction in pearl millet consumption was highest in rural Tamil Nadu (-95.5 per cent) followed by rural Andhra Pradesh (-80.95 per cent). In Finger millet there is a slight decline of 9 per cent in consumption in urban Karnataka, whereas in Rural Karnataka the decline in consumption was four fold (36 per cent) than in urban consumption. The Andhra Pradesh rural consumption of finger millet also declined drastically (-67.27 per cent).

Causal Factors for the Decline in Consumption of Nutritious Cereals

1. Easy availability of fine cereals like rice and wheat at a cheaper price to that of nutritious cereals under Public Distribution System (PDS).
2. Rapid urbanization.
3. Nutritious cereals are socially less valued necessitated the decline in their consumption on rising of capita income.
4. Tedious and time consuming preparation of nutritious cereals foods.
5. Preference to cash payment by laborers rather than taking food grain in lieu of work performed.
6. Change in food habit particularly preference to fast food by younger generation.

Measures for Increasing the Consumption/Demand of Nutritious Cereals

1. Creating public awareness about the nutritional superiority of millets.
2. Supply of nutritious cereals through PDS.
3. Value-addition in nutritious cereals.
4. Provisions for supply of millet foods under Mid Day Meal/Other Food Supplement programmes being arranged by various States.
5. Utilization of millet foods under larger Scheme like Adolescent Girl Nutrition Scheme, under consideration of Ministry of Women and Child Welfare.

Utilization of Millets for Alternative Uses

Millets are traditionally used for preparation of fermented foods and beverages.

Grains with high starch (65-72 per cent) and low protein (8-12 per cent) could be utilized for production of alcohol. The utilization of *Kharif* sorghum grain as a raw material in various industries is increasing, given the limited prospects of rainy season *Kharif* sorghum for human consumption. The important traditional/ fast food and industrial usages of millets are given in the Table 1.4.

Most recently pearl millet is being used as a low cost substitute for maize in poultry and dairy feed, which is gaining momentum. In addition, there are other niches for pearl millet:

- Due to its rich protein content, the grain can be used for preparation of protein-rich foods.
- The bran is rich in oil, hence it can complement rice bran in oil extraction.
- Pearl millet can also be used for brewing in those favorable areas with marketed surplus where the factories are located.

Table 1.4: Important Food and Industrial Uses of Millets

Millets	*Food Products*	*Industrial Products*
Sorghum	Roti, ugali, popped sorghum, malt food, snack/roasted mix grains.	Malting, high fructose syrup, starch, Jaggery, bakery, value added products for diabetics, poultry and animal feed.
Pearl millet	Roti, ugali, fermented food products, pizza, roasted mix grains.	Brewing/malting, starch, bakery, poultry and animal feed.
Finger millet	Roti, dumpling, popped millet, malt-food.	Malting/brewing, baby foods, bakery and food for diabetics.
Other millets	Roti, cooked cereals	Value added food for devotees (Barnyard millet), Feed, value added food products for diabetics.

* The high fiber content in the grain can be exploited to prepare health foods. Diabetic food products can also be prepared.

The main industries currently using sorghum and millets in India are the poultry feed, animal feed, potable alcohol distilleries, etc. At present poultry feed sector is using approximately 2.0 million tonnes sorghum annually; animal feed sector uses about 0.60 million tonnes sorghum followed by alcohol distillers (about 0.49 million tonnes).

1. **Poultry feed/Animal Feed:** Poultry feed industry in the country is progressing well. Several independent feeding trials conducted on layer and broiler birds could conclusively demonstrate that sorghum and pearl millet as good as maize although, the latter is marginally superior in total energy value. Sorghum and millets based feed in layer birds has to be supplemented with carotene to ensure yellow colour to egg yolk. In the case of cattle feed, sorghum based feed is found satisfactory. Moulded *Kharif* grain is found as an acceptable grain component in the feed of goat and swine too. Sorghum is used when maize is in short supply and priced up to 20 per cent higher than sorghum.
2. **Starch:** Sorghum is rarely used by the starch industry in view of processing problem and starch yield as the starch extraction machineries are adapted to the bigger grain size of maize. While there is scope to improve starch recovery from sorghum by modified extraction, selective breeding and use of specifically suited machineries, starch production from the present sorghum cultivars be economically viable only if the grain price is 20 per cent lesser than that of maize. Sorghum grain is found more suitable for production of glucose and liquid glucose.

3. **Potable alcohol:** The sorghum grain offers an excellent ethanol and feed stock with a high alcohol recovery of 422-448litres per tonne. This provides scope for utilization of rain-damaged blackened and less market value sorghum grain for utilization as raw material for malting and production of quality potable alcohol. Currently more than 10 distilleries are using sorghum grain (mainly blackened) as raw material for potable alcohol. Government of Maharashtra has announced incentives for those sorghum grain based distilleries which are established as alternative to sugarcane molasses. The setting of such industry in the potential areas would benefit the farmers in getting assured market with remunerative price through contract farming. Like sorghum, pearl millet is also being used for production of potable alcohol.
4. **Malt:** Another potential area could be the use of ragi and sorghum malt as an adjunct in brewing industry. Malted sorghum is used for brewing beer in Ghana and Nigeria. Sorghum's comparative advantage would be its low output price, especially in production regions of rainy sorghum in India over that of existing raw material.
5. **Bio-ethanol:** Sweet stalked and high energy sorghum has the potential to emerge as a major bio-energy crop for production of industrial alcohol, gasohol and even electricity. While the national trials on sweet sorghum had shown its potential productivity as 50-60 t stalk and 2-3 t grain/ha in 125-130 days, reports from China indicate a far high productivity to the tune of 90 t stalk and 6 t grain/ha from the crop in 140-150 days. This productivity potential places the crop at an advantage over sugarcane or molasses at their ruling prices for ethanol production. As a bio-energy crop sweet sorghum is more efficient than sugarcane because of its high productivity with relatively low levels of fertilization and irrigation. Sweet sorghum, in this context, could be effectively exploited as an energy crop. IIMR and ICRISAT and many R and D organizations were actively involved in fine-tuning technologies related to sweet sorghum to project it as an alternative feed stock for bioethanol production. However, government policy support in declaring high ethanol pricing will determine its current utilization and future demand.
6. **Production of jaggery**: Production of jaggery and brown or colourless syrup or high fructose syrup are other opportunities from sweet sorghum. These diversifications could be done at village or farm level as small scale enterprises.

Limitations with Storage and Consumption of Millets

The awareness about the high nutritive values of millets is very low both among the producers and consumers. Besides being small seeded and low priced commodity, the produce is not properly cleaned, graded and dried before they are brought to the markets, fetching low price to the farmers and poses storage problems. Besides, rains at the time of maturity/harvest also lead to poor quality of *Kharif* produce. The high fat, fiber, inert matters coupled with high moisture content/high humidity leads to infestation of insect/fungus during storage. Grains stored with high moisture content often develop Mycotoxins/Aflatoxin which is highly poisonous both for human and live stock. The shelf life of flour of millets is also low because of high fat content in the grains. The dark seed coat color of some millet such as ragi and bajra also possess problem in acceptance of food products of these commodities. Thus, enhancement of shelf life could be an important challenge to process grain to products and market them sustainability.

Government Policies that Determine Millets Production and Consumption in the Country

Minimum Support Price (MSP)

Sorghum, pearl millet and finger millet are covered under MSP while small millets are not covered so far. An analysis of MSP announced for millets vis-a.-vis wheat and rice during 1977-78 to 2012-13 reveals an increase in MSP of millets at par with wheat and rice and of late even higher. However, MSP could not help in sustainability of millets because of Lack of procurement system for millets through public agencies and Lack of provision for supply of millets under PDS like wheat and rice. Millets are included in the new Food security bill, thus it is expected the millets are going to be procured, however, it needs to be seen how this is going to be operationalized. There is no MSP for small millets, some states have suggested for coverage of small millets under MSP and supply of millet grains under PDS for regional food security.

The Government of India has increased the Minimum Support Price (MSP) of bajra or pearl millet of Fair Average Quality (FAQ) from Rs. 980 per quintal (100 Kilograms) to Rs 1,175 per quintal of the crop year 2012-2013 and marketing season 2013-2014. This is an increase of Rs. 195 a quintal or 19.89 percent over last year's Minimum Support Price fixed for bajra. Similarly, highest increase is witnessed in sorghum.

Public Distribution System (PDS)

PDS provided a substantive captive market for rice and wheat produced in Green Revolution areas. PDS has had other important effect-on dietary habits of Indians. Effectively, PDS has made rice and wheat the staple diet across the country, replacing millets and other cereal.

The GoI production policy for coarse grains/millets is similar to that for wheat and rice, but there is less focus on these crops. Also, the government's support for coarse grain via the procurement and PDS operation is very limited and ineffective compared to rice and wheat. Unlike wheat and rice, the government does not typically maintain a buffer stock of coarse grains to keep process in check. GoI also announce Minimum Support Price for coarse grains and there has been sharp increase in MSP.

Decentralized procurement, storage and distribution by elected bodies such as Panchayats or even state governments remains a dream despite the obvious economic benefits. If one were to assume that decentralized procurement will provide communities with the choice to procure produce of their liking, then the possibility of coarse grains and cereals finding a way back into the dietary habits of people will exist. Alas, the political support for this is lacking.

The obvious choice, then, is to hope that the markets can do the trick. Ask children what they would like in their mid-day meal and the chances are that noodles will figure in the top five choices.

The millets may be targeted in public funded welfare programmes such as Mid-day meals in those states where millets are traditionally grown and consumed, at least 2 times a week to begin with.

Development Programmes MMA

Developmental programmes of millets are being supported through Centrally Sponsored Scheme of Macro Management in Agriculture (MMA). The Scheme provides flexibility to the States to develop and pursue activities on the basis of their location specific needs. The States are also free to include new/innovative interventions in their Work Plan provided; it is not covered under any other Scheme of the Central Government or is not part of any ongoing State Schemes. Schemes also have a provision for utilization of at least 33 per cent funds for small, marginal and women farmers, who have larger stake in millets cultivation. Assistance to the States/UTs is provided as 100 per cent Grant-in-aid. All States are eligible to receive assistance under MMA Scheme.

INSIMP

Initiative for Nutrition Security through Intensive Millets Promotion (INSIMP), a sub-scheme of Rashtriya Krishi Vikas Yojna (RKVY) which aims to increase production of millets especially Sorghum, Pearl millet, Finger millet and Small millets in the country by demonstrating post-harvest technologies in an integrated manner. This scheme is also expected to generate consumer demand for millet based food products through processing and value addition techniques. Government of India allotted Rs. 300 Crore in 2011-12 under RKVY for promotion of millets as Nutri-cereals and is being continued till date. It has components of support of on farm production, seed production, establishment of processing clusters, popularization and promotion of millets and marketing linkages. The INSIMP established 3 Centers of Excellence (COE) (a) on sorghum processing at IIMR, Hyderabad, (b) on pearl millet at CCSHAU, Hisar and (c) on Small millets processing at UAS, Bangalore. The major objectives of CoE includes refinement of technology, retrofitting of machinery and their demonstration and Providing entrepreneurship development and training on sorghum post-harvest technologies and value addition. In XII plan the activities of INSIMP are going to be merged with National Food Security Mission (NFSM).

Alcoholic Beverages

Millets are traditionally important grains used in brewing millet beer in some cultures, for instance by the Tao people of Orchid Island and in Taiwan. Various peoples in East Africa brew a drink from millet or sorghum known as *ajono*, a traditional brew of the Teso. The fermented millet is prepared in a large pot with hot water and people share the drink by sipping it through long straws. Millet is also the base ingredient for the distilled liquor *rakshi* in Nepal and the indigenous alcoholic drink of the Sherpa, Tamang, Rai and Limbu people, *tongba*, in eastern Nepal. In Balkan countries, especially Romania and Bulgaria, millet is used to prepare the fermented drink *boza*.

Figure 1.2: *Tongba*: A Millet-based Alcoholic Brew found in the Far Eastern Mountainous Region of Nepal and Sikkim, India.

As a Food Source

Millets are major food sources in arid and semi-arid regions of the world and feature in the traditional cuisine of many others. In western India, sorghum (called *jowar*, *jola*, *jwaarie*, or *jondhahlaa* in Gujarati, Kannada, Hindi and Marathi languages, respectively; *mutthaari*, *kora*, or *pangapullu* in Malayalam; or *cholam* in Tamil) has been commonly used with millet flour (called *bajari* in western India) for hundreds of years to make the local staple, hand-rolled (that is, made without a rolling pin) flat bread (*rotla* in Gujarati, *bhakri* in Marathi, or *roti* in other languages). Another cereal grain popularly used in rural areas and by poor people to consume as a staple in the form of *roti* or other forms is called *ragi* in Karnataka, *naachanie* in Maharashtra, or *kezhvaragu* in Tamil, with the popular *ragi rotti* in Kannada. *Ragi mudde* is a popular meal in southern India. In Telugu, it is called *jonnalu*. *Jonna* is dark in colour like rye, but rougher in texture. Millet porridge is a traditional food in Russian, German and Chinese ñuisines. In Russia, it is eaten sweet (with milk and sugar added at the end of the cooking process) or savoury with meat or vegetable stews. In China, it is eaten without milk or sugar, frequently with beans, sweet potato, and/or various types of squash. In Germany, it is also eaten sweet, boiled in water with apples added during the boiling process and honey added during the cooling process.

Per capita consumption of millets as food varies in different parts of the world. It is highest in western Africa. In the Sahel region, millet is estimated to account for about 35 percent of total cereal food consumption in Burkina Faso, Chad and the Gambia. In Mali and Senegal, millets constitute roughly 40 percent of total cereal food consumption per capita; while in Niger and arid Namibia it is over 65 percent (see *mahangu*). Other countries in Africa where millets are a significant food source include Ethiopia, Nigeria and Uganda. Millet is also an important food item for the population living in the drier parts of many other countries, especially in eastern and central Africa and in the northern coastal countries of western Africa. In developing countries outside Africa, millet has local significance as a food in parts of some countries, such as China, India, Burma and North Korea. The use of millets as food fell between the 1970s and the 2000s, both in urban and rural areas, as developing countries such as India have experienced rapid economic growth and witnessed a significant increase in per capita consumption of other cereals. People with coeliac disease can replace certain gluten-containing cereals in their diets with millet. Millets are also used as bird and animal feed.

Figure 1.3: *Awaokoshi*, **Candied Millet Puffs, are a Specialty of Osaka, Japan. This millet confection tradition began when it was presented to Sugawara no Michizanewhen he stopped in Naniwa during the early Heian period, about 1000 years ago.**

Composite Flours

Composite flour technology initially referred to the process of mixing wheat flour with cereal and legume flours for making bread and biscuits. However, the term can also be used in regard to mixing of non-wheat flours, roots and tubers, legumes or other raw materials (Dendy, 1992). One example is the mixture of sorghum and maize flour for tortillas.

Diluting wheat flour with locally available cereals and root crops was found to be desirable to encourage the agricultural sector and reduce wheat imports in many developing countries. In Africa there has been an ever-increasing demand for wheat products such as bread. Africa is not a major wheat-growing region, but it produces large quantities of other cereals such as sorghum and millets. It has been reported that replacing wheat with 20 per cent non-wheat flour for the manufacture of bakery products would result in an estimated savings in foreign currency of US$320 million annually (FAO, 1982). At 30 per cent substitution the savings would be US$480 million annually. Thus composite flour technology holds excellent promise for developing countries. Although actual consumer trials have been rare, products made with composite flour have been well accepted in Colombia, Kenya, Nigeria, Senegal, Sri Lanka and the Sudan (Dendy, 1992).

When sorghum or millets are used for bread-making, addition of bread improvers or modification of the bread-making process is needed. A higher level of substitution is possible with hard than with soft wheat flour (United Nations Economic Commission for Africa, 1985). For the production of biscuits from composite flours, the fat content of the non-wheat flour should be kept as low as possible to promote a longer shelf-life.

Crabtree and Dendy (1979) reported that bread could be produced from composite flour made by co-milling wheat with pearl, prove, barnyard or finger millets. The proportion of millet in the flour can be up to 15 per cent. Potassium bromate treatment of the dough tends to improve the loaf volume. Bread containing 10 per cent pearl millet flour had an excellent texture and flavour similar to that of whole-wheat bread (Bad), Hoseney *et al.*, 1987; Perten, 1972).

Sorghum flour milled at 80 per cent extraction rate could be blended with white wheat flour for bread-making without any adverse effect (Rao and Shurpalekar, 1976). Acceptability studies conducted at the Food Research Centre in Khartoum, the Sudan, indicated that breads made with composite flour of 70 per cent wheat and 30 per cent sorghum was acceptable. Milling at 72 to 75 per cent extraction rate yielded fine sorghum flour that is more suitable for bread-making. Consumer acceptance trials in Nigeria indicated that breads made with 30 per cent sorghum flour were comparable to 100 per cent wheat bread (Aluko and Olugbemi, 1989; Olatunji, Adesina and Koleoso, 1989). The protein content of composite flour was lower than that of wheat flour, while its crude fiber was higher. Addition of pentosan improved the quality of bread made with composite flour. The Institute of Food Technology in Dakar, Senegal, prepared bread consisting of 30 per cent millet and 70 per cent wheat using the popular millet varieties Souna and Sanio (Thiam, 1981). Bread, called panble, was prepared with 15 per cent millet and 85 per cent wheat. Bread with 30 per cent sorghum and 70 per cent wheat was also prepared in Senegal (Thiam and Ndoye, 1977). Breads with up to 15 per cent prove millet were acceptable and comparable to white wheat bread (Lorenz and Dilsaver, 1980).

A combination of 80 per cent non-wheat cereal and 20 per cent wheat can be used to produce biscuits with acceptable quality. Sorghum and pearl millet flour blended with wheat flour can be used to make biscuits (Badi and Hoseney, 1976, 1977). Olatunji, Adesina and Koleoso (1989) reported that a proportion of 55 per cent sorghum could be used for biscuits without adversely affecting biscuit quality. Proso millet was found suitable for making biscuits; biscuit spread and quality score increased with increased levels of prove millet flour because of its high fat content (Lorenz and Dilsaver, 1980). Millet flour imparted a slight grittiness, however. Pearl millet could replace 50 per cent of the wheat for cake and 80 per cent for biscuits (Thiam, 1981). In Senegal, traditional foods such as faux, conus and beignets (fritters) are prepared by mixing millet flour with rice, maize or wheat flour (Thiam, 1981).

Alternative Uses of Sorghum and Millet

Sorghum and pearl millet production has considerably increased in several countries during the past few years. With the simultaneous increase in the production of wheat and rice and the available surplus in storage, millets face competition from the utilization point of view. There is already an increasing trend of using wheat or rice in place of sorghum even in those regions where sorghum has been the traditional staple grain in the past.

Sorghum and millets will continue to be major food crops in several countries, especially in Africa (and in particular in Nigeria and the Sudan, which together account for about 63 per cent of Africa's sorghum production). These grains will be used for traditional as well as novel foods. However, there is a need to look into the possibilities of alternative uses. Though sorghum and millets have good potential for industrial uses, they have to compete with wheat, rice and maize. Sorghum in particular could be in great demand in the future if the technology for specific industrial end uses is developed. Although pearl millet has some potential for industrial use, other millets have limited potential because of their small grain size and the associated difficulties of adopting a suitable dehulling technology. However, they can be considered for animal and poultry feed. There is a need to compare their performance as feed in comparison with maize.

Sorghum and millets can be adopted for other food products by using appropriate processing methods. Dehulling and milling practices to improve the quality of foods made from sorghum and millets have been described in Chapter 3. It may be possible to select grain types with improved milling quality that will make these crops competitive with other cereals in terms of utilization. Wheat milling technology with suitable modification can be effectively used for grinding sorghum and millets. Although bread can be produced from whole sorghum flour, the quality of the bread can be improved by using sorghum flour from which the bran fraction has been removed by passage through sieves (Caster *et al.*, 1977). Kulkarni, Parlikar and Bhagwat (1987) reported that sorghum malt could be used to make biscuits, weaning foods and beer wort. Addition of up to 40 per cent sorghum malt in biscuits caused reduction in stack height and increase in spread because of increased water absorption.

The use of sorghum in common foods such as *idli* (a steamed product), *dosa* (a leavened product) *endponganum* (a shallow-fat-fried product) can be popularized for wider use in sorghum-growing areas (Subramanian and Jambunathan, 1980). A few important sun-dried or extruded and sun-dried products from sorghum are *papad, badi* and *kurdigai*. These products usually have a shelf-life of over one year. They can be popularized through marketing channels similar to those used for rice products. A wide range of bakery and snack products prepared with dehulled sorghum were accepted by consumers in Andhra Pradesh, India (Andhra Pradesh Agricultural University, 1991). It was indicated that these foods should be marketed commercially to make them reach more people and become more popular.

Grazing Millet

In addition to being used for seed, millet is also used as a grazing forage crop. Instead of letting the plant reach maturity it can be grazed by stock and is commonly used for sheep and cattle. Millet is a C4 plant which means it has good water efficiency and utilizes high temperature and is therefore a summer crop. A C4 plant uses a different enzyme in photosynthesis from C3 plants and this is why it improves water efficiency. In southern Australia millet is used as a summer quality pasture, utilizing warm temperatures and summer storms. Millet is frost sensitive and is sown after the frost period, once soil temperature has stabilised at 14 °C or more. It is sown at a shallow depth. Millet grows rapidly and can be grazed 5–7 weeks after sowing, when it is 20–30 cm high. The highest feed value is from the young green leaf and shoots. The plant can quickly come to head, so it must be managed accordingly because as the plant matures the value and palatability of feed reduces. The Japanese millets (*Echinochloa esculenta*) are considered the best for grazing and in particular Shirohie, a new variety of Japanese millet, is the best suited variety for grazing. This is due to a number of factors: it gives better regrowth and is later to mature compared to other Japanese millets; it is cheap cost of seed is $2–$3 per kg and sowing rates are around 10 kg per hectare for dryland production; it is quick to establish; it can be grazed early; and it is suitable for both sheep and cattle.

Compared to forage sorghum, which is grown as alternative grazing forage, animals gain weight faster on millet and it has better hay or silage potential, although it produces less dry matter. Lambs do better on millet compared to sorghum. Millet does not contain prussic acid which can be in sorghum. Prussic acid poisons animals by inhibiting oxygen utilisation by the cells and is transported in the blood around the body ultimately the animal will die from asphyxia. There is no need for additional feed supplements such as sulphur or salt blocks with millet. The rapid growth of millet as a grazing crop allows flexibility in its use. Farmers can wait until sufficient late spring / summer moisture is present and then makes use of it. It is ideally suited to irrigation where livestock finishing is required.

Nutrition

Millets, like sorghum, are predominantly starchy. The protein content is comparable to that of wheat and maize. Pearl and little millet are higher in fat, while finger millet contains the lowest fat. Barnyard millet has the lowest carbohydrate content and energy value. Millets are also relatively rich in iron and phosphorus (Table 1.5). The bran layers of millets are good sources of B-complex vitamins. However, millets also feature high fiber content and poor digestibility of nutrients, which severely limit their value in nutrition and influence their consumer acceptability. Finger millet has the highest calcium content among all the food grains, but it is not highly assimilable. The protein content in millet is very close to that of wheat; both provide about 11 per cent protein by weight, on a dry matter basis. Millets are rich in B vitamins (especially niacin, B_6 and folic acid), calcium, iron, potassium, magnesium and zinc (Table 1.6). Millets contain no gluten, so they are not suitable for raised bread. When combined with wheat (or xanthan gum for those who have celiac disease) they can be used for raised bread. Alone, they are suited for flatbread.

Table 1.5: Nutrient Content in Various Millets per 100g

Component	*Cassava*	*Wheat*	*Rice*	*Sweet Corn*	*Potato*	*Sorghum Millet*	*Proso Millet*
Water (g)	60	13.1	12	76	82	9.2	8.7
Energy (kJ)	667	1368	1527	360	288	1418	1582
Protein (g)	1.4	12.6	7	3	1.7	11.3	11
Fat (g)	0.3	1.5	1	1	0.1	3.3	4.2
Carbohydrates (g)	38	71.2	79	19	16	75	73
Fiber (g)	1.8	12.2	1	3	2.4	6.3	8.5
Sugars (g)	1.7	0.4	>0.1	3	1.2	1.9	–
Iron (mg)	0.27	3.2	0.8	0.5	0.5	4.4	3
Manganese (mg)	0.4	3.9	1.1	0.2	0.1	<0.1	1.6
Calcium (mg)	16	29	28	2	9	28	8
Magnesium (mg)	21	126	25	37	21	<120	114
Phosphorus (mg)	27	288	115	89	62	287	285
Potassium (mg)	271	363	115	270	407	350	195
Zinc (mg)	0.3	2.6	1.1	0.5	0.3	<1	1.7
Pantothenic acid (mg)	0.1	0.9	1.0	0.7	0.3	<0.9	0.8
Vitb6 (mg)	0.1	0.3	0.2	0.1	0.2	<0.3	0.4
Folate (µg)	27	38	8	42	18	<25	85
Thiamin (mg)	0.1	0.38	0.1	0.2	0.1	0.2	0.4
Riboflavin (mg)	<0.1	0.1	>0.1	0.1	>0.1	0.1	0.3
Niacin (mg)	0.9	5.5	1.6	1.8	1.1	2.9	4.7

As none of the millets are closely related to wheat, they are appropriate foods for those with celiac disease or other forms of allergies/intolerance of wheat. However, millets are also a mild thyroid peroxidase inhibitor and probably should not be consumed in great quantities by those with thyroid disease.

Health Benefits gets from the Sorghum and Millets Consumption

1. Sorghum and millets food provides non-glutinous flours which are useful for avoiding damage to the lining of the intestine and make easy for nutrients to be absorbed.
2. They avoid celiac disease problems.
3. They also avoid dermatitis herpetiforms disease.
4. Sorghum and millets are non-acid forming food.
5. Sorghum and millet foods are a soothing and easy to digest.
6. Sorghum foods are least allergenic and mostly digestible.
7. They are good source of dietary fibers.
8. They also good source of vitamins and minerals.
9. They also provide antioxidants for controlling cancers.
10. Malted sorghum and millet flours can be used for anti-constipation drinks.
11. They provide sufficient amount of dietary fiber in human diet.
12. Selected sorghum and millet genotypes can be providing enough amount of lysine to meet human requirement.
13. Sorghum and millet diet is very good to the people suffering from the Jaundice.
14. Sorghum and millets has very low glycimic index so it helps to human health.
15. Sorghum and millets helps in loose and control of body weight.
16. They increase body's sensitivity to insulin.
17. They also help in diabetes control.
18. They reduce risk of heart diseases.
19. They also reduce blood cholesterol levels.
20. They also reduce hunger and keep fuller for longer time.
21. They also help to manage polycystic ovary syndrome in women.

Comparison with other Major Staple Foods

The following table shows the nutrient content of millet compared to major staple foods in a raw form. Raw forms, however, are not edible and cannot be fully digested. These must be prepared and cooked as appropriate for human consumption. In processed and cooked form, the relative nutritional and antinutritional contents of each of these grains are remarkably different from that of raw forms reported in this Table 1.5 and Table 1.6. The nutritional value in the cooked form depends on the cooking method.

Table 1.6: Nutrient Content of Various Millets with comparison to Rice and Wheat g/100g

Crop/Nutrient	*Protein (g)*	*Fiber (g)*	*Minerals (g)*	*Iron (mg)*	*Calcium (mg)*
Pearl millet	10.6	1.3	2.3	16.9	38
Finger millet	7.3	3.6	2.7	3.9	344
Foxtail millet	12.3	8	3.3	2.8	31
Proso millet	12.5	2.2	1.9	0.8	14
Kodo millet	8.3	9	2.6	0.5	27
Little millet	7.7	7.6	1.5	9.3	17
Barnyard millet	11.2	10.1	4.4	15.2	11
Rice	6.8	0.2	0.6	0.7	10
Wheat	11.8	1.2	1.5	5.3	41

Source: Millet Network of India, http://www.milletindia.org.

Processing Technology in Cereals

Physico-chemical Properties of Grains

A grain is a living biological product which germinates and respires also. The respiration process in the grain is externally manifested by the decrease in dry weight, utilization of oxygen, evolution of carbon dioxide and release of heat. The rate of respiration is dependent upon moisture content and temperature of the grain. 'The rate of respiration of paddy increases sharply (at 25° C) at 14 to 15 per cent moisture content which is called the critical point. On the other hand the rate of respiration increases with the increase of temperature to 40° C. Above this temperature the viability of the grain as well as the rate of respiration decreases significantly.

Structure

Wheat and rye consist mainly of pericarp, seed coat, aleurone layer, germ and endosperm whereas oats, bailey, paddy, pulses and some other crops consist not only of the above five parts but an outer husk cover also. The husk consists of strongly lignified floral integuments. The husk reduces the rate of drying significantly. The embryo or germ is the principal part of the seed. All tissues of the germ consist of living cells which are very sensitive to heat. The endosperm, which fills the whole inner part of the seed, consists of thin-walled cells, filled with proto-plasm and starch granules and serves as a kind of receptacle for reserve foodstuff for the developing embryo.

Chemical Composition

The grain is composed of both organic and inorganic substances, such as carbohydrates, proteins, vitamins, fats, ash, water, mineral salts and enzymes. Paddy, corn, wheat, buck wheat seeds are especially rich in carbohydrates whereas legumes are rich in proteins and oil seeds rich in oils. Generally, percarp (and floral integuments also) contains cellulose, pentosan and ash, the aleurone layer contains mainly albumin and fat. The endosperm contains the highest amount of carbohydrate in the form of starch, small amount of reserve protein and a very little amount of ash and cellulose whereas the germ contains the highest amount of fat, protein and a small amount of carbohydrate in the form of sugars and a large amount of enzyme.

Effects of Temperature on the Quality of Grain

Proteins

The proteins present in cereal grains and in flour are hydro-phillic colloids. The capacity of flour proteins to swell plays an important role in the preparation of dough. At temperatures above 50° C denaturation and even coagulation of proteins take place. As a result, the water absorbing capacity of the proteins and their capacity for swelling decrease.

The second major component of sorghum and millet grains is protein. Both genetic and environmental factors affect the protein content of sorghum and millets. In sorghum the variability is large, probably because the crop is grown under diverse agroclimatic conditions which affect the grain composition (Burleson, Cowley and Otey, 1956; Waggle, Deyoe and Smith 1967; Deosthale, Nagarajan and Visweswar Rao, 1972). Fluctuations in the protein content of the grain are generally accompanied by changes in the amino acid composition of the protein (Waggle and Deyoe, 1966).

The quality of a protein is primarily a function of its essential amino acid composition. To assess the protein quality, Block and Mitchell (1946) introduced the concept of an amino acid or chemical score, in which the amount of the essential amino acid that is in greatest deficit is expressed as a percentage of the amount present in a standard or reference protein. Egg and human milk proteins, for their very high biological value, have been considered as reference standards. Sorghum and millet proteins differed in their essential amino acid profile (Table 1.7). However, the most common feature was that lysine was always found to be the most limiting amino acid. The highest deficit of lysine was in the protein of barnyard millet (chemical score 31), closely followed by little millet (chemical score 33). Sorghum protein, with a chemical score of 37, did not differ very much in quality from the proteins of barnyard and little millet.

The primary function of dietary protein is to satisfy the body's needs for nitrogen and essential amino acids. According to the World Health Organization (1985), the chemical score of a protein if calculated in relation to the essential amino acid requirement pattern as reference would be more realistic and indicative of the capacity of the protein to meet human requirements. Such data on chemical score relative to amino acid requirement for different age groups (Table 1.8) suggested that the inherent capacity of the existing varieties commonly consumed was not adequate to meet the growth requirements of infants and young children, though all of them except sorghum may be able to meet the maintenance requirements in adults.

Table 1.7: Essential Amino Acid Composition (mg/g) and Chemical Score of Sorghum and Millet Proteins

Grain	*Isoleucine*	*Leucine*	*Lysine*	*Methionine*	*Cystine*	*Phenyl-alanine*	*Tyrosine*	*Threonine*	*Tryptophan*	*Valine*	*Chemical Score*
Sorghum	245	832	126	87	94	306	167	189	63	313	37
Pearl millet	256	598	214	154	148	301	203	241	122	345	63
Finger millet	275	594	181	194	163	325	–	263	191	413	52
Foxtail millet	475	1 044	138	175	–	419	–	194	61	431	41
Common millet	405	762	189	160	–	307	–	147	49	407	56
Little millet	416	679	114	142	–	297	–	212	35	379	33
Barnyard millet	288	725	106	133	175	362	150	231	63	388	31
Kodo millet	188	419	188	94	–	375	213	194	38	238	55

Table 1.8: Lysine Amino Acid Scores for different Age Groups Based on the 1985 WHO Index

Grain	*Infant (<1 year)*	*Preschool Child (2-5 Years)*	*School Child (10-12 Years)*	*Adult*
Wheat	43	46	62	100+
Rice (hushed)	57	61	82	100+
Maize	41	43	58	100+
Sorghum	17-51	18-55	25-74	71-100+
Pearl millet	26-69	28-74	38-100+	100+
Foxtail millet	28-38	30-40	40-55	100+
Finger millet	39-63	41-68	56-91	100+
Kodo millet	46-52	48-55	65-74	100+
Barnyard millet	26	27	37	100+
Proso millet	23-72	24-74	32-98	93-100+

Sources: WHO, 1985; Hulse, Laing and Pearson, 1980.

Grain proteins are broadly classified into four fractions according to their solubility characteristics: albumin (water soluble), globulin (soluble in dilute salt solution), prolamin (soluble in alcohol) and glutelin (extractable in dilute alkali or acid solutions). In solubility fractionation studies with sorghum and pearl, finger and foxtail millets, five protein fractions were obtained (Table 1.9) and digestibility also mentioned in Tables 1.10 and 1.11. The levels of albumin plus globulin were higher in pearl millet varieties than in sorghum, while amounts of the cross-linked prolamin, ß-prolamin, were higher in sorghum than in pearl millet.

Pearl Millet

Pearl millet, like sorghum, is generally 9 to 13 per cent protein, but large variations in protein content, from 6 to 21 per cent, have been observed (SernaSaldivar, McDonough and Rooney, 1991). Lysine is the first limiting amino acid of pearl millet protein. A significant inverse correlation has been reported between the level of protein in the grain and the lysine content of the protein (Deosthale *et al.*, 1971). In high-protein varieties of pearl millet with protein content ranging from 14.4 to 27.1 per cent, significant inverse correlations have also been observed between protein and threonine, methionine and tryptophan. The essential amino acid profile shows more lysine, threonine, methionine and cystine in pearl millet protein than in proteins of sorghum and other millets. Its tryptophan content is also higher (Table 1.7).

Wide variation is observed in the lysine content of pearl millet protein, with values ranging from 1.59 to 3.8 g per 100 g protein. From chemical scores calculated in relation to amino acid requirements for different age groups it was apparent that pearl millet has greater potential to meet the lysine requirements of growing children than most other cereals (Table 1.8). Pushpamma, Parrish and Deyoe (1972) observed in rat feeding trials a PER of 1.84 for pearl millet as against 1.74 for finger

Table 1.9: Distribution of Protein Fractions in Sorghum and Millet Grains (Percentage of total protein)

Fraction	*Sorghum*		*Pearl Millet*		*Finger Millet*		*Foxtail Millet*	
	Range	*Mean*	*Range*	*Mean*	*Range*	*Mean*	*Range*	*Mean*
Albumin+globulin	17.1-17.8	17.4	22.6-26.6	25.0	17.3-27.6	22.4	11.6-29.6	17.1
Prolamin	5.2-8.4	6.4	22.8-31.7	28.4	24.6-36.2	32.3	47.6-63.4	56.1
Cross-linked prolamin	18.2-19 5	18.8	1.8-3.4	2.7	2.5-3.3	2.78	6.4-17.6	8.9
Glutelin-like	3.4-4.4	4.0	4.7-7.2	5.5	–	–	5.2-11.9	9.2
Glutelin	33.7-38.3	35.7	16.4-19.2	18.4	12.4-28.2	21.2	–	6.7
Residue	10.4-10.7	10.6	3.3-5.1	3.9	16.1-25.3	21.3	–	2.0
Total	91.2-94.0	92.9	78.6-87.5	83.9	74.7-83.9	78.7	–	98.0

millet, 1.46 for sorghum and 1.36 for maize. This has supported the view that the protein quality of pearl millet ranks quite high in comparison with that of other cereals. On fortification of a pearl millet diet with 0.3 per cent lysine hydrochloride, the growth response of rats was enhanced and nearly equaled that of controls fed a casein diet (Howe and Gilfillan, 1970).

Table 1.10: Protein Quality and Digestible Energy in Dehulled Millets (per cent)

Grain	*True Digestibility*	*Biological Value*	*Net Protein Utilization*	*Digestible Energy*
Pearl millet (low protein)	95.9	65.6	62.9	89.9
Pearl millet (high protein)	94.6	58.8	55.7	85.3
Foxtail millet	95.0	48.4	46.3	96.1
Common millet	99.3	52.4	52.0	96.6
Little millet	97.7	53.0	51.8	96.1
Barnyard millet	95.3	54.8	52.2	95.6
Kodo millet	96.6	56.5	54 5	95.7

Table 1.11: PER of Diets Based on Millets or Blends of Millet and Chickpea[a]

Protein Source	*PER*
Foxtail millet	0.80
Proso millet	1.10
Proso millet + chickpea	1.80
Pearl millet	1.60
Pearl millet + chickpea	2.16
Finger millet	2.00
Finger millet + chickpea	2.10
Rice	2.09
Whole wheat	1.30

Protein content in diet: 10 per cent. As a supplemental protein source. Chickpea provided 40 per cent of protein; *Source*: Casey and Lorenz, 1977.

Protein quality is associated with the distribution pattern of protein fractions in the grain (Table 1.9). Sawhney and Naik (1969) observed large variability in the protein fractions of pearl millet varieties. Albumin ranged from 6.1 to 26.5 per cent (mean 15.1 per cent), globulin from 3.5 to 14.7 per cent (mean 8.7 per cent), prolamin from 21.3 to 38.0 per cent (mean 30.2 per cent) and glutelin from 23.8 to 37.7 per cent (mean 30.3 per cent). As in other cereals, albumin and globulin are rich in lysine as well as the other basic amino acids arginine and histidine. The globulin fraction appeared to be very rich in sulphur amino acids. The prolamin fraction is characterized by high glutamic acid proline and leucine and was also shown to be rich in tryptophan, whereas glutelin was found to contain more lysine and less tryptophan.

True protein digestibility in rats fed pearl millet varied little, from 94 to 97 per cent (Singh *et al.*, 1987) and it was not affected by the protein content of the grain (Table 1.10). The digestible energy content was lower in high-protein types because of their high prolamin content. In high-protein genotypes, the lysine content of the protein was low and this was reflected in low biological value and low net protein utilization. But the net utilizable protein (per cent protein x NPU) from the high-protein genotypes was two to three times higher than that from normal millets. Rats fed raw pearl millet flour exhibited higher digestibility of protein and energy than rats fed raw wheat flour (Dassenko, 1980). However, the digestibility and PER were lower when the millet was fed as chapatti, probably because the longer cooking time required for millet chapatti resulted in heat damage to the protein. In nitrogen balance studies in 11 - to 12-year-old boys, the apparent protein digestibility of a pearl millet-based diet was 52.9 per cent and the nitrogen balance was positive (Kurien, Swaminathan and Subrahmanyan, 1961).

Finger Millet

Finger millet is poor in protein content compared with other common cereals (Table 1.7). Wide variability in the composition of the grain, including its protein content, was reported (Hulse, Laing and Pearson, 1980). Both genetic and environmental factors appear to have an important role in determining the protein content of finger millet (Pore and Magar, 1977; Virupaksha, Ramachandra and Nagaraju, 1975). Prolamin is the major protein fraction in finger millet (Table 1.10). The high protein of white-grain varieties was attributed to the higher prolamin content of the grain, while the lysine content and hence the protein quality of these varieties are low (Virupaksha, Ramachandra and Nagaraju, 1975). Differences in amino acid composition in different varieties of finger millet are large and as in other cereals both the lysine content and the methionine content of the protein are inversely correlated with the protein content of the grain. The protein tractions also showed wide variation in their amino acid composition. While the albumin and globulin fraction was found to contain a good complement of essential amino acids, the prolamin fraction contained higher proportions of glutamic acid, proline, valine, isoleucine, leucine and phenylalanine but low lysine arginine and glycine. The amino acid composition of prolamin was almost the same as that of endosperm protein.

In vitro studies showed that proteins of finger millet and kodo millet were resistant to pepsin digestion unless the millet was first cooked in an autoclave for 15 minutes or boiled for at least two hours in water. Digestibility of the protein was found to be adversely affected by tannin in the grain, which was as high as 3.42 per cent in some of the finger millet varieties studied (Ramachandra, Virupaksha and Shadaksharaswamy, 1977). A finger millet diet was found to be adequate to maintain a positive nitrogen balance in adults (Subrahmanyan *et al.*, 1955). The subjects also showed positive calcium and phosphorus balances and the digestibility of finger millet protein was found to be 50 per cent. Supplementation of a finger millet protein diet with lysine or with leaf protein in addition to lysine significantly improved nitrogen retention in young boys. They also showed greater increase in height and weight (Doraiswamy, Singh and Daniel, 1969). The use of finger millet in child and infant feeding, however, appeared to be limited because of its poor digestibility and the large quantity required meeting energy requirements. The growth performance of growing rats fed sprouted finger millet was better than that of animals fed raw grain. However, the protein quality as judged by PER was unaltered by sprouting (Hemanalini *et al.*, 1980). Further processing of germinated finger millet grains by drying, roasting and filtering through a cloth gave a product low in fiber. Animals fed this product as a source of protein showed improved calcium retention, probably because of the low fiber content of the flour.

Malted grains of finger millet have a significantly higher saccharifying enzyme activity useful in brewing. This activity is higher than that of malted sorghum, pearl millet or maize (Rao and Mushonga, 1985). A weaning food with low hot paste viscosity and high energy density was developed in which malted finger millet from which the vegetative portion had been removed was combined with green gram. Amix of 70 parts malted finger millet grains and 30 parts green gram, combined with 10 per cent skimmed milk powder, had a PER of 2.7 and NPU of 63 per cent (Malleshi and Desikachar, 1982). An extruded cooked product prepared from a blend of rice (42.5 parts), finger millet (42.5 parts) and defatted soy flour (15 parts) exhibited a significant improvement in protein quality over the unprocessed blend (Dublish, Chauhan and Bains, 1988). The PER values after extrusion had increased from 1.92 to 2.41, while the trypsin inhibitory activity was reduced by about 70 to 100 per cent, the tannin content was below measurable amounts and the phytin phosphorus as percentage of total phosphorus had decreased by about 4 to 13 per cent. These changes obviously might have contributed to the improved protein quality. A blend of finger millet and defatted soy flour (85:15) on extrusion had a PER of 1.81 before processing and 2.23 after extrusion (Table 1.11).

Foxtail Millet

The protein in foxtail millet is also deficient in lysine its amino acid score (Table 1.8) is comparable to that of maize (Baghel, Netke and Bajpai, 1985). Monteiro, Virupaksha and Rajagopol Rao (1982) observed large variation in the grain protein content and in the distribution pattern of different solubility fractions. Prolamin constituted the major storage protein (Table 1.9) and showed a positive correlation with total protein in the grain. Evaluations of the amino acid composition of the protein fractions and of total protein in different varieties have confirmed that lysine is the first limiting amino acid, followed by tryptophan and sulphur amino acids.

With increase in grain protein, the lysine content of the protein decreased. The protein was found to be high in leucine. Naren and Virupaksha (1990) observed that the prolamin was relatively rich in the sulphur amino acid methionine and that the sulphur status of the soil affected the synthesis of prolamin in the grain. In studies on in vitro protein digestibility, 90.5 to 96.9 per cent of the protein in foxtail millet was digestible by pepsin and 89.7 to 95.6 per cent by papain (Monteiro *et al.*, 1988). Poor digestibility with trypsin (21.6 to 36.9 per cent) was improved by prior treatment with acid. The protein quality of dehusked grain (Table 1.10) was the lowest among the minor millets tested (Geervani and Eggum, 1989). Heat treatment or lysine supplementation improved the protein quality (Geervani and Eggum, 1989). In growing rats fed 10 per cent protein, the nitrogen balance improved from 19 to 31 per cent when the foxtail millet diet was fortified with lysine protein digestibility and biological value were also enhanced (Ganapathy, Chitra and Gokhale, 1957). Supplementation of dehusked millet with chickpea rose the PER from 0.5 to 2.2.

Common Millet

Though the range of protein content in common millet can be very wide, the values appear to lie most frequently in a narrow range of 11.3 to 12.7 per cent, with a mean of 11.6 per cent, on a dry-matter basis (Serna-Saldivar, McDonough and Rooney, 1991). The protein of common millet is deficient in lysine as well as threonine and its tryptophan content is also marginal (Chung and Pomeranz, 1985). Studies on the protein solubility fractions of common millet showed that more than 50 per cent of the grain protein was prolamin and the next predominant fraction was glutelin, about 28 per cent. The prolamin fraction was very poor in lysine arginine and glycine compared to the albumin and globulin fraction and had more alanine, methionine and leucine (Jones *et al.*, 1970). When ted as the sole source of protein (8.4 per cent) in the diet, common millet had a PER of 0.95. According to data presented by Kuppuswamy, Srinivasan and Subramanian (1985), a common millet diet containing 9 to 11 per cent protein had a PER of 1.2 and biological value of 56. A protein isolated (84.8 per cent) by alkali extraction of common millet (Tashiro and Maki, 1977) was compared for its protein quality with casein and gluten. In a 21 -day feeding trial in young rats with 10 per cent protein in the diet, the PER of the protein isolate of common millet was 3.1 while that of casein was 2.8. Animals fed whole millet flour as a source of protein failed to grow. In adult rats the biological value of millet flour was higher than that of the other protein sources. In vitro studies, the isolated protein was digestible by pepsin and by pepsin-pancreatin but not by trypsin.

Other Millets

Kodo, barnyard and little millets have been investigated less from the nutritional point of view. Kodo millet grains are enclosed in a hard, corneous husk which is difficult to remove. The fiber content of the whole grain is very high. Kodo millet has around 11 per cent protein and the nutritional value of the protein has been found to be slightly better than that of foxtail millet but comparable to that of other minor millets (Table 1.9). Apart from lysine the protein of kodo millet is deficient in tryptophan (Chung and Pomeranz, 1985). As with other foodgrains, the nutritive value of kodo millet protein was improved by supplementation with legume protein (Rajalakshmi and Mujumdar, 1966). The PER of kodo millet on supplementation with chickpea and amaranth leaves was increased from 0.9 to 1.9 (Patwardhan, 1961).

Barnyard and little millets are comparable to common millet in their protein and fat content (Geervani and Eggum, 1989) and both are very high in fiber. With lysine amino acid scores of 31 and 33, little millet and barnyard millet have the poorest quality proteins among the millets. Barnyard and little millets are comparable in their protein digestibility, biological value, net protein utilization and digestible energy content (Table 1.10) and therefore in their overall nutritive value.

Starch

Starch is insoluble in cold water. It swells in hot water. Up to a temperature of 60° C, the quality of starch does not change appreciably. With a further increase in temperature, particularly above 70° C and especially in the presence of high moisture in the grain, gelatinization and- partial conversion of starch to dextrin take place. In addition, a partial caramelisation of sugars with the formation of caramel may take place which causes deterioration in colour of the product. These effects have been discussed in detail in Section II on Parboiling.

Fats

Fats are insoluble in water. Compared to alburmns and starch, fats are more heat resistant. But at temperatures above 70° C, fats may also undergo a partial decomposition resulting in an increase of acid numbers. In the range of temperatures from 40 to 45° C, the rate of enzymatic activity on fats increases with the increase of moisture and temperature. With a further rise of temperature the enzymatic activity begins to decrease and at temperatures between 80 and 100° C the enzymes are completely inactivated *Vitamins.* The heats sensitive B-vitamins present in the germ and aleurone layer are destroyed at high temperature.

Physical Properties

The knowledge of important physical properties such as shape, size, volume, surface area, density, porosity, colour, etc., of different grains is necessary for the design of various separating and handling, storing and drying systems. The density and specific gravity values are also used for the calculation of thermal diffusivity and Reynolds number. A few important physical properties have been discussed here.

Sphericity

Sphericity is defined as the ratio of surface area of sphere having same volume as that of the particle to the surface area of the particle.

Porosity

It is defined as the percentage of volume of inter-grain space *to* the total volume of grain bulk. The per cent void of different grains in bulk is often needed in drying, air flow and heat flow studies of grains. Porosity depends on: (a) shape, (b) dimensions and (c) roughness of the grain surface. Porosity of some crops is tabulated as follows:

Grain	*Porosity Per cent*
Com	40-45
Wheat	50-55
Paddy	48-50
Oats	65-70

Coefficient of Friction and Angle of Repose

Angle, of repose and frictional properties of grains play an important role in selection of design features of hoppers, chutes, dryers, storage bins and other equipment for grain flow.

Coefficient of Friction

The coefficient of friction between granular materials is equal to the tangent of the angle of internal friction for the material. The frictional coefficient depends on(a) grain shape (b) surface characteristics and (c) moisture content.

Angle of Repose

The flowing capacities of different grains are different. It is characterized by the angle of natural slope (angle of repose). The angle of repose is the angle between the basic and the slope of the cone formed on a free vertical fall of the grain mass to a horizontal plane. The angle of repose for a few important grains is tabulated as follows:

Grain	*Angle of Repose (degree)*
Wheat	23-28
Corn	30-40
Millets	20-25
Rye	23-28
Oats	31-44
Barley	28-40
Paddy	30-45

Thermal Properties

The raw foods are subjected to various types of thermal treatment namely, heating, cooling, drying, freezing, etc., for processing. The change of temperature depends on the thermal properties of the product. Therefore, knowledge of thermal properties, namely, specific heat, thermal conductivity, thermal diffusivity, is essential for the design of different thermal equipments and for solving various problems on heat transfer operation.

Specific Heat

The specific heat of a substance is defined as the amount of heat required to raise the temperature of unit mass through 1°C. The specific heat of wet grain may be considered as the sum of specific heat of bone dry grain and of its moisture content.

General Grain Milling Operations

Food grains are naturally endowed with outer protective husk/bran layers composed of rough, fibrous, pigmented and waxy substances which are undesirable for edible purposes. It also consists of oily germs which are undesirable for storage purposes. Removal of these parts constitutes the most fundamental prerequisite in grain milling or flour milling technology of cereals, In grain milling the outer husk/ bran layers are removed from the grain with its shape retained whereas in flour milling, flour without or with negligible bran content is prepared without the grain shape. In general, milling refers to the size reduction and separation operations used for processing of food grains into edible form by removing and separating the inedible and undesirable portions from them. Milling may involve cleaning/ separation, husking, sorting, whitening, polishing, grinding, sifting, etc. To increase the milling quality: of the food grains or to improve the quality and quantity of their end products or to facilitate milling operations for the desired products, food grains are sometimes subjected to hydrothermal treatment prior to milling called conditioning. Basic milling operations and hydrothermal treatment involved in cereal milling technology have been discussed.

Classification of Separation Methods

Any mixture of solid materials can be separated into different fractions according to their difference in length, width, thickness, density, roughness and drag in moving air, electrical conductivity, colour and other physical properties.

Each of the various types of separators employed in flour and grain milling is designed on the basis of the difference in the following physical' characteristics of grain: (1) Width and thickness of the grain for sieves, screen cleaners, sifters, thickness graders, grading reels, inclined sifters, etc.; (2) Length of the grain for indented type or disc type pocket separators; (3) Aerodynamic properties for husk aspirators, cyclone separators. (4) Form and state of the surface for separators for coarse grain, spiral separators, belt type separators; (5) Specific gravity and coefficient of friction for separating tables, stone separator; (6) Ferromagnetic properties for magnetic and electromagnetic separators; (7) Electrical properties for electrostatic separators and (8) Colour for electronic separators.

Separation According to Aerodynamic Properties

The pneumatic separation is based on the difference in aerodynamic properties of the different 'components. The aerodynamic properties of a particle depend on the shape, dimensions and weight of the particle, the state and position of the particle with respect to the air current. The aerodynamic properties have been discussed in detail in other section.

Separation According to Specific Gravity

If the components of a mixture are different in densities and subjected to a reciprocating movement of an inclined table or screen then the components of the mixture are readily separated into different fractions according to their densities. This type of separator works on the principle of self sorting or stratification. The heavier particles of the mixture sink to the bottom by the to and fro movement of the inclined table and are then separated by any suitable method. In a mixture of components of same density, the finer particles sink while in a mixture of components of equal dimensions but of different densities, the heavier particles sink to the bottom. This is the principle of operation of stone separators. Composite stone separators of pneumatic grading-table type are used to enhance settling of stone. The stone separator has been discussed in detail in Rice Milling Chapter. The effectiveness of operation of a stone separator depends on many factors. Of them kinematic parameters are most important. Continuous and uniform feeding of mixture is also important.

Separation According to Magnetic Properties

Metallic impurities in the grain accelerate wear and tear of different parts of the milling machinery. Moreover, even minute quantity of meta-Ilk impurities present in the milling products can make them unfit for human consumption. The magnetic impurities like steel, pig iron, nickel and cobalt particles present in a grain mixture can be separated on the basis of their differences in magnetic properties. Since effectiveness of removal of metallic impurities depends on the force of; attraction of the magnet, an electromagnet is preferred to an ordinary permanent magnet as the force of attraction of the former can be increased by increasing the strength of electric current only. Magnetic separation consists of three steps: (a) distribution of feed over the magnet, (b) collection and retention of the ferromagnetic impurities by the magnet and (c) cleaning of the magnet from the impurities.

Separation According to Electric Properties

When different particles are charged with statically electricity and are passed through another electric field, then the action of the outer electric held on the electric field of the charged particles- produces some mechanical work which is used for separation. The electrical separation consists of two stages: Preliminary charging of the particles with electricity and separation of the charged particles by electrostatic forces in accordance with the magnitude and nature of the charges on the particles. Magnitude of the preliminary charge is determined by the following factors: electrical conductivity, dielectric constant and other properties of the particles such as particle size and shape, specific gravity and design of the separators also.

Separation According to Colour

Difference in colour can be used for separation. With the help of an electronic separator some fruits, vegetables and cereal grains can be sorted from the discoloured or defective ones in accordance with their differences in colour. In an electronic separator the seeds are uniformly fed to the optical chamber. Two photo cells are set at a certain angle in order to direct both beams to one point of the parabolic trajectory of the seeds. A needle connected to a high voltage source is placed on the other side. When a beam through photoelectric cells falls on a dark object, a current is generated on the needle. The end of the needle receives a charge and imparts *it* to the dark seeds. The grains are then allowed to pass between two electrodes with a high potential difference between them. As a result two fractions of the mixture are separated according to difference in colours. Electronic separators have been used in various industries for a long time. Recent electronic sorters are being used for the separation of discoloured grains in some advanced countries. But the sorting capacity of these units is limited resulting in very high cost of separation.

Frictional Separations

Separation according to surface properties: The frictional properties can be utilized for the separation of a mixture of grains of almost same size. Sizes of oats and hulled oats, millets and bind weed are almost the same. These mixtures can be separated by frictional separators consisting of an inclined plane surface; the operation of the separators is based on the differences between the friction angles of two types of grain, when grains are allowed to move along an inclined plane, frictional forces of different magnitudes act upon these particles. Therefore, different particles move on inclined surface at different velocities. In this case also heavier particles will sink at the lower layer and move at a lower velocity, while lighter particles will float at the top and move downward at a higher velocity.

Husking/Scouring/Hulling of Grain

In general, husking refers to the removal of outer seed coat from the grain kernel. The terms hulling and scouring are also used in cereal milling. In grain milling, husk is removed from the grain retaining its original shape whereas in flour milling bran is removed from the grain to produce flour without any emphasis on its shape. Husking and scouring are the most important operations in grain milling or flour milling technology of cereals. Different types of machines are employed for husking and scouring operations because of the differences in anatomical structure, type of bonds and strength properties among husk, bran and kernel of different grains.

Methods of Husking

The operation of husking and scouring machines can be divided according to the following three basic principles: (1) Compression and shear: Compression and shear can compress, split and strip off the husk from grain. Concave type of husking machine, rubber roll husker, etc., are designed on the basis of this principle. (2) Abrasion and friction: Hollanders are based on the friction of grain on an abrasive surface (emery). (3) Impact and friction: Husk can be stripped off by the action of impact and frictional force. Centrifugal type paddy Sheller comes under this group.

Concave Type Husker

This machine consists of a horizontal rotating cylinder, called roll and a stationary cylindrical surface known as concave. The husk/bran layer can be removed from buck wheat and millets keeping the original shape of the kernel by applying mild shear and compression. These types of machines are fairly efficient and require low power. On feeding into the mill the grain is first caught up by the rolls and passed through the husking zone between the roll and the concave where it is subjected to shear and compression simultaneously. One part of the husk is shared by the rotating roll while the other part is pressed against the stationary concave and subjected to breaking forces. The minimum clearance between the roll and the concave must be greater than the dimensions, of the grain kernel; otherwise the kernel will be crushed. The radius of curvature of the concave is usually the same as the radius of the roll. Composition of roll and concave varies with the type of grain to be husked. As for example roll made of abrasive material and concave made of commercial rubber are used for millets. Usually the following specifications of rolls and concaves are used:

Grain	*Angle of Repose (degree)*
Diameter of concave	50 to 60 cm
Length of concave	19 to 30 cm
Angle of contact	40 to 70º
Peripheral speed	10 to 15 m/sec

Husking by the Action of Rubber Rolls

In case of paddy, deformation caused by shear and compression of the two rotating rubber surfaces are sufficient to split and separate the husk from the grains.

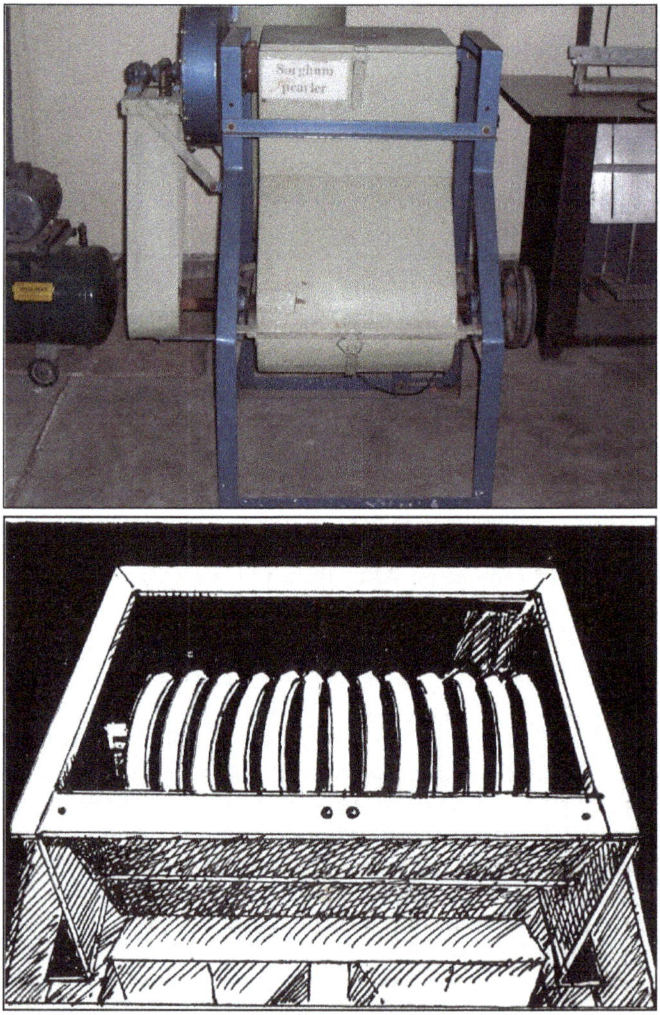

Figure 1.4: Dehulling or Pearling Machine.

The paddy is passed through the clearance between two rubber rolls, rotating in opposite directions at different speed. The clearance between them is smaller than the mean thickness of the paddy. One part of the husk is subjected to shearing forces whereas the other part in contact with the slower roll IS under compression and is thus subjected to breaking forces. Husking is done by the action of these forces.

Grinding

Cereal grinding system can be divided into two groups: plain grinding and selective grinding. In plain grinding hard bodies are grounds to a free flowing material consisting of particles of sufficiently uniform size. This material is either the final product or a product ready for further processing. In selective grinding, the grinding operation is carried out in a number of stages successively using differences in structural and mechanical properties of the components of the body. It should be noted that the power consumption for grinding is about 5O to 80 per cent of the total power required for all operations. Therefore, the following important points are to be considered for power design of the grinding system: rational utilization of the raw material, quality of the products, size and colour of the products, efficiency of the grinders, specific power consumption and costs of production.

The characteristics of grinding operation are affected by the £following grain parameters:

(i) Type of cereal grain, (ii) Variety, (iii) moisture content, (iv) extent of hydrothermal treatment given to the grain, (*v*) mechanical properties. Each type of grain has optimum moisture content for highest efficiency of grinding. The maximum formation of new surface for each type of wheat is related to the optimum moisture content. Moreover, power consumption is also dependent on moisture content. The starch granules are separated from the proteins due to deformation of hard wheat. Hardness of rice and barley endosperms is 11.5 to 14.5 kg/mm^2 and 6.0 to 9.0 kg/mm^2 respectively.

Effectiveness of Grinding

The main criteria for evaluation of effectiveness of grinding of any solid body, including grain are: degree of grinding, specific power consumption and the specific load of the initial product on the working tool of the grinder.

Machinery Used in Cereal Grinding

Depending on the objective of grinding and mechanical properties of cereal grain, effects of the following are utilized for grinding: Compression and simultaneous shear of the material, impact followed by crushing of the material etc.

Grinding of Grain in Roller Mills

The roller mill consists of two cylindrical steel rolls revolving in opposite directions at different speeds. In the roller mill the grain or its parts are ruptured in a space, which is narrowed towards the bottom. Certain degree of grain rupturing starts at above the line connecting the centers of the rolls. The slow roll holds the grain during the action of the fast Toll. In the grinding zone, the grain or its parts

Figure 1.5: Flourmills.
1. Vertical mill, 2. Horizontal mill, 3. Pin mill.

are simultaneously subjected to compression and shear resulting in deformation of grain. Rupturing of endosperm without much grinding the bran of wheat under grinding conditions is the characteristic of the first breaking system (Figures 1.6 and 1.7).

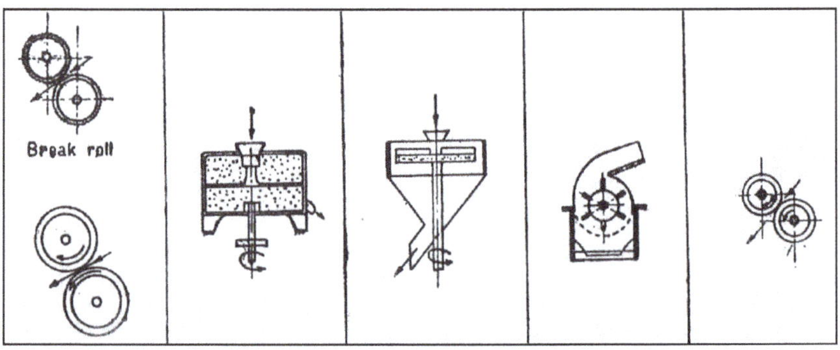

Grinding Machines	Roller Mill	Burr Mill	Attrition Mill	Hammer Nill	Flatting mill
Grinding mechanism	Compression and shear	Compression and shear	Impact and friction	Impact and crushing	Compression

Figure 1.6: Classification of Grinding Machinery

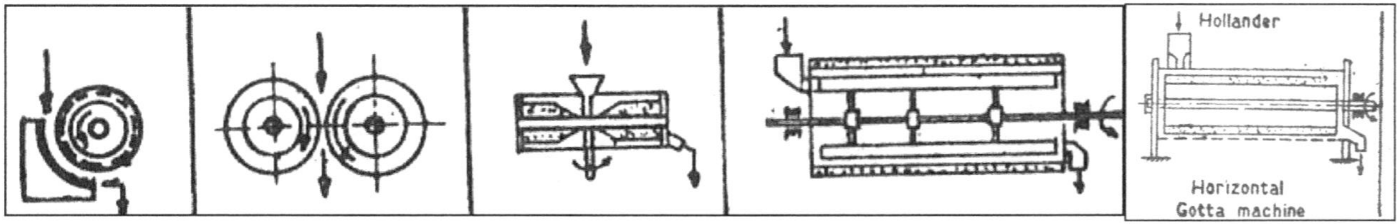

Characteristics of hull	For millets with free hulls around no brittle kernels	For paddy with free husk around brittle kernels	For oats with hulls around elastic kernels	For barley tightly joined hulls over strong kernels	For legumes tightly held hulls with kernels
Hulling method and machine	Concave huller	Rubber roll husker	Under runner disc sheller	Blade type emery scourer	Horizontal gotta machine
Hulling mechanism	Shear and compression	Shear and compression	Shear, compression and friction	Impact, abrasion and friction	Friction and abrasion

Figure 1.7: Classification of Husking/Scouring/Hulling Methods.

Factors Affecting the Effectiveness of Roller Mills

Clearance between Rolls

Even a small variation in rollers' clearance leads to considerable variations in the products.

Geometrical Parameters of Rolls

Conditions of rupture of the particles to be ground depend upon the roll diameter, clearance and initial size of the particle. Shape, number, slope, mutual position and shape of the cross-section of corrugations of the rolls have significant effects on the quality and yield of flour, total output and specific power consumption of the roller mills. Position of the corrugation edges has also the same effects on the product.

Kinematic Parameters of the Rolls

The important kinematic parameters of the rolls are speed of the fast *roll and* ratio of the speeds between fast and slow rolls. The efficiency of grinding depends upon the kinematic parameters.

Capacity of the Roller Mills and Power Consumption for their Operation

The capacity of a roller mill is the amount of product in kilograms, ground per unit time.

Grinding Grain in Hammer Mills

Grinding in a hammer mill involves impact on the material followed by crushing. The main working tools of this type of mill are hammers made of high quality steel, screens and metal lines. The output of hammer mill depends on the peripheral speeds of the hammer, the clearance between the screen and hammers, the area of the screen, the size of the screen openings and the structural-mechanical properties of the material.

Post-harvest Processing of Millets for Value Addition

Millet is a generic term used for the grains from the heterogeneous group of forage grasses. They are small sized grains and are grouped along with maize and sorghum as 'coarse cereals' perhaps because of their typical grain texture, which makes them difficult to process as well as cook in convenience form similar to rice and wheat. They assume significance for food and nutritional security in most of the Asian and African countries because of their hard nature and ability to grow in rain-fed lands with very little agricultural inputs as compared to most of the cereals. The annual production of the millets worldwide is about 32 million tonnes, of which a little more than half the quantity is produced in India. These are cultivated in India in almost all the states and the major millet producing states are Karnataka, Rajasthan, Gujarat, Haryana and Maharashtra.

Bajra or pearl millet (*Pennisetum americanum*), *ragi* or finger millet (*Eleusine coracana*), *navane* or foxtail millet (*Setaria italica*), *samai* or little millet (*Panicum miliare*), *haraka* or kodo millet (*Paspalum scrobiculatum*), *panivaragu* or proso millet (*Panicum *miliaceum*), banti or barnyard millet (*Echinochloa frumentacea*) are the important millets cultivated largely in the Asian and African countries. Fonio (*Digitaria exilis*) and Tef (*Eragrostic tef*) are specific to Nigeria and Ethiopia respectively. Most of the millets are grown in different regions of world from east to west, which is evident from their colloquial names, namely, foxtail millet as Italian millet, proso as French millet, barnyard as Japanese barnyard millet, indicating their specific regionality. Proso millet is cultivated in US as well as in Russia and is popular as a bird feed in both the countries, whereas, foxtail millet is largely harvested in China also besides in India. These are predominantly used for food purposes in India as well as in other major developing countries but are mainly used as feed ingredients in developed countries. Absence of appropriate primary processing technologies to prepare ready-to-use or ready-to-cook products and also secondary as well as tertiary processing to prepare ready-to-eat value added products have been the major limiting factors for their diversified food uses and better economic status. However, studies towards processing of millets have demonstrated their successful utilization for various traditional foods and also in newer convenience health oriented products. Hence, efforts are needed for popularization and wider adoption of these technologies to promote them for diversified utilization among the non-traditional urban population.

Millets are not only comparable to major cereals with respect to their nutritional features but are very good sources of carbohydrates, micronutrients and phytochemicals with nutraceutical properties. The millets contain 7-12 per cent protein, 2-5 per cent fat, 65-75 per cent carbohydrates and 15-20 per cent dietary fiber. Among them, pearl millet contains considerably high proportion of proteins (12-16 per cent) as well as lipids (4-6 per cent) whereas; finger millet contains lower levels of protein (6-8 per cent) and fat (1.5-2 per cent). The essential amino acid profiles of the millet proteins is better than maize. The niacin content in pearl millet is higher than all other cereals whereas, finger millet proteins are unique because of the sulphur rich amino acid contents. Similar to cereal proteins, the millet proteins are poor sources of lysine, but they complement well with lysine-rich vegetable (leguminous) and animal proteins and form nutritionally balanced composites of high biological value. There are no reports in the Indian sub-continent of any adverse effects on regular consumption of the millets. On the other hand, incidence of *diabetes mellitus* and gastro-intestinal tract related disorders are minimal among the population using these grains as staple. Even then, the millets, although nutritious, still form staple only for the under-privileged section of the society. The pearl millet products have very poor storage characteristics, which is also one of the reasons for it's under utilization.

Carbohydrates of the millets comprise of large proportion of starch (65-70 per cent) and a good amount (16-20 per cent) of non-starchy polysaccharides (NSP). The NSP, which constitute nearly 95 per cent of the dietary fiber content of the grain, are derived not only from bran portion but also from the endosperm cell walls.

A variety of foods from the millets are traditionally prepared and consumed, over the centuries, in Indian subcontinent, Africa and Central America. Majority of the conventional products are made from flour prepared either from the whole grain or from partially decorticated grains. These include unleavened pancakes from fermented or unfermented dough, stiff porridge or *mudde*, snack foods, deep fried products, sweet or sour opaque beer, non-alcoholic beverages and decorticated grains, boiled similar to rice.

Millets are known for their health benefits from time immemorial. Each of the millets has its nutritional and nutraceutical specialities and probably because of this, traditionally, different millets are consumed specifically in different seasons of the year. For example, pearl millet *roti* is normally recommended during the winter season probably due to its high energy content whereas finger millet and other small millets like foxtail and little millets are recommended for consumption in the form of "*nucchu*" (semolina based food product) mostly in summer season, since they are considered as cool foods. Even though, the scientific reasoning for these have not been investigated, one can attribute the health benefits of the millets to the special features of their biochemical constituents such as carbohydrates, proteins, dietary fiber, vitamins, minerals and also for the phytochemical contents with nutraceutical characteristics. Almost all of these millets are good source of both soluble and insoluble dietary fibers, which is contributed by both seed coat as well as endosperm cell walls, unlike in rice, wherein mostly the bran layers accounts for its dietary fiber content. Since, the dietary fiber component of the food offers several physiological benefits including the health of gastrointestinal tract, the millets could be gainfully utilized in the preparation of functional and health foods for target population.

Among the millets, finger millet is better source of calcium and several micronutrients. Similarly, pearl millet is a very good source of niacin and pyridoxine. Apart from this, these are good source of B-group vitamins. The seed coat of almost all the millets is coloured and the endosperm of some of the millets also contain light coloured pigments. The pigments are generally constituents of carotene and because of that, some of the millets contain beta-carotene. Polyphenols are important phytochemicals with strong anti-oxidants potential and because of this they offer protective action against several non-communicable diseases. The other prominent phytochemical in the millet is phytate, which also exerts some health benefits due to its metal chelating properties, especially with 'Fe' chelation and thereby hindering free radical formation. The gelatinization temperature of the millet starch is slightly higher compared to rice and wheat. It also differs from other cereal starches with respect to the crystallinity and the amylose and amylopectin organization and because of this; the millet based foods may digest slowly compared to other cereal foods.

The term milling refers to processing of the millets similar to rice as well as wheat. Decortication or debranning of foxtail, little, kodo, barnyard and proso millets is carried out to use them similar to that of rice. The milled grains are also pulverized to prepare flour and semolina based foods. Incipient moist conditioning, milling and grinding is followed to prepare refined flours from the millets. This treatment minimises pulverization of seed coat and hence the major portion of the seed coat could be separated by sifting or aspiration. In the case of pearl and finger millets, the glumes, whereas, husk in the case of other millets are non-edible and are separated by just abrasion or abrasion followed by mild attrition. The contributions of CFTRI, Mysore, towards milling or decortication of the millets are unmatched in the world. Whether it is refining or pearling, the Institute has explored all the possible avenues to prepare ready-to-cook millets. It has developed process to decorticate even finger millet, which was not at all possible hitherto due to its peculiar textural features.

Cereal flakes are popular snacks and are largely prepared from rice or corn. Flaking of millets has been successfully attempted adapting the normal cereal flaking methods using roller flaker. Millet flaking would be a new avenue for wide spread utilization, because of the inactivation of lipase and stabilization of the oil leading to better shelf-life to the flakes. Besides conventional flakes, fiber regulated flakes could also be prepared from the millets. The flakes could be toasted or expanded by hot air or sand which may serve as snack or supplementary food for the obese and calorie conscious people. Since, millets are small seeded grains; the flakes from these will hydrate easily and may serve as quick cooking cereals. After toasting they could be conveniently used as ingredients of muesli and such other products. Also, the bran separated as a by-product of the flaking industry could serve as a source of the edible oil similar to that of rice bran oil. Deoiled bran from the millets would have less ash and silica content as compared to that of deoiled rice bran and could be efficiently used as a source of dietary fiber as the millet contains high proportion of soluble dietary fiber and could be tapped for development of hypocholestrolemic and hypoglycaemic foods.

Popping of finger millet is very popular among the millets as it develops highly desirable aroma with crunchy texture. Popping of other millets is also practiced to some extent. Normally, the expansion volume of the popped millet is about 7-10 ml/g. Popped millets are good sources of dietary fiber and carbohydrates. The varieties with hard endosperm and medium thick pericarp exhibit superior popping quality. The lipolytic enzymes are denatured during the process of popping and hence, popped product including pearl and foxtail millets also will have good shelf-life. Popped millet could be consumed as snacks after seasoning with salt and spice and also for preparation of sweets such as *laddu* or *sattu* and *chikki*. Popped millet flour blended with popped or toasted legumes, such as puffed Bengal gram and jaggery or sugar makes delicious and nutritionally balanced convenience foods for growing children and lactating mothers. The product has the advantage of being low cost food permits addition of other flavouring and ingredients like groundnut, dried coconut etc., for improving the taste and nutritional quality. Since millets are rich source of micro-nutrients and phytochemicals, such products may score over similar products from rice and wheat.

Extrusion cooking is one of the very popular contemporary food processing technologies and is largely followed for corn and rice but the millets could also be extruded to prepare ready-to-eat (RTE) products. Since, the seed coat or the bran affects the expansion ratio and also the eating quality of the extrudates, it is desirable to use refined grits and flour, preferably of less than 40 mesh size. Equilibrating the flour to about 18 per cent moisture content and extruding in a single- or twin-screw extruder at about 150°C and 200 rpm, gives extrudates of highly desirable food qualities with an expansion of 1.5 - 2 times. The products will have crunchy texture and can be coated with traditional ingredients to prepare sweet or savoury snacks. Alternately, the grits could be mixed with spices and condiments prior to extrusion to obtain RTE snacks of desirable taste.

The flour from the millets could be roller dried to prepare a ready-to-eat food, most suitable as a thickener in soup or porridge. The roller dried millet may also find usage as a component of edible films.

Malting is one of the very early biotechnological processes adopted for cereal processing for food and brewing. Although, barley has the place of pride in malting, finger millet malting is also practiced at house hold and cottage industrial scale in India and Africa, especially for preparation of specialty food products and also for lactic acid fermented brew. Pearl millet and other minor millets have very limited scope for malting because of poor keeping quality of the malt. The ratio of alpha-to-beta amylases, developed during the germination of finger millet, is lower than that of barley and hence the millet malt requires supplementation of amylases from external source to prepare yeast fermented lager beers. In view of this, some of the millet malt is used as an adjunct in industrial brewing based on barley. Malted finger millet has better nutritional characteristics than the native millet. The malt is not a ready–to–eat product and needs further processing for various food uses. The malt flour is a good source of nutrients, besides, serves as a source of amylases. Hence, it is also termed as "Amylase Rich Food" (ARF). It can be mixed with powdered sugar, milk powder and flavouring agents such as cardamom to use as milk based beverage, which is popularly sold as *"ragi malt"* in South India. Since, the malt flour contains hardly 3-5 per cent protein; it can be blended with vegetable or animal protein sources such as grain legumes, milk powder, egg powder, etc., to prepare supplementary nutritious food for children. Now-a-days, about 5 per cent finger malt is invariably blended with the energy food to improve its texture. This food is produced on bulk and supplied to the weaning children. The process for preparation of weaning food based on malted millet (two parts) blended with malted green gram (one part) has been developed at CFTRI, Mysore and the food is popularly termed as "Malted Weaning Food" (MWF). The malt flour as a substitute to maltodextrin can be blended with milk and spray dried to prepare the infant food also. The special feature of the malt flour to form nutrient dense free flowing slurry consistency (about 1 kcal/ml) has been utilized towards development of enteral foods or the liquid foods suitable for feeding the needy patients. For the purpose, the malt flour is blended with other ingredients such as milk powder, sugar, soya flour, legume flours, vegetable oils and the blend is fortified with essential vitamins and minerals. The blend can be cooked to use as low cost enteral food or can be spray dried to prepare ready-to-eat enteral foods. The enteral foods prepared using the millet malt were found to be cost effective and clinically efficient in improving the nutritional status of patients and reducing the hospitalization period.

Papad is one of the important adjuncts in the Indian dietary. Even though, all the millets could be exploited for the preparation of *papads,* finger millet is more suitable. For preparation of *papads* the flour from the millets is cooked in appropriate quantity of water to completely gelatinize the starch, kneaded thoroughly and flattened to desired thinness and size, using roller pins and finally dried in shade. The *papads* from finger millet appear dark and unattractive, they turn to light coloured product with crisp texture and characteristic aroma on deep frying or toasting or microwaving with improved consumer appeal.

Utilization of millets for vermicelli-noodles is very rare, but is gaining popularity. Millet noodles are unique with respect to taste and aroma and provide dietary fibre. The noodles, on cooking in water retain the texture of their strands and firmness without disintegration with less than 6 per cent solid loss. The noodles from millets could be used as savoury as well as sweet products. Since the present trend of the towards low fat and high fibre foods, noodles from these minor cereals will have a great future, not only in the domestic market but also for export to Gulf countries, Europe and Australia.

Parboiling or hydrothermal treatment to minor millets seems to be highly promising because of the nutritional benefits, improved milling and culinary characteristics. However, no concerted efforts to standardize the methodology of parboiling of millet have been reported. Only recently, efforts made at CFTRI, Mysore to this effect have yielded desirable results. Since, the minor millet kernels resemble rice in their morphological features, the husk provides protection to the grains during steaming and prevents burst opening of the kernel. Normally the grains can be soaked at ambient to higher temperature (70°C). Soaking the grains at elevated temperature reduces the time of hydration to hardly few hours. The grains soaked to their equilibrium moisture content (35 per cent) can be steamed at atmospheric or at higher pressure to gelatinize the starch. The millets can also be parboiled adopting shella or dry heating methodology. The steamed grains can be dried in a conventional manner and milled in rice milling machinery similar to parboiled rice. This offers considerable advantages with respect to milling characteristics, namely, reduced breakage during milling and higher yield of head grains. Similar to rice, probably the parboiled millets will have better retention of vitamins, especially thiamine and also enhanced storage life. Parboiled grains on cooking form less sticky food products and the cooked material can be utilized similar to rice along with other adjuncts or can be seasoned with spice and condiments to prepare *chitranna*, tamarind rice and such other products. Besides, the parboiled millets can be processed to prepare a ready-to-eat product similar to expanded rice. However, by incipient germination and hydrothermal treatment to the millets, the expansion characteristics will be superior to the normal parboiled material. The expanded millets posses all the desirable qualities for preparation of snacks and also as adjuncts in specialty health products such as health bars and museli. Now days, there is a growing demand for ready-to-eat high fiber products and hence, the expanded millets will have very high economic potential in this direction.

Millets could be conveniently incorporated in multigrain products like snack or supplementary foods or health bars either in their flour form or as expanded grains. The parboiled and decorticated millets can be subjected to secondary processing like 'high temperature and short time treatment' to prepared expanded products similar to popped cereals. These expanded grains are almost free from adhering husk or seed coat of the kernel and will have better consumer acceptability. They can be incorporated into specialty foods like multi-grain cereal bars. On the other hand, the flour from the millets could be suitably blended with other ingredients like legumes etc and could be sheeted and cut into products of required shape and size and toasted either in oil or using hot air to prepare ready-to-eat multi-grain snack products.

Use of millets in bakery items like biscuits or bread as a composite flour is gaining popularity in recent years. The flour from the millets is blended in 15-40 per cent with the refined wheat flour to prepare biscuits. These biscuits possess the nutritional benefits associated with the millets and also provide the consumer with improved taste and flavour.

The knowledge base on the millet grain quality, processing and value addition is now increasing with recent R and D work across the world. Special features of the millets, in comparison to rice and wheat have been established. The information could form the basis for exploitation of the millet for value addition and also for widening the scope of their utilization. Feasibility for production of non-conventional products such as extrusion cooked and expanded ready-to-eat snacks and supplementary foods, noodles, malt foods, *papad* and such other food adjuncts, decorticated ready-to-cook cereals and flakes have been established which would pave the way for their commercial-scale processing and their utilization even by the non-millet consumers. Hence, there is a great scope for utilization of the millets in variety of foods and value added products in the years to come.

Traditional Processing Methods

Processing Untreated Grains

Flour made by grinding whole grain is occasionally used, particularly with the smaller millets, but in most places where sorghum and millets are consumed the grain is partially separated into its constituents before food is prepared from it. The first objective of processing is usually to remove some of the hull or bran - the fibrous outer layers of the grain. This is usually done by pounding followed by winnowing or sieving. The grain may first be moistened with about 10 per cent water or soaked overnight. When hard grains are pounded, the endosperm remains relatively intact and can be separated from the heavy grits by winnowing. With soft grains, the endosperm breaks into small particles and the pericarp can be separated by winnowing and screening.

When suitably prepared grain is pounded, the bran fraction contains most of the pericarp, along with some germ and endosperm. This traction is usually ted to domestic animals. The other fraction, containing most of the endosperm and much of the germ along with some pericarp, is retained for human consumption. Retaining the germ in the flour will improve aspects of its nutritional quality, but at the same time it will increase the rate at which the flour will become rancid. This is particularly important in the case of pearl millet.

Dry, moistened or wet grain is normally pounded with a wooden pestle in a wooden or stone mortar. Moistening the grain by adding about 10 per cent water facilitates not only the removal of the fibrous bran, but also separation of the germ and the endosperm, if desired. Although this practice produces slightly moist flour, many people temper the grain in this way before they pound it. Pounding moist or dry grain by hand is very laborious, time consuming and inefficient. A woman working hard with a pestle and mortar can at best only decorticate 1.5 kg per hour (Perter, 1983). Pounding gives a non-uniform product that has poor keeping qualities.

Many pearl millet grains have an irregular indentation in the pericarp. This makes it more difficult to decorticate pearl millet than it is to decorticate most other cereal grains (Kent, 1983).

The particle size of the endosperm fraction can be reduced by crushing or grinding to produce coarse grits or fine flour. This unpleasantly hard work is almost always done by women. Traditional grinding stones used to grind whole or decorticated grain to flour usually consist of a small stone which is held in the hand and a larger flat stone which is placed on the ground (Subramanian and Jambunathan, 1980; Vogel and Graham, 1979). Grain, which should be fairly dry, is crushed and pulverized by the backward and forward movement of the hand-held stone on the lower stone. The work is very laborious and it is hard work for anyone to grind more than 2 kg of flour in an hour. In a traditional process used in many countries of Africa and Asia, decorticated grain is crushed to coarse flour either with a pestle and mortar or between stones. Grain is also ground to coarse or fine flour in mechanized disk mills now located in many villages.

In wet milling, the sorghum or millet is soaked in water overnight (and sometimes longer) and then ground to a batter by hand, often between two stones. Soaking makes the endosperm very soft and the pericarp quite tough and makes grinding much easier, but it gives a batter or paste instead of flour.

Processing Malted Grains

Malting involves germinating grain and allowing it to sprout. Typically the grain is soaked for 16 to 24 hours, which allows it to absorb sufficient moisture for germination and for sprouts to appear. However, germinated sorghum rootless and sprouts contain very large amounts of dhurrin, a cyanogenic glucoside, which on hydrolysis produces a potent toxin variously known as prussic acid, hydrocyanic acid (HCN) and cyanide (Panasiuk and Bills, 1984). The fresh shoots and rootless of germinated sorghum and their extracts must therefore never be consumed, either by people or by animals, except in very small quantities (*e.g.* when the germinated grain is used just as a source of enzymes). Dada and Dendy (1988) showed that the removal of shoots and roots and subsequent processing reduced the HCN content by more than 90 per cent.

Malted sorghum has traditionally been used in several countries in Africa, but always after careful removal of the toxic parts. Hullu-murr is an important traditional food prepared from malted sorghum in the Sudan (Bureng, Badi and Monawar, 1987). Alcoholic beverages and dumplings are prepared in Kenya from germinated sorghum and millet.

In the germination process, the grain produces a-amylase, an enzyme that converts insoluble starch to soluble sugars. This has the effect of thinning paste made by heating slurry of starch in water, in turn allowing a higher caloric density in paste of a given viscosity, since as much as three times more flour can be used when the grain has been germinated. The energy that young children can consume is often limited by the bulk that they can consume. Thus using germinated grain can make food more suitable for certain categories of young children. Flour from malted grain is consequently used quite widely in the production of children's food, but when such foods are made from sorghum, great care must always be taken to ensure that the level of cyanide is adequately low, as children are particularly vulnerable to cyanide.

In India, malted finger millet is common and is considered to be superior to malted sorghum and malted maize. Studies have shown that finger millet develops higher amylase activity than sorghum and other millets (Seenappa, 1988). Germination of grain is reported to change the amino acid composition, convert starch into sugars and improve the availability of fat, vitamins and minerals.

Pal, Wagle and Sheorain (1976) measured the changes in the constituents of sorghum and various millets (finger, pearl, prove, kodo and barnyard) during malting. The malting losses for finger millet and foxtail millet were high. Pearl millet had the highest a-amylase activity. Amylolytic and proteolytic enzyme levels in malted pearl millet were comparable to those in malted barley.

The use of only 5 per cent malted sorghum or finger millet was found to reduce the viscosity of weaning foods (Mosha and Svanberg, 1983; Seenappa, 1988).

Processing Grain Treated with Alkali

To produce a particular type of tortilla that is popular in Mexico, sorghum grains are cooked in lime water for a short time and steeped overnight, washed to remove the excess alkali and then ground to a paste (Rizley and Suter, 1977).

Wood ash is used in traditional treatments to reduce the level of tannin in brown sorghums and improve the nutritional quality. Muindi and Thomke (1981) reported the use of wood ash in the United Republic of Tanzania. Mukuru (1992) described a tannin-reducing technique used in parts of eastern and central Africa where, because of grain-eating birds, only "high-tannin" sorghums are grown. The sorghum is first soaked overnight in slurry of wood ash in water. After draining it is left for three or four days to germinate. The germinated grains are sun-dried and pounded to loosen the adhering wood ash and to remove the sprouts, with their high levels of cyanide. The grain is then ground and used to prepare either a non-alcoholic beverage called obushara or an alcoholic drink containing about 3 per cent alcohol called omuramba.

Processing Parboiled Grain

Parboiling is reported to help in dehusking kodo millet (Shrestha, 1972) and to eliminate the stickiness in cooked finger millet porridge (Desikachar, 1975).

Industrial Processing

While there are many machines available for processing hard white sorghum, there is unfortunately no well-proven industrial process available that is entirely satisfactory for making white products from coloured sorghums and millets.

Cereal grains can be milled wet, in the form of thin aqueous slurry, usually to produce starch, or in an essentially dry form (often suitably dampened or "tempered") which usually produces meal (coarse or fine flour). A factory in Texas, United States, for wet milling sorghum operated intermittently from the 1940s to the 1970s (Rooney, 1992) but is now closed. No millets have ever been wet milled commercially to produce starch. The following technologies are all for dry and semi-wet milling.

In industrial processing, once the grain has been cleaned, the first operation is usually the separation of offal (the portion not normally used for human consumption) from the edible portion. The offal consists of the pericarp and sometimes the germ. Offal removal is frequently called decortication or dehulling.

Following the removal of offal, the edible portion is often milled to reduce the particle size of the edible fraction. There is usually a choice of techniques and mills that may be used for particle size reduction if a finer product is desired. Some of the earliest research and development work on milling technology for pearl millet and sorghum was promoted by FAO in 1964, initially on a laboratory scale in Senegal and later on a semi-industrial scale in the Sudan. The conclusion was reached that the technology for milling wheat is not optimal for milling sorghum and millet (Perter, 1977).

Most industrial operations that can be carried out on untreated grain can also be used with grain that has been prepared in some way, for example grain that has been germinated and then suitably dried. Three types of processors can be used to mill sorghum and millets on a commercial scale: abrasive decorticators, which abrade the pericarp away, *i.e.* progressively remove offal from the outside; machines that rub (rather than abrade) the pericarp off the endosperm; and roller mills, which cut the endosperm from the inside of the pericarp.

Abrasive Decortications

Abrasive decorticators work by abrading away the fibrous pericarp. Obviously, the outer layers of the seed-coat are abraded away first and the innermost layers, which in many varieties contain those factors that most need to be removed, are the last to be abraded away. If all parts of all grains could be abraded away at the same rate, abrasive decortication would be an efficient way of removing the pericarp. However, different parts of individual grains are abraded away at very different rates and there is some loss of endosperm (particularly from damaged grains) even when the grain is only lightly abraded. Also, non-spherical seeds, *e.g.* pearl millet grains, tend to abrade away much more quickly at some points than at others.

When hard white sorghum grains, uncontaminated with seeds with a red testa, are decorticated in an abrasive decorticator, any pericarp left on the grain is hard to see and when the pearled grain is milled, the presence of pericarp goes largely unnoticed. However, the ability of abrasive decorticators to produce an adequately white product falls sharply with increasing levels of contamination from seeds with a coloured seed-coat. When the contaminating seeds have a red testa (which is deeply coloured and is practically the last layer to be abraded away) a decorticator's ability to produce an acceptably white product falls even more sharply. The problem is compounded by the fact that many contaminating seeds are comparatively soft and their exposed endosperm is ground away quickly. As a result, milling yields often fall to unacceptably low levels.

Decorticators produce what is visually a very acceptable product in a good yield from grain well suited to abrasive decortication. However, if the grain to be ground is not always going to consist of a very high proportion of hard, white, spherical seeds of fairly regular size, a very careful analysis of the economics of operating an abrasive decorticator should be made on the basis of recovery rates derived from trial runs.

Even though decorticators are well suited to small-scale operations, these machines have often proved to be too large for the system into which they were introduced. In many cases they have been introduced less successfully than originally hoped, either because of a lack of supplies of the high-quality grain that is needed for them to work properly or because of insufficient local demand for the product. Very small units are likely to be run less efficiently than larger ones.

Most decorticators are based on a prototype put out by the Prairie Regional Laboratory (PRL) in Canada. This type of decorticator has the enormous advantages of being relatively inexpensive to install and relatively simple to maintain and operate. Bassey and Schmidt (1989) described the development of this type of decorticator and its use in Africa. More recently it has been introduced in India.

In 1976, a prototype decorticator was established in Maiduguri, Nigeria. A larger unit to process 5 to 10 tonnes of sorghum per day was installed at Pitsane in southern Botswana in 1978 but the demand for the product was inadequate to keep the equipment running at full capacity. The Centre national de recherches agronomiques (CNRA) in Bambey, Senegal, began to use a PRL decorticator to decorticate sorghum and millet in 1979. The capacity of this decorticator also exceeded the demand for the product.

FAO supplied the Food Research Centre (FRC) in the Sudan with a pilot plant including a decorticator manufactured in Germany after FRC had compared decorticators made by several different manufacturers. FRC is currently decorticating white sorghum for a local urban market. The centre has also produced pearled sorghum as a substitute for rice (Bad), Perten, 1983); although the product has to be cooked much longer than rice, it was well accepted. Of the five most popular varieties of sorghum grown in the Sudan, two (Feterita and Mayo) are unsuitable for abrasive decortication.

James and Nyambati (1987) described the industrial preparation of pearled brown and white sorghum in Kenya using a decorticator that could mill sorghum in batches or continuously, but they found it was difficult to obtain sufficient sorghum suitable for processing. The product was sold at 60 per cent of the price of rice and consumer acceptance was very good. Flour was also produced from the pearled grain.

Various modifications have been made to the PRL design to suit specific conditions. A variant of the PRL decorticator was developed in the early 1980s by Palyi and tested in Canada. The Palyi-Hanson BR 001-2 can mill 3 tonnes per hour. Under local management in the Gambia a PRL decorticator processed 50 tonnes of pearl millet over a one-year test period, after which modifications were made to the design. In 1986 the Rural Industrial Innovation Centre (RIIC) introduced a modification that enabled the machine to handle small quantities of grain (Bassey and Schmidt, 1989). By 1989, about 35 RIIC decorticators had been installed in Botswana, but for one reason or another, not all of these machines are still being used for milling sorghum or millet. In turn, local agencies in some of the main countries to which the RIIC design has been exported (*e.g.* Zimbabwe, Senegal) have deemed it necessary to modify the RIIC design for improved operation for local grain.

In Zimbabwe, decorticators were placed in five rural locations for evaluation. A local research group, Environment Development Activities, produced a modified version that can process one tonne of grain in eight hours. In Senegal, a local modification was evaluated in ten villages. Decorticators based on a second local design (called the mini-SISMAR/ISRA), which can mill about 600 kg of grain in eight hours, were then introduced.

Equipment of RIIC design was introduced at Morogoro, United Republic of Tanzania, in 1982. Although the first unit was unsuccessful, four pilot systems were established locally for evaluation. In 1982, a mill with an RIIC decorticator was established in Ethiopia, but the supplies of grain for it were inadequate because of the drought.

There has also been an intensive effort to introduce RIIC decorticators in Andhra Pradesh. Decortication improved the quality of the flour from sorghum and millets so that it could be used in new ways (Geervani and Vimala, 1993). High-yielding sorghums introduced in Mali were soft and could not be decorticated in PRL-type decorticators (Scheuring *et al.*, 1983).

A number of large decorticators have been installed around the world with capacity ranging from 1 to 2.5 tonnes per hour. Typically, they are vertical axis units with abrasive disks that have been carefully selected for the optimal degree of abrasion. The grain is first cleaned to separate sand, dust, coarse material and other impurities. An aspirator removes the abraded bran through a screen. The bran is sometimes further separated into fine bran (mostly pericarp) and a mixture of germ, broken grain and coarse bran. An l-t/hour decorticator manufactured in Switzerland was run for several years in Zimbabwe, preparing coarse sorghum flour that was introduced into a wheat flour mill. A 2.5-l/hour unit manufactured in Germany was installed in the Sudan. Other large units are reportedly in operation in Nigeria. As with small units, high-quality sorghum is needed to produce an acceptably white product in these larger decorticators. Sufficient quantities of high-quality sorghum to keep large mills running at full capacity are not often available.

Rubbing Techniques

Munck, Bach Knudsen and Axtell (1982) described a new industrial milling process developed in Denmark, which does not involve abrasive milling. Decortication is achieved by a steel rotor rotating the grain mass within a generally cylindrical chamber. When the grain is properly tempered, the pericarp is rubbed off by the movement of one seed against another. However, when the grain is too dry, as was the case in a factory in the Sudan, abrasion of the internal components of the mill becomes severe. The hulls and the endosperm fragments are separated in a cyclone and the endosperm particles are milled in a proprietary mill. These units have a capacity of 2 tonnes of sorghum per hour. The system was reported to yield 80 per cent flour with whiteness comparable to traditional milling, but to do this it requires grain with specifications similar to those required for efficient abrasive decortication.

Roller Mills

Most wheat is milled in a type of mill called a roller mill. Roller mills are the most efficient mills for separating the constituents of cereals. Two types of rollers are used: rollers with axial grooves, which cut the endosperm from the pericarp (effectively cutting it away from the inside) and smooth rollers, which progressively crush the endosperm pieces into finer and finer flour. Normally the grain is passed through a number of roller mills, often 20 or more. Wheat milling technology is suitable for milling large quantities of grain, but it requires a large investment and experience in operating and maintaining the equipment. For all these reasons, it is therefore not suitable for milling sorghum and millets in very small-scale operations. However, roller mills are very efficient in separating the edible portion of cereals from the offal and can do so with sorghum and millets regardless of the physical characteristics of the grain; it does not matter if the grain is soft, coloured or broken. Roller milling might therefore have a place where high-quality products are required from comparatively large quantities of grain of poor or indifferent quality, particularly where there is spare capacity in an existing wheat mill.

To withstand the stresses of roller milling, the pericarp of sorghum and millets has to be much moister than that of wheat. Early efforts to roller-mill sorghum and millets always ended in failure because the grain was dry when it was milled. It would shatter, the pericarp breaking into small pieces that were too brittle to allow separation of the endosperm. Using conventional tempering techniques, Perten (1983) was unable to achieve efficient separation of the offal of either sorghum or millets from the endosperm. He concluded that sorghum and millets are more difficult to grind than wheat and those they produce coarser and much darker flour which contains high levels of fat and ash.

The use of moisture levels much higher than those used for milling wheat was first reported by Abdelrahman, Hoseney and Varriano-Marston (1983) for milling pearl millet and by Cecil (1986, 1992) for milling other millets and sorghum. The term semi-wet milling was adopted for this technique. For millets, about 10 per cent water must be equilibrated in the grain for four hours before it is ready for milling; for sorghum, about 20 per cent moisture must be added and the grain conditioned for six hours. The damp material flows almost as easily as normally tempered wheat products do and no holdup problems were encountered in several hours of running 2 tonnes per hour of red sorghum in a commercial mill. In early experiments, comparatively low yields of fine flour were obtained, but subsequent work produced low-fiber, low-tannin grits from red sorghum in a commercial mill with six roller passes with a yield of 72 per cent (compared with typical wheat recovery of 70 per cent). In a laboratory mill with three milling passes, 84 per cent yield of grits was obtained from commercial white sorghum from Botswana and 83 per cent from white sorghum from Lesotho. All the grits contained very low levels of fiber.

Semi-wet milling has several advantages, including the excellent separation of the offal from the edible portion and the opportunity for using well-tested existing commercial wheat-milling equipment without the need for any changes in the set-up of the mills. White flour with practically no tannin, which tastes better, looks better and is nutritionally better than flour that contains tannin, can be produced from high-tannin coloured varieties. Mixtures of sorghum or millet varieties, soft varieties, misshapen seeds and mixtures of sorghum with other grains (including wheat) can all be milled together if necessary. Moistening the endosperm softens it to such an extent that very little energy is needed to mill it. Semi-wet milling of pearl millet, unlike abrasive decortication, may also help eliminate substances that cause goitre (Klopfenstein, Leipold and Cecil, 1991).

Redundant or underutilized wheat mills can be used with minimal additions and the mill can be reverted to milling wheat within a few minutes. Alternatively, any type of sorghum can be milled together with wheat. For a period of about five days, 0.6 tonnes of red sorghum and 14 tonnes of wheat per hour were milled together without difficulty in a commercial mill in Zimbabwe. Semi-wet milling has some disadvantages. Although it would not be difficult or very expensive in a commercial system to dry the products of semi-wet milling, they are usually too damp for long-term storage. In semi-wet milling, microbiological growth might be more vigorous than in conventional milling of wheat, but reasonable attention to hygiene will minimize this problem. Semi-wet milling is not suitable for very small operations. Finally, although it has been shown that sorghum can be milled semi-wet without any difficulty in commercial equipment, the technique has not yet been proved over an extended period of operation.

Size Reduction

Many mills can be used to reduce the size of the particles obtained by decortication, but the type that is usually used (and is also probably the simplest to use and the cheapest to install) is the hammer mill. Hammer mills are available in all sizes. They consist of blunt blades rotating rapidly in an enclosed cylinder with an outlet covered by a screen. The size of the holes in the screen determines the size of the particles of flour, but small holes will reduce the throughput of the mill and if they are too small overheating may result.

If roller mills are used for separating the endosperm from the offal, the particle size is usually reduced in roller mills with smooth rollers.

Processing of Millets mostly based on Grains and their Structure

Kernels of sorghum and millets show considerable diversity in colour, shape, size and certain anatomical components (Table 1.12). The basic kernel structure is similar in sorghum and different millets. The principal anatomical components are

Table 1.12: Structural Features of Kernels of Sorghum and some Millet

Grain	*Type*	*Shape*	*Colour*	*1000-kenel weight (g)*
Sorghum	Caryopsis	Spherical	White, yellow, red, brown	25-30
Pearl millet	Caryopsis	Ovoid, hexagonal, globose	Grey, white, yellow, brown, purple	2.5-14
Finger millet	Utricle	Globose	Yellow, white, red, brown, violet	2.6
Proso millet	Utricle			4.7-7.2
Foxtail millet	Utricle			1.86

Grain	*Seed-coat*			*Alcurone*	
	Number of Layers	*Pigmented*	*Thickness (pm)*	*Number of Layers*	*Cell Size (pm)*
Sorghum	1	Sometimes	0.4	1	
Pearl millet	1	Sometimes	0.4	1	16-30 x 5-15
Finger millet	5	Yes	10.8-24.2	1	18 x 7.6
Proso millet	1	No	0.2-0.4	1	12 x 6
Foxtail millet	1			1	

Grain	*Starch Granules*				*Protein Bodies*		
	Diameter (μm)	*Peripheral Zone (μm)*	*Corneous Zone (μm)*	*Floury Zone (μm)*	*Type*	*Size (μm)*	*Location*
Sorghum	20-30				Simple	0.3-3	All areas
Pearl millet	10-12	6.4	6.4	7.6	Simple	0.6-0.7	All areas
Finger millet	3-21	8-16.5	3-19	11-21	Simple/ compound	2.0	Peripheral/ corneous
Proso millet	2-10	3.9	4.1	4.1	Simple	0.5-1.7	Peripheral
Foxtail millet	10						

Grain	*Germ*	
	Size (μm)	*Endosperm : Germ*
Ratio		
Sorghum		8.4:1
Pearl millet	1 420 x 620	4.5:1
Finger millet	980 x 270	11:1
Proso millet	1 100 x 310	12:1
Foxtail millet		12:1

pericarp, germ or embryo and endosperm. In finger, prove and foxtail millets the pericarp is like a sack, loosely attached to the endosperm at only one point. In these utricle-type kernels the pericarp easily breaks away, leaving the seed-coat or testa to protect the inner endosperm. The kernels of sorghum and pearl millet are of the caryopsis type, in which the pericarp is completely fused to the endosperm. The relative distribution of the three main kernel components varies. In the sorghum kernel the distribution by weight is pericarp 6 percent, endosperm 84 percent and germ 10 percent (Hubbard, Hall and Earle, 1950). In pearl millet, it is pericarp 8.4 percent, endosperm 75 percent and germ 16.5 percent (Abdelrahman, Hoseney and Varriano-Marston, 1984). The ratio of endosperm to germ in pearl millet is 4.5:1, while in the sorghum kernel it is 8.4:1. In finger and prove millets the germ is very small and therefore the endosperm-togerm ratio, 11:1 to 12:1, is much higher than in sorghum.

Pericarp

Pericarp is the outermost structural component of the caryopsis and is composed of three sublayers, namely epicarp, mesocarp and endocarp. The epicarp is further divided into epidermis and hypodermic. In the sorghum caryopsis, the epidermis is composed of thick, elongated, rectangular cells which have a coating of cutin on the outer surface. Often a pigment is present in the epidermis. The hypodermic is composed of slightly smaller cells than the epidermis and is one to three cell layers in thickness. The mesocarp, the middle part, is the thickest layer of the sorghum pericarp, but its thickness varies widely among genotypes. Mould resistance in sorghum is associated with thin mesocarp. Grains with thick mesocarp on a hard endosperm are preferred for dehulling by hand-pounding. The endocarp, the innermost sublayer of the pericarp, consists of cross cells and a layer of tube cells which transport moisture into the kernel. During dry milling of sorghum, the breakage occurs at the cross and tube cell layers.

The pericarp of the pearl millet caryopsis consists of an epicarp with one or two cell layers, a mesocarp that varies in thickness because of genetic factors and an endocarp made up of cross and tube cells. The mesocarp layer of pearl millet does not contain starch granules; these are found only in sorghum mesocarp. During decortication or milling, the pericarp of pearl millet breaks at the cross and tube cell layers and fragments of endocarp may remain with the endosperm.

Seed-Coat or Testa

Just underneath the endocarp is the testa layer or seed-coat. In some sorghum genotypes the testa is highly pigmented. The presence of pigment and the colour are a genetic character. The thickness of the testa layer is not uniform. It is thick near the crown area of the kernel and thin near the embryo portion. In some genotypes there is a partial testa, while in others it is not apparent or is absent. In pearl millet the testa layer is thin and sometimes pigmented. In other millets the testa is always pigmented and is only a single layer thick. In finger millet the testa is very thick, with five cell layers and is also pigmented.

Endosperm

The largest component of the cereal kernel is the endosperm, which is a major storage tissue. It is composed of an aleurone layer and peripheral corneous and floury zones. In all the millets and sorghum, the aleurone layer is a single layer of cells which lies just below the seed-coat or testa. The aleurone cells are rich in minerals, B-complex vitamins and oil and contain some hydrolyzing enzymes. The peripheral endosperm is distinguished by long rectangular cells which are densely packed and contain starch granules and protein bodies enmeshed in the protein matrix. The starch in these cells is therefore not easily available for enzyme digestion, unless the protein associated with it is also reduced (Chandrashekar and Kirleis, 1988). The matrix protein in general is alkalisoluble glutelin and the protein bodies are alcohol-soluble prolamins which account for the largest proportion of total protein in the kernel.

The protein bodies in the endosperm of sorghum and millets are spherical and differ in size among species and also within the endosperm of a single kernel. In sorghum the number of protein bodies decreases as the starch content increases from the peripheral zone to the central core where the floury endosperm is located. In pearl millet the protein bodies are more numerous in the floury than in the corneous zone. Adams, Novellie and Liebenberg (1976) have reported the presence of several enzymes, *e.g.* protease, 3-glucosidases, 3galactosidase and phosphatases, in the protein bodies of sorghum. The protein bodies of sorghum, pearl millet and finger millet also contain phosphorus, calcium, potassium and magnesium.

The starch granules of corneous endosperm are polyhedral and differ in size in different millet species. In floury endosperm the starch granules are spherical and bigger than the starch granules of the corneous zone. The starch in the floury zone is more amenable to enzyme digestion. In pearl and finger millets, the starch granules of the floury endosperm are spherical and big. The starch in pearl millet is hydrolyzed more slowly than that of sorghum by hog pancreatic amylase (Sullies and Rooney, 1977).

The proportions of corneous and floury endosperm determine the texture of the millet kernel. In soft-textured kernels there is more floury than corneous endosperm. In hard-textured kernels, on the other hand, there is more densely packed corneous endosperm than floury endosperm. Foxtail millet contains very little floury endosperm and is of a hard, corneous texture. Finger and prove millet kernels, with the endosperm evenly divided between the corneous and floury zones, are of intermediate texture. In pearl millet and sorghum the kernel texture varies widely, from all floury, very soft endosperm to all corneous, very hard or vitreous endosperm.

Grain texture is one of the most important determinants of the processing and food quality of sorghum and millets (Rooney, Kirleis and Murty, 1986). Hardendosperm sorghum when decorticated gives fewer brokers and more full grains than softer-endosperm sorghum (Desikachar, 1982). In dry milling, the flour yield is higher in corneous than in soft floury types. On the other hand, in wet milling the starch yield is higher in soft-endosperm genotypes. In the preparation of thick porridge, varieties with a higher proportion of vitreous endosperm are preferred. Such varieties are also suitable for popping (Chandrashekar and Desikachar, 1986; Murty, Patil and House, 1982). For preparation of bread, fermented or unfermented, the flour of soft-endosperm sorghum is highly preferred (Rooney, Kirleis and Murty, 1986).

Germ

The embryonic axis and the scutellum are the two major parts of the germ. The scutellum is a storage tissue rich in lipids, protein, enzymes and minerals. In pearl millet the ratio of germ to endosperm is larger than in sorghum and other millet kernels. The oil in the sorghum germ is rich in polyunsaturated fatty acids and is similar to corn oil (Rooney, 1978).

Conclusion

Millets are very nutritious and it has medicinal values which is very essential for human nutrition. Therefore, millets need to be processed properly and use in daily diet. It gives various photochemical which can prevent several disorders in human. It provides dietary fibers, minerals, antioxidants and neutraceutical properties. Therefore, it is necessary to give more attention on the production and processing of millets for utilization in human diet by various processed products.

References

Abdelrahman, A., Hoseney, R.C. and Varriano-Marston, E. 1983. Milling process to produce low-fat grits from pearl millet. *Cereal Chem.* 60: 189-191.

Adams, C.A., Novellie, L. and Liebenberg, N.V.D.W. 1976. Biochemical properties and ultra structure of protein bodies isolated from selected cereals. *Cereal Chem.*, 53: 1-12.

Aluko, R.E and Olugbemi, L.B. 1989. Sorghum as a raw material in the baking iindustry. Paper presented at the Symposium on the Current Status and Potential of Industrial Uses of Sorghum in Nigeria, Kano. Nigéria, 4-6 décembre.

Andhra Pradesh Agricultural University. 1991. Rural food enterprises for production of low cost supplementury foods using millets and legumes. Terminal report of the IDRC-funded project. Hyderabad. Inde. 135 p.

Bad, S.M. and Hoseney, R.C. 1976. Use of sorghum and pearl millet flours in cockies. *Cereal Chem.*, 53: 733-738.

Badi, S.M. and Hoseney, R.C. 1977. Use of pearl millet and sorghum flours in bread and cockies. In D.A.V. Dendy, éd. *Proceedings of symposiumon sorghum and millets for human food*, Vienne, 1 l-12mai 1976, p. 37-39. Londres, Institut des produits tropicaux.

Baghel, R.P.S., Netke, S.P. and Bajpai, L.D. 1985. Nutritive value of kangni. *Poult. Guide* 22: 28-29.

Bassey, M.W. and Schmidt, O.G. 1989. Ahrasive-disk dehullers in Africa -from research to dissemination Ottawa, Canada, Centre de recherche pour le développement international. 97 p.

Block, R.J. and Mitchell, M.M. 1946. The correlation of the aminoacid composition of proteins with their nutritive value. *Nutr. Abstr. Rev.* 16: 249-278.

Bureng, P.L., Badi, S.E.N.M. and Monawar, L.Y. 1987. *Hullu-murr production.* Khartoum. Soudan, Food Research Centre. 10 p.

Burleson, C.A., Cowley, W.R. and Otey, G. 1956. Effect of nitrogen fertilization on yield and protein content of grain sorghum in the lower Rio Grande valley of Texas. *Agron. J.*, 48: 524-525.

Casey, P. and Lorenz, K. 1977. Millet - functional and nutritional properties. Baker's Dig., 51: 45-51.

Casier, J.P.J., de Paepe, G., Willems, H., Goffings, G. and Noppen, H. 1977. Bread from starchy tropical crops. II. Bread production from pure millet and sorghum flours, using cereal endosperm-cellwall-pentosan as a universal baking factor. In D.A.V. Dendy, éd. Proceedings of a symposium on sorghum and millets for human food, Vienne, 11 - 12 mai 1976, p. 127- 131. Londres, Institut des produits tropicaux.

Cecil,.*I.E.* 1986. Roller milling of sorghum and millet grain using a semi-wet process. Londres, Tropical Development Research Institute. 25 p.

Cecil, *I.E.* 1992. Semi-wet milling of red sorghum. In M.l. Gomez, L.R. House, L.W. Rooney and D.A.V. Dendy, (Eds.) *Utilisation of sorghum and millets.* p. 23-26. Patancheru, Inde, ICRISAT

Chandrashekar, A. and Desikachar, H.S.R. 1986 Sorghum quality studies Part 11. Suitability for making dumpling (mudde). *J. Food Sci. Technol.*, 23: 7- 10.

Chandrashekar, A. and Kirleis, A.W. 1988 Influence of protein on starch gelatinization in sorghum. *Cereal Chem.* 65: 457-462.

Chung, O.K. and Pomeranz, Y. 1985 Amino acids in cereal proteins and protein fractions. In J.W Finley and D.T Hopkins, éds *Digestibility and Amino Acid avaiability in Cereals and Oilseeds* p. 65- 107 St Paul, Minnesota, Etats-Unis, American Association of Cereal Chemists.

Crabtree, J. and Dendy, D.A.V. 1979. Comilling of four species of millet with wheat to produce composite flours. *Cereal Foods World*, 24: 103- 107.

Crawford, Gary W. (1983). *Paleoethnobotany of the Kameda Peninsula.* Ann Arbor: Museum of Anthropology, University of Michigan. ISBN 0-932206-95-6.

Crawford, Gary W. (1992). "Prehistoric Plant Domestication in East Asia". In Cowan C.W., Watson P.J. *The Origins of Agriculture: An International Perspective.* Washington: Smithsonian Institution Press. pp. 117–132. ISBN 0-87474-990-5.

Crawford, Gary W. and Gyoung-Ah Lee (2003). "Agricultural Origins in the Korean Peninsula". *Antiquity* 77 (295): 87–95. Doi: 10.1017/s0003598x00061378.

Dada, I, O. and Dendy, D.A.V. 1988. Cyanide content of germinated cereals and influence of processing techniques. In D. Alnwick, S. Moses and O.G. Schmidt, (Eds.) *Improving young child feeding in eastern and southern Africa* p. 359-365. Nairobi, Kenya, Centre de recherche pours le développement international.

Dassenko, S. 1980. Effect of milling, fermentation and cooking on nutritive value of pearl millet (*Pennisetum americanum* (L.) Leeke). Thèse de doctorat. Kansas State University, Manhattan, Kansas, Etats-Unis.

Dendy, D.A.V. 1992. Composite flour - past, present and the future: a review with special emphasis on the place of composite flour in the semi-arid zones. In M.l. Gomez, L.R. House, L.W. Rooney and D.A.V. Dendy, (Eds.) *Utilizution of songhum millets*, p. 67-73. Patancheru, Inde, ICRISAT.

Deosthale, Y.G., Nagarajan, V. and Visweswar Rao, K. 1972. Some factors influencing the nutrient composition of sorghum grain. *Indian J. Agric. Sci.* 42: 100-108.

Deosthale, Y.G., Visweswara Rao, K., Nagarajan, V. and Pant, K.C. 1971. Varietal differences in protein and amino acids of grain bajra (*Pennisetum typhoides*). *Indian J. Nutr. Diet.* 8: 301-308.

Desikachar, H.S.R. 1975. Processing of maize, sorghum and millets for food uses. *J. Sci. Ind. Res.* 34: 231 -237.

Desikachar, H.S.R. 1982. Pearling and milling studies on sorghum. In L.W. Rooney and D.S. Murty, Eds. *Proceedings of the International Symposium on Sorghum Grain Quality*, Hyderabad, Inde, 28-31 October 1981, p. 194-199. Patancheru, Inde, ICRISAT.

Doraiswamy, T.R., Singh, N. and Daniel, V.A. 1969. Effects of supplementing ragi (Eleusine coracana) diets with lysine or leaf protein on the growth and nitrogen metabolism of children. *Br. J. Nutr.* 23: 737-743.

Dublish, R.K., Chauhan, G.S. and Bains, G.S. 1988. Nutritional quality of extruded rice, ragi and defatted soy flour blends. *J. Food Sci. Technol.* 25: 35-8.

FAO. 1982. The potential for industrial processing of sorghum for baking and allied food *industries in Africa. Report of the Regional Workshop on Composite Flours*, Shambat, Soudan, 7 décembre 1981. Rome. 32 p.

Ganapathy, N.S., Chitra, R.G. and Gokhale, S.K. 1957. The effect of the protein of Italian millet (*Setaia italica*) on nitrogen retention in albino rats. *Indian J. Med. Res.* 45: 395-399.

Geervani, P. and Eggum, B.O. 1989. Nutrient composition and protein quality of minor millets. *Plant Foods Hum Nutr.* 39: 201-208.

Geervani, P. and Vimala, V. 1993. Operational research on development of sorghum food enterprises for alternative uses and supplementary feeding. Hyderabad, Inde, College of Home Science, Andhra Pradesh Agricultural University.

Hemanalini, G., Umapathy, K.P., Rao, J.R. and Saraswathi, G.1980. Nutritional evaluation of sprouted ragi. *Nun. Rep. Int.*, 22: 271 -277.

Hoseney, R.C., Andrews, D.J. and Clark, H. 1987. Sorghum and pearl millet. In R.A. Olsen and K.J. Frey, (Eds.) *Nutritional quality of cereal grains: genelic and agronomic improvement*, p. 397-456. Madison, Wisconsin, Etats-Unis, American Society of Agronomy.

Howe, E.E. and Gilfillan, E.W. 1970. Limiting nutrients in Cereal grains for the growth of the laboratory rat. *Indian J. Nutr. Diet.* 7: 17-20.

Hubbard, J.E., Hall, H.H. and Earle, F.R. 1950. Composition of the component parts of the sorghum kernel. *Cereal Chem.* 27: 415-420.

Hulse, J.H., Laing, E.M. and Pearson, O.E. 1980. *Sorghum and the millets: their composition and nutritive value.* New York, Academic Press. 997 p.

James, A.W. and Nyambati, G.M. 1987. New food uses for sorghum in Kenya. *Approp. Technol.*, 14: 24-26.

Jones, R.W., Beckwith, A.C., Khoo, U. and Inglett, G.F. 1970. Protein composition of proso millet. *J. Agric. Food Chem.*, 18: 37-39.

Kent, N.L. 1983. Technology of cereals. Oxford, Royaume-Uni, Pergamon Press.

Klopfenstein, C.F., Leipold, H.W. and Cecil, J.E. 1991. Semiwet milling of pearl millet for reduced goitrogenicity. *Cereal Chem.* 68: 177-179.

Kulkarni, K.D., Parlikar, D. and Bhagwat, S.M. 1987. Product application studies on sorghum malt: cookies, weaning food and beer wort. In U.M. Ingle, D.N. Kulkarni and S.S. Thorat, (Eds.) *Proceedings of the National Seminar and Technology and Applications for Alternative Uses of Sorghum* p. 130- 143. Parbhani, Maharashtra, Inde, Marathwada Agricultural University.

Kuppuswamy, S., Srinivasan, M. and Subramanian, V. 1958. *Protein in foods.* Special Report Series No. 33. New Delhi, Inde, Indian Council of Medical Research.

Kurien, P.P., Swaminathan, M. and Subrahmanyan, V. 1961. The metabolism of nitrogen, calcium and phosphorus in children on a poor Indian diet based on bajra (*Pennisetum typhoideum*). *Ann. Biochem. Exp. Med.* 21: 41 -46.

Lorenz, K. and Dilsaver, W. 1980. Rheological properties and food applications of proso millet flours. *Cereal Chem.* 57: 21 -24.

Malleshi, N.G. and Desikachar, H.S.R. 1982. Formulation of a weaning food with low hot paste viscosity based on malted ragi (*Eleusine coracana*) and green gram (*Phaseolus radiatus*). *J. Food Sci. Technol.* 19: 193- 197.

Monteiro, P.V., Gopal, D.N., Virupaksha, T.K. and Ramachandra, G. 1988. Chemical composition and *in vitro* protein digestibility of Italian millet (*Setaria italica*). *Food Chem.* 29: 19-26.

Monteiro, P.V., Virupaksha, T.K. and Rajagopol Rao, D. 1982. Proteins of Italian millet: amino acid composition, solubility fractionation and electrophoresis of protein fractions. *J.Sci. Food Agric.* 33: 1072-1079.

Mosha, A.C. and Svanberg, U. 1983. Preparation of weaning foods with high nutrient density using flour of germinated cereals. *Food Nutr. Bull.*, 5: 10- 14.

Muindi, P.J. and Thomke, S. 198 1. The nutritive value for rats of high- and low-tannin sorghums treated with Magadi soda. *J. Sci. Food Agric.*, 32: 139- 145.

Mukuru, S.Z. 1992. Traditional technologies in small grain processing. In M.I. Gomez, L.R. House, L.W. Rooney and D.A.V. Dendy, (Eds.) *Utilization of sorghum and millets*, p. 47-56. Patancheru, Inde, ICRISAT.

Munck, L., Bach Knudsen, K.E. and Axtell, J.D. 1982. Milling processes and products as related to kernel morphology. In L.W. Rooney and D.S. Murty, (Eds.) *Proceedings of the International Symposium on Sorghum Grain Quality*, Hyderabad, Inde, 28-31 octobre 1981, p. 200-210. Patancheru, Inde. ICRISAT.

Murty, D.S., Patil, H.D. and House, L.R. 1982. Sankati quality evaluation of sorghum cultivars. In L.W. Rooney and D.S. Murty, Eds. *Proceedings of the International Symposium on Sorghum Grain Quality*, Hyderabad, Inde, 2831 October 1981, p. 36-38. Patancheru, Inde, ICRISAT.

Naren, A.P. and Virupaksha, T.K. 1990. Effect of sulfur deficiency on the synthesis of ocsetarin, a methionine-rich protein of Italian millet. *Cereal Chem.* 67: 136-138.

Olatunji, O., Adesina, A.A. and Koleoso, O.A.1989. Use of sorghum as compasite flour in baking Paper presented at the *Symposium on the Current Status and Potential of Industrial Uses of Sorghum in Nigeria*, Kano, Nigéria, 4-6 décembre.

Pal, A., Wagle, D.S. and Sheorain, V.S. 1976. Some enzymatic studies on bajra (*Pennisetum typhoides*) and barley (*Hordeum vulgare*) during malting. *J. Food Sci. Technol.*, 13: 75-78.

Panasiuk, O. and Bills, D.D. 1984. Cyanide content of sorghum sprouts. *J. Food Sci.*, 49: 791-793.

Patwardhan, V.N. 1961. Nutritive value of cereal and pulse proteins. In: *Progress in Meeting Protein needs of Infants and Preschool Children*, p. 201-210. Publication No. 843. Washington, DC, National Academy of Sciences, National Research Council.

Perten, H. 1972. Camposite flour studies in Senegal. Presented at a meeting on the production and marketing of composite flour bakery products and pasta goods, Bogota, Colombie, octobre.

Perten, H. 1977. Specific characteristics of millet and sorghum milling. In D.A.V. Dendy, (Ed.) *Proceedings of a Symposium on Sorghum and Millets for Human Food*, Vienne, 11 12 mai 1976, p. 47-51. Londres, Institut des produits tropicaux.

Perten, H. 1983. Practical experience in processing and use of millet and sorghum in Senegal and Sudan. *Cereal Foods World* 28: 680-683.

Pore, M.S. and Magar, N.G. 1977. Nutritive value of hybrid varieties of finger millet. *Indian J. Agric. Sci.* 47: 220-228.

Pushpamma, S., Parrish, D.B. and Deyoe, C.W. 1972. Improving protein quality of millet, sorghum and maize diets by supplementation. *Nutr. Rep. Int.* 5: 93100.

Rajalakshmi, R. and Mujumdar, N. 1966. Effect of different legume supplements on the nutritive value of maize and jowar to albino rats. *Baroda J. Nutr.*: 2: 2130.

Ramachandra, G., Virupaksha, T.K. and Shadaksharaswamy, M. 1977. Relationship between tannin levels and in vitro protein digestibility in finger millet (*Eleusine caracana* Gaertn.). *J. Agric. Food Chem.* 25: 1101 - 1104.

Rao, H. and Shurpalekar, S.R. 1976. Utilization of milo in bakery products. *J. Food Sci. Technol.* 13: 293-299.

Rao, S.A. and Mushonga, J.N. 1985. Traditional food crops in Zimbabwe. l. Finger millet. *Zimbabuwe Agric. J.*, 82: 101-104.

Rizley, N.F. and Suter, D.A. 1977. Sorghum tortillas: process and product attributes. *J. Food Sci.* 42: 1435-1438.

Rooney, L.W. 1978. Sorghum and pearl millet lipids. *Cereal Chem.* 55: 584-590.

Rooney, L.W. 1992. Wet milling nixtamalization and micronization of sorghum. In M.I. Gomez. L.R. House, L.W. Rooney and D.A.V. Dendy, (Eds.) *Utilization of sorghum and millets* p. 19-21. Patancheru, Inde, ICRISAT.

Rooney, L.W., Kirleis, A.W. and Murty, D.S. 1986. Traditional foods from sorghum: their production, evaluation and nutritional value. *Adv. Cereal Sci. Technol.* 8: 317-353.

Sawhney, S.K. and Naik, M.S. 1969. Amino acid composition of protein fractions of pearl millet and the effect of nitrogen fertilization on its proteins. *Indian J. Genet. Plant Breed.* 29: 395-406.

Scheuring, J.F., Sidibe, S., Rooney, L.W. and Earp, C.F. 1983. Sorghum pericarp thickness and its relation to decortication in a wooden mortar and pestle. *Cereal Chem.* 60: 86-89.

Seenappa, M. 1988. Sorghum and millets in East Africa with reference to their use in weaning foods. In D. Alnwick, S. Moses and O.G. Schmidt, (Eds.) *Improring young child feeding in eastern and southern Africa*, p. 39-54. Nairobi, Kenya, Centre de recherche pour le développement international.

Serna-Saldivar, S.D., McDonough, C.M. and Rooney, L.W. 1991. The millets. In K.J. Lorenz and K. Kulp, (Eds.) *Handbook of cereal science and technology*, p. 271-300. New York, Marcel Dekker.

Shrestha, K.B. 1972. Dehusking of varagu and its ulilization for edible purposes. Thèse de maîtrise. University of Mysore, Karnataka, Inde.

Singh, P., Singh, U., Eggum, B.O., Kumar, K.A. and Andrews, D.J. 1987. Nutritional evaluation of high protein genotypes of pearl millet (*Pennisetum americanum* (L.) Leeke). *J. Sci. Food Agric.*, 38: 41 -48.

Subrahmanyan, V., Narayana Rao, M., Rama Rao, G. and Swaminathan, M. 1955. The metabolism of nitrogen, calcium and phosphorus in human adults on a poor vegetarian diet containing ragi (*Eleusine coracana*). *Br. J. Nutr.*, 9: 350357.

Subramanian, V. and Jambunathan, R. 1980. Traditional methods of processing sorghum (*Sorghum bicolor* L. Moench) and pearl millet (*Pennisetum americanum* L.) grains in India. *Rep. Int. Assoc. Cereal Chem.*, 10: 115-118.

Sullins, R.D. and Rooney, L.W. 1977. Pericarp and endosperm structure of pearl millet (*Pennisetum typhoides*). In D.A.V. Dendy, Ed. *Proceedings of a sympo sium on sorghum and millets for human food Vienne,* 11 -12 mai 1976 p. 79. Londres, Institut des produits tropicaux.

Tashiro, M. and Maki, Z. 1977. An evaluation of the nutritive value of proso-millet protein [résumé en anglais]. Kyoto Furitsu Daigaku Gakujutsu Hokoku: Rigaku. Seikatsu Kagaku, 28: 23-30.

Thiam, A.A. 1981. Contribution of ITA to the development of composite flours. Paper presented at the *Regional Workshop on Composite Flours for Industrial Processing of Sorghum for Bakery and Allied Food Industries in Africa*, Institute of Tropical Agriculture, Khartoum, Soudan, 7- 12 décembre.

Thiam, A.A. and Ndoye, A. 1977. Bread from millet. *League Int. Food Educ. Newsl.* August.

Virupaksha, T.K., Ramachandra, G. and Nagaraju, D. 1975. Seed proteins of finger millet and their amino acid composition. *J. Sci. Food Agric.* 26: 1237-1246.

Vogel, S. and Graham, M., (Eds.) 1 979. *Sorghum and millet: food production and use. Report of a workshop*, Nairobi, Kenya, 4-7 juillet 1978. Ottawa, Canada, Centre de recherche pour le développement international.

Waggle, D.H. and Deyoe, C.W. 1966. Relationship between protein level and amino acid composition of sorghum grain. *Feedstuffs* 38: 1819.

Waggle, D.H., Deyoe, C.W. and Smith, F.W. 1967. Effect of nitrogen fertilization on the amino acid composition and distribution in sorghum grain. *Crop Sci.*, 7: 367-368.

Chapter 2

Barnyard Millet (*Echinochloa* spp.)

Millets are group of highly variable small-seeded grasses, widely grown around the world as cereal crops or grains for both human food and fodder. It's not just a type of bird seed. It can be consumed by humans in the form of pilafs, snacks and cookies too. It has a sweet nutty flavor. Millets are widely grown in Africa and northern china. It is gluten free and rich in calcium, fiber, protein and minerals. It has high nutritional values when compared with rice and wheat. Millets are called as miracle grains or wonder grains. It is a great food for people looking for weight loss, diabetic and cardiac patients.

Millet provides a host of nutrients, has a sweet nutty flavor and is considered to be one of the most digestible and non-allergenic grains available. It is one of the few grains that are alkalizing to the body. Millet has always been a favorite grains.

Barnyard millet (*Echinochloa* spp.) is a very widespread genus of plants in the grass family. Some of the species are known by the common names barnyard grass or cockspur grass. Some of whose members of the genus are millets grown as cereal or fodder crops. The most notable of these are Japanese millet (*E. esculenta*) in East Asia, Indian barnyard millet (*E. frumentacea*) in South Asia and burgu millet (*E. stagnina*) in West Africa. Collectively, the members of this genus are called barnyard grasses (though this may also refer to *E. crus-galli* specifically), barnyard millets or billion-dollar grasses. When not grown on purpose, these grasses may become a nuisance to farmers. In particular, common barnyard grass (*E. crus-galli*) is notorious as a weed. It is not easily suppressed with living mulches such as velvet bean (*Mucuna pruriens* var. *utilis*). Early barnyard grass (*E. oryzoides*) is a well-known example of Vavilovian mimicry: the plants have evolved to resemble rice (*Oryza*), enabling them to escape weeding more easily. Among the plant pathogens that affect this genus are the sac fungus *Cochliobolus sativus*, which has been noted on common barnyard grass and rice hoja blanca virus. Both affect many other grass species, in particular most important cereals and *Echinochloa* weeds may serve as a reservoir. The fungi *Drechslera monoceras* and *Exserohilum monoceras* have been evaluated with some success as potential bio-control agents of common barnyard grass in rice fields. More research is necessary, however, because they may not be host specific enough to be of practical use.

Health Benefits

1. Millet is alkaline and it digests easily.
2. The Hunzas who live in a remote area of the Himalayan foothills and are known for their excellent health and longevity enjoy millet as a staple in their diet.
3. Millet will hydrate your colon to keep you from being constipated.
4. Millet acts as a prebiotic feeding microflora in your inner ecosystem.
5. The serotonin in millet is calming to your moods.
6. Millet is a smart carb with lots of fiber and low simple sugars. Because of this it has a relatively low glycemic index and has been shown to produce lower blood sugar levels than wheat or rice.
7. Magnesium in millet can help reduce the effects of migraines and heart attacks.
8. Niacin (vitamin B_3) in millet can help lower cholesterol.
9. Millet consumption decreases triglycerides and C-reactive protein. Scientists in Seoul, South Korea concluded that millet may be useful in preventing cardiovascular disease.
10. All millet varieties show high antioxidant activity. A team of biochemists analyzed the antioxidant activity; all varieties showed high antioxidant activity.
11. Millet is gluten-free and non-allergenic. Great grain millets are for sensitive individuals.
12. Millet's high protein content (15 percent) makes is a substantial addition to a vegetarian diet.

How to Buy and Store

- ☆ When purchasing from bins in the bulk section, make sure the bins are covered and that there is a good product turnover ensuring it is fresh. Also, make sure that there is no moisture.
- ☆ Store millet in an airtight container in a cool, dry and dark place; it will keep for several months.

Tips for Eating or Cooking

- ☆ Instead of rice or potatoes, serve millet.
- ☆ Add millet to your favorite chopped vegetables; make a stir fry.
- ☆ Add millet to your salad (I like warm millet).

- ☆ Make breakfast porridge with cooked millet; add your favorite nuts and fruits.
- ☆ Add ground millet to bread and muffin recipes.
- ☆ The Hunzas use millet as a cereal, in soups and for making dense, whole grain bread.
- ☆ In Indian flat bread called *roti* are often made from millet flour.
- ☆ In Eastern Europe, millet is used in porridge and kasha, or is fermented into a beverage.
- ☆ In Africa it is used to make baby food and as breakfast porridge.
- ☆ In some countries it is used as a stuffing ingredient for cabbage rolls.

Food and Safety

Millet contains small amounts of goiterogenic substances that limit uptake of iodine to the thyroid. The "thyroid function inhibitors" can cause goiter when consumed in large quantities; this may explain the correlation between millet consumption and goiter incidence in developing countries where millet constitutes a significant part of the diet. Also, in some of these countries there may be a lack of sufficient dietary iodine.

Barnyard millet can produce ripe grain in 45 days from the sowing time under optimal weather conditions. Barnyard millet does not suffer from major fungal diseases and is a valuable fodder plant. Small seeds of barnyard millet are processed on groats that are used for preparation of different types of porridges. It has been variously known as barn grass and barnyard millet. Barnyard grass is distributed throughout much of North America as a weed in moist cultivated areas. The other Indian names for barnyard millet are Vari, Varai, Sanwa, Samwa and Swang.

Nutritional Value

Millets are predominantly starchy. The protein content is comparable to that of wheat and maize. Barnyard millet has the lowest carbohydrate content and energy value amongst all the varieties of Millets. The bran layers of millets are good sources of B-complex vitamins. Barnyard millet is an appropriate food for patients intolerant to gluten causing celiac disease.

Processing of Barnyard Millets into Various Products

1. Indian Barnyard Millet/Varai Pulao

Ingredients

440 gm (2 cups) barnyard millets or varai; 2 carrots, cut into stripes; 2 bell peppers, cut into cubes; 1/2 cup cabbage, shredded (optional); 1 potato, cut into cubes (optional).

Dry Masala

1 tsp cumin seeds; 6 cloves; 6 green cardamom seeds; 8 to 10 peppercorns; 1" cinnamon.

Seasoning

4 to 5 tbsp ghee or vegetable oil; 1 tsp grated fresh ginger; 1 green chili, smashed with a mortar and pestle; 1 liter (4 cups) water; 1/4 cup to 1/2 cup extra water if required; 2 1/2 tsp salt or to taste.

Method

Dry roast the millets on a low flame until they acquire a nutty flavour, about 8 to 10 minutes, while ensuring that they do not turn brown. This step is crucial to get cooked millets which are fluffy and do not stick to each other.

Prepare the Vegetables and keep aside

In a small iron skillet, put in the whole spices and roast until aromatic, about 3 to 4 minutes. Powder these spices with a mortar and pestle or a coffee grinder. In a large pan or a cocotte, put in the ghee or vegetable oil and heat over a medium flame until the ghee melts if using ghee or until the oil is hot but not smoking. Lower the heat and put in the ginger and chili, stir fry for a minute. Put in the vegetables and cook covered for about 3 to 4 minutes on low heat.

Pour in the water over the vegetables and bring to a boil. Lower the heat. Stir in the powdered garam masala and salt. Wash the millets and put them into the water, just as it comes to a boil. Squeeze in the lime juice. Cover and cook till done on the lowest heat possible, for about 15 to 20 minutes. After 15 minutes, if you find that the top layer is uncooked, then sprinkle 1/4 cup to 1/2 cup water over it, cover and cook for 5 minutes more. Serve with a cooling *raita* and a vegetable curry.

2. Barnyard Millet Upma Recipe

Ingredients

Barnyard millet - 1/2 cup; Water - 1 and 1/2 cups; Onion chopped-1; carrot, finely chopped - 2 tblsp; Beans, finely chopped – 4; Ginger, very finely chopped - 1 tsp; Green chilli – 2; Salt - as needed.

To Temper

Coconut oil/or any cooking oil - 1 tblsp; Mustard - 3/4 tsp; Urad dal - 1 tsp; Channa dal - 1 tblsp; Curry leaves - 1 sprig.

Barnyard Rice

Barnyard Chhiki

Barnyard Idali

Barnyard Laddoo

Barnyard Biscuits

Barnyard Puri

Method

Wash the millet once or twice and drain water completely. Heat a kadai with oil, temper with the items given under 'To temper' table in order. Followed by onion, green chilli and ginger. Fry till onion turns transparent. Add veggies and fry for 2 minutes in low flame. Add water and salt. Bring to boil and add the millet. Mix well. Cook covered in low flame for 8 to 10 minutes or until all water dries up.

3. Other Various Food Products also Prepared as per the Standard Procedures of Rice Products

Some of the species of barnyard millet

1. *Echinochloa brevipedicellata*
2. *Echinochloa callopus*
3. *Echinochloa chacoensis*
4. *Echinochloa colona* – shama grass, samo
5. *Echinochloa crus-galli* – common barnyard grass, cockspur grass
6. *Echinochloa crus-pavonis* – gulf cockspur grass - southwestern + south-central USA; Mexico, Honduras
7. *Echinochloa elliptica*
8. *Echinochloa esculenta* – Japanese millet
9. *Echinochloa frumentacea* – Indian barnyard millet
10. *Echinochloa glabrescens*
11. *Echinochloa haploclada*
12. *Echinochloa helodes*
13. *Echinochloa holciformis*
14. *Echinochloa inundata*
15. *Echinochloa jaliscana* - Jalisco
16. *Echinochloa jubata*
17. *Echinochloa kimberleyensis*
18. *Echinochloa lacunaria*
19. *Echinochloa macrandra*
20. *Echinochloa muricata* – rough barnyard grass, American barnyard grass - widespread in Canada, USA, northern Mexico
21. *Echinochloa obtusiflora*
22. *Echinochloa oplismenoides* - Mexico, Guatemala, Arizona
23. *Echinochloa oryzoides* – early barnyard grass
24. *Echinochloa paludigena* - Florida, Cuba, Puerto Rico
25. *Echinochloa picta*
26. *Echinochloa pithopus*
27. *Echinochloa polystachya* - from Texas, Louisiana, + West Indies to Argentina
28. *Echinochloa praestans*
29. *Echinochloa pyramidalis*
30. *Echinochloa rotundiflora*
31. *Echinochloa stagnina* – burgu millet
32. *Echinochloa telmatophila*
33. *Echinochloa turneriana*

34. *Echinochloa ugandensis*
35. *Echinochloa walteri* – Walter's barnyard grass - Quebec, Ontario, USA (Northeast, Southeast, Midwest, South-central, California); much of Latin America + West Indies

References

Barrett, S. (1983): Mimicry in Plants. *Scientific American* 257 (3): 76-83.

Biota of North America Program 2013. County distribution maps.

Caamal-Maldonado, J.A.; Jimenez, J.J.; Torres, A. and Anaya, A. (2001): The use of allelopathic legume cover and mulch species for weed control in cropping systems. *Agronomy Journal* 93 (1): 27-36.

Gould, F. W., M. A. Ali and D. E. Fairbrothers. 1972. A revision of *Echinochloa* in the United States. *American Midland Naturalist* 87(1): 36–59

Huang, S.W.; Watson, A.K.; Duan, G.F. and Yu, L.Q. (2001): Preliminary evaluation of potential pathogenic fungi as bioherbicides of barnyardgrass (*Echinochloa crus-galli*) in China. International Rice Research Institute Notes 26 (2): 36-37.

Pheng, S.; Khiev, B.; Pol, C. and Jahn, G.C. (2001): Response of two rice cultivars to the competition of *Echinochloa crus-galli*. International Rice Research Institute Notes 26 (2): 36-37.

Chapter 3

Brown Top Millet (*Urochloa* spp.)

Brown Top Millet Seed: Browntop millet is a warm season annual seed crop. Browntop Millet performs or grows best in well drained soils. Browntop Millet seed is a great choice for dove fields, duck plots, quail, turkey, pheasant, deer, food plots and exotic bird feed. Browntop millet can sprout within five days of moisture. Browntop is also used as a cover crop or nurse grass for permanent or perennial grass plantings. When planting Browntop millet as a nurse crop or grass is sure not to plant more than 5-10 pounds per acre. If it is planted to heavy it can shade out the perennial grasses. Keep the millet mowed at 4-6 inches for best results when using for cover.

Browntop Millet Food Plot Seed Quick Facts

- ☆ *Type of crop:* Browntop Millet is a warm season annual grass that is a heavy seed producer.
- ☆ *Origin:* Southeast Asia
- ☆ *Soil adaptations:* Well-drained soils of Southern US in slightly acidic soils with pH of 5.0 to 7.0. Will grow in soils of low fertility.
- ☆ *Uses:* Browntop is a leafy annual grass that grows from 2-3 feet tall producing heavy seed yields in about 50-60 days after emergence. Use for wild game food plots, pasture, hay and seed production and erosion

Brown Top Millet Pakoda

Brown Top Millet Cake

Brown Top Millet Pizza

Brown Top Millet Chhiki

Brown Top Millet Cookies

Brown Top Millet Cakes

control. Great food plot crop for game birds and may be flooded to attract ducks.

- ☆ ***Benefits:*** Very tolerant of acidic soils and low fertility. Nutritious food plot plant that establishes relatively quick.
- ☆ ***Nutrition:*** Produces good quality forage when young shatters abundant seed for game birds. 60 - 70 days to maturity.

Planting Browntop Millet Seed

- ☆ ***Plant Rate Drilled:*** 20-30 pounds of seed per acre
- ☆ ***Plant Rate Broadcast:*** 30 to 40 lb./acre or 2 lb./1000 sq. ft.
- ☆ ***Planting Time:*** May - August
- ☆ ***Planting Method:*** Well-prepared, firm seed bed 1/4 to1/2 inch deep
- ☆ ***Grazing:*** Can be grazed or cut for hay but is less productive than hybrid Pearl Millet or Sorghum Sudan grass when used for livestock grazing or hay production
- ☆ ***Food Plot Management:*** To provide food plots for game birds throughout the summer and into the fall make sure to stagger your plantings by three to four weeks.
- ☆ ***Use in Food Plots For:*** Deer, Turkey, Duck, Dove, Pheasant, Rabbit and Quail

Browntop millet seed produces fast growing millet that in turn produces large quantities of seeds. These millet seeds are use in food plots for game birds that are highly attracted to the nutritious seed. This millet seed is grown in a variety of soils and climates. Browntop Millet is one of the few types of millet that can be planted and flooded for ducks OR planted in dry areas for deer, quail, dove, turkey and other wildlife. Browntop is an early producer and can supply seeds in about 60 to 70 days.

Browntop millet can easily be over-seeded in erosion sites, harvested grain fields and added to conservation sites. It is used as a companion crop with many grassing seeds for early planting and forage. Plant in early April through August depending on the climate zone at the rate of 25-45 lb. per acre or in divided food plots alone or in combination with other crops. Brown top is a non-aggressive/ non-invasive plant.

Browntop millet is an annual grass, growing from 2 up to 5 feet tall. Browntop is used as a wildlife food plot crop, livestock summer grazing crop, for erosion control, hay production. It makes an abundance of seed which is attractive to birds such as dove and quail. When planted in combinations with corn or alone and then flooded in the fall, it is one of the best food attractants for wild ducks available.

Browntop millet is generally less productive for livestock grazing and hay production than Sudan grass or Cattail Millet but is more rigidly stemmed. Browntop has much less mass in its growth producing better quality hay. Browntop will volunteer or reseed itself somewhat if the fields are not flooded - Generally it is best to reseed each year to be sure of an adequate stand.

References

Black Bean and Millet Salad" found at http://www.ichef.com, March 23, 1999, Bumgarner, Marlene Anne, *The New Book of Whole Grains* New York, First St. Martin's Griffin, 1997

Cook It Right; http://www.phys.com/b_nutrition/02solutions/06database/grains/millet.html March 25, 1999, Crowley, Marilyn, "Not Just for the Birds", *Chatelaine*; October, 1995

Haas, Elson M., M.D., *Staying Healthy With Nutrition*, Berkley, California; Celestial Arts, 1992

Hulled Millet at http://www.glness.com/agway/millet.html, March 23, 1999

Millet at http://encarta.msn.com/index/conciseindex/25/02545000.htm, March 25, 1999

Millet at http://www.orst.edu/food-resource/g/millet.html, March 25, 1999

Millets at http://teach1cses.vt.edu/cses3444/3444lec9.html, March 28, 1999

Pearl Millet at http://www.znfu.org.zm/technical/pearl.html, March 28, 1999

Raloff, Janet, "Goiter, Do you eat millet?" *Science News*, May 3, 1986

Roehl, Evelyn, *Whole Food Facts*, Rochester, Vermont; Healing Arts Press, 1996

Strauss, Karen, "Mighty Millet: This Nutrition-packed, Mild-flavored Grain Isn't Just for the Birds" *Vegetarian Times*, February 1997

Taylor, Renée, *Hunza Health Secrets*, New Canaan, Connecticut; Pivot Health Books, 1978.

Chapter 4

Finger Millet (Ragi/Nachani) (*Eleusine coracana*)

Finger millet (*Eleusine coracana*) is an annual plant widely grown as a cereal in the arid areas of Africa and Asia. Earliest Karnataka civilisation shows it was grown in Hallur in the later Iron Age. It remains one of the main ingredients of the staple diet in Karnataka. It is commonly known as finger millet, African finger millet, red millet, caracan millet, koracan and ragi Kannada: *E. coracana* is native to the Ethiopian Highlands. It is very adaptable to higher elevations and is grown in the Himalaya up to 2,300 meters in elevation.

History

Finger millet has been cultivated in India from as far back as 4000 years ago. Now the Indian states of Karnataka, Andhra Pradesh, Tamil Nadu, Bihar and few others produce ragi. Karnataka is the top producer of Ragi and has 58 per cent share in India's export of this crop.

Cultivation

Eleusine coracana is often intercropped with legumes such as peanuts (*Arachis hypogea*), cowpeas (*Vigna sinensis*) and pigeon peas (*Cajanus cajan*), or other plants such as Niger seeds (*Guizotia abyssinica*). Although statistics on individual millet species are confused and are sometimes combined with sorghum, it is estimated that finger millet is grown on approximately 38,000 km^2. India is a major cultivator of finger millet with a total cultivated area of 15870 km^2. The state of Karnataka is the leading producer of finger millet, known as Ragi in the region, accounting for 58 per cent of India's Ragi production.

Storage

Once harvested, the seeds keep extremely well and are seldom attacked by insects or moulds. The long storage capacity makes finger millet an important crop in risk-avoidance strategies for poorer farming communities.

Nutrition

Finger millet is especially valuable as it contains the amino acid methionine, which is lacking in the diets of hundreds of millions of the poor who live on starchy staples such as cassava, plantain, polished rice, or maize meal (Table 4.1). Finger millet can be ground and cooked into cakes, puddings or porridge. The grain is made into a fermented drink (or beer) in Nepal and in many parts of Africa. The straw from finger millet is used as animal fodder. It is also used for a flavored drink in festivals.

Table 4.1.: Nutritional Value of Finger Millet per 100g

Parameter	*Content*
Protein (per cent)	7.6
Fat (per cent)	1.5
Fiber (per cent)	3.0
Carbohydrate (per cent)	88
Calcium (mg)	370
Vitamin A (mg)	0.48
Thiamine (B_1; mg)	0.33
Riboflavin (B_2; mg)	0.11
Niacin (B_3; mg)	1.2

Growing Finger Millet to Improve Nutrition

The International Crops Research Institute for the Semi-Arid Tropics (ICRISAT), a member of the CGIAR consortium, partners with farmers, governments, researchers and NGOs to help farmers grow nutritious crops, including finger millet. This helps their communities have more balanced diets and become more resilient to pests and drought. For example, the Harnessing Opportunities for Productivity Enhancement of Sorghum and Millets in Sub-Saharan Africa and South Asia (HOPE) project is increasing yields of finger millet in Tanzania by encouraging farmers to grow improved varieties. Finger millet is very high in calcium, rich in iron and fiber and has better energy content than other cereals. These characteristics make it ideal for feeding to infants and the elderly.

Preparation as Food

In India

In India, finger millet (locally called by various name Including *ragi*, *kezhvaragu* and *nachani*) is mostly grown and consumed in Karnataka, Rajasthan, Andhra

Pradesh, Tamil Nadu, Odisha, Maharashtra, Garhwal and Kumaon (Uttarakhand) and Goa. Ragi flour is made into flatbreads, including thin, leavened dosa and thicker, unleavened roti. Ragi grain is malted and the grains are ground. This ground flour is consumed mixed with milk, boiled water or yoghurt. In India, Ragi recipes are hundreds in number and even common food stuffs such as dosa, idly and laddu are made out of ragi.

Western and Konkan Region

In Goa ragi is very popular and *satva, pole* (dosa), *bhakri, ambil* (a sour porridge) are very common preparations. Nachani Ladus are common in some families. In Maharashtra, bhakri a type of flat bread is prepared using finger millet (ragi) flour. Bhakri is called *ragi rotti* in Kannada) in Northern districts of Karnataka.

In South India

In Tamil Nadu, ragi is called *kezhvaragu* or just *keppai*. Ragi is dried, powdered and boiled to form a thick mass that is allowed to cool. This is the famed *kali* or *keppai kali*. This is made into large balls to quantify the intake. It is taken with sambar or thick spicy lentil soups flavored with tamarind extracts. For children, ragi is also fed with milk and sugar (malt). It is also made in the form of pancakes with chopped onions and tomatoes. *Kezhvaragu* is used to make puttu with jaggery/sugar and adai (by making a thick paste (sweet or salt is used) and tapping it flat on a hot skillet). Apart from that, ragi has its medicinal uses for sinus and severe cold by applying boiled *kezhvaragu* flour cooled to skin bearable warmth on the forehead.

In Karnataka, ragi is generally consumed in the form of ragi mudde (balls) *ragi mudde* in Kannada). It is the staple diet of majority of Kannadigas in South Karnataka, especially in the rural areas. Mudde is prepared by cooking the ragi flour with water to achieve a dough-like consistency. This is then rolled into 'balls' of desired size and consumed with huLi, Saaru, or curries. Mudde is cut into biteable sizes using fingers, dipped in the curry preparation and swallowed wholly without biting into the pieces. Kannadigas also use ragi to make roti, idli, dosa and conjee. In the malnad region of Karnataka, the whole ragi grain is soaked and the milk is extracted to make a dessert known as "Keelsa". In Telangana, *ambali* - is consumed in the morning as a breakfast helps them sustain through the whole day and it keeps the body cool.

In Andhra Pradesh, *ragi sankati* or *ragi muddha* - ragi balls - are eaten in the morning with a chilli, onions, sambar (dish) or meat curry and helps them sustain through the whole day and it keeps the body cool. In Kerala, putu a traditional breakfast dish, is usually made with rice powder with grated coconut and steamed in a cylindrical steamer. The preparation is also made with ragi powder, which is more nutritive.

Central and Northern India

In the tribal and western hilly regions of Odisha, ragi or *mandiaa* is a staple food. The porridge and pitha made of ragi are more popular among village folk. In the Garhwal region of Uttarakhand, "koda" or *maddua* is made into thick rotis (served with ghee) and also made as a dish – *badi* - similar to halwa but without sugar. In the Kumaon region of northern India, ragi is traditionally fed to women after child birth.

In South and Far East Asia

In Nepal

In Nepal, thick dough made of millet flour is cooked and eaten by hand. Fermented millet is used to make a beer and the mash is distilled to make liquor.

In Sri Lanka

In Sri Lanka, finger millet is called *kurakkan* and is made into *kurakkan roti* - an earthy brown thick roti with coconut and *thallapa* - thick dough made of ragi by boiling it with water and some salt until like a dough ball. It is then eaten with a very spicy meat curry and is usually swallowed in small balls, rather than chewing. It is also eaten as a soup (kurrakan kendra) and as a sweet called 'Halape'.

In Vietnam

In northwest Vietnam, finger millet is used as a medicine for women at childbirth. A minority use finger millet flour to make alcohol (*bacha* alcohol is a good drink of the H'mong minority).

As Health Food

In southern parts of India, pediatricians recommend finger-millet-based food for infants of six months and above because of its high nutritional content, especially iron and calcium. Home-made ragi malt is a popular infant food.

Holy Deity Food

In Tamil Nadu, ragi is considered to be the holy food of Amman, otherwise known as "Goddess Kali". Every small or large festival of this goddess is celebrated with women making ragi porridge in the temples and distributing it to the poor and needy. This porridge is called *Koozh* - a staple diet in farming communities, alongside raw onions and green chillies.

As Beverage

Ragi malt porridge is made from finger millet which is soaked and shadow dried, then roasted and ground. This preparation is boiled in water and given to children, patients, adults, etc. This is a good substitute for milk powder-based beverages.

Finger millet (*Eleusine coracana* L.) also known as Ragi in India is one of the important cereals which occupies the highest area under cultivation among the small millets. The state of Karnataka is the largest producer of ragi in India. Ragi is a crop which can withstand severe drought conditions and can be easily grown throughout the year. Nutritionally, when ragi is used as a whole grain, it is higher in protein and minerals in comparison to all other cereals and millets. It is a remarkable source of protein, making it perfect for vegetarian diets.

Finger millet contains important amino acids *viz.*, isoleucine, leucine, methionine and phenyl alanine which are not present in other starchy meals. It has the highest amount of calcium (344 mg per cent) and potassium (408 mg per cent). Ragi is a great source of iron making it beneficial for individuals with low hemoglobin levels.

Millets also contains B vitamins, especially niacin, B_6 and folic acid. Some of the health benefits of ragi are attributed to its polyphenol and dietary fiber contents. Due to its high content of polyphenols and dietary fiber ragi exhibits anti-diabetic and antioxidant and antimicrobial properties; it protects against tumors and atherosclerosis (narrowing and hardening of blood vessels). Being low in fat and gluten free, ragi is easy to digest. It is therefore, given as first foods to babies in the form of *ragi porridge.*

- ☆ **Ragi for weight loss**: Ragi contains an amino acid, Tryptophan which reduces appetite. It has a much higher amount of dietary fiber compared to white rice and other grains. It is also a low fat cereal and most of the fats are in the unsaturated form. This makes it the best choice in grains for people trying to lose weight. The bulkiness of the fibers and the slower digestion rate makes one feel fuller on fewer calories and therefore helps prevent excess caloric consumption.
- ☆ **Ragi for bones:** Ragi is the richest source of calcium among plant foods. Calcium along with vitamin D helps to strengthen bones. It is an excellent source of natural calcium for children and aging people. It helps in development of bones in growing children and in maintenance of bone health in adults. Thus Ragi helps to keep bone problems at bay and could reduce risk of fractures too.
- ☆ **Ragi for diabetics:** Regular consumption of finger millet is known to reduce the risk of diabetes mellitus and gastrointestinal tract disorders and these properties are attributed to its high polyphenol and dietary fiber content. According to research Finger Millet based diets help diabetics as they contain higher amounts of fiber over other cereals and millets. Diets based on whole finger millet have a lower glycemic response *i.e.* lower ability to increase blood sugar level. This has been attributed to presence of certain factors in ragi flour which lower digestibility and absorption of starch.
- ☆ **Ragi for lowering high cholesterol levels:** Ragi contains amino acids Lecithin and Methionine which help in bringing down **cholesterol** levels by eliminating excess fat from Liver. Finger Millet also contains the amino acid threonine which hinders fat formation in the liver and helps to bring down the cholesterol levels. The high fiber content of ragi also helps to manage cholesterol problems.
- ☆ **Ragi in anemia:** Ragi is an excellent plant source of natural Iron. Its consumption helps in Anemia. Vitamin C increases iron absorption. Sprouted ragi develops vitamin C in the process of sprouting, therefore the iron in ragi becomes more bioavailable when consumed as sprouted ragi flour or ragi malt.

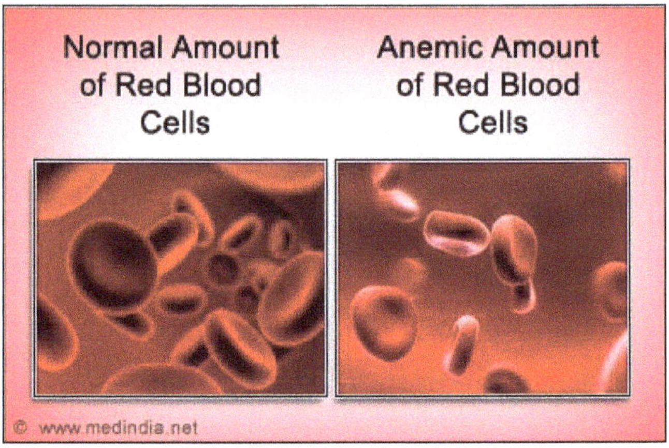

- **Ragi for relaxation:** One amazing benefit of Ragi is that it helps in relaxing the body naturally. It is found to be beneficial in conditions of anxiety, depression, migraine and insomnia.
- **Ragi as a source of essential amino acids:** Ragi is rich in Amino Acids vital for the body. Finger Millet contains amino acids namely Tryptophan, Threonine, Valine, Isoleucinc and Methionoine. Isoleucine helps in muscle repair, blood formation, bone formation and improves skin health. Valine is an essential amino acid which facilitates metabolism and repair of body tissues. Another essential amino acid, not found in most cereals, is Methionine. It is found in Ragi and therefore makes it a complete protein food unlike other cereals and millets. The amino acid Methionine is useful in various body processes, helps in eliminating fat from the body and is main provider of sulfur in body. Sulfur is essential for production of Glutathione which is the body's natural and most important antioxidant.

Ragi Food Preparations

Finger millet is normally consumed in the form of flour-based foods such as *roti* (unleavened pancake), mudde (stiff porridge/dumpling) and ambli (thin porridge). Ragi can be used to make porridge, upma, cakes and biscuits. Ragi flour is used to make various Indian preparations like dosas, idlis and ladoos.

'Puttu' is a traditional breakfast of Kerala, usually made with rice powder together with coconut and steamed in a cylindrical steamer. The same preparation is also made with ragi powder, which makes it more nutritious.

Multigrain flour preparation by combining wheat and finger millet in the ratio of 7:3 (wheat: finger millet) is one of the simplest ways of incorporating ragi in the daily diet as no Indian meal is complete without the Indian style bread or roti. In the proposed blend, though the gluten content is reduced significantly, the making of chapatti is not much affected. However, the color turns a little dark. Fortification of finger millet in chapattis is very helpful in controlling glucose levels in diabetic patients efficiently.

Malted ragi grains are ground and consumed, mixed with milk, boiled water or yogurt. In southern parts of India, it is a recommended food, by doctors, for infants of six months and above because of its high nutritional content. Home-made ragi malt is one of the most popular infant foods till date.

Malting characteristics of finger millet are superior to other millets. On malting the vitamin-C is elaborated, phosphorus availability is increased, digestion is easier and amino acids are synthesized. In south India, the malted ragi flour is extensively used in preparation of weaning foods, instant mixes and beverages.

Despite finger millet's rich nutrient profile (Table 4.2); low cost and easy availability recent studies indicate its lower consumption in general by urban Indians. Obesity has become a matter of health concern in India. Unhealthy foods have increasingly become a part of the food choices made by youth. Large populations of children in the country are malnourished and are deficient in calcium and protein. The millet ragi then could be the answer to all the above problems relating to nutrient deficiencies. Ragi is truly a wonder cereal grain and should be consciously incorporated in the diets in one way or the other.

Table 4.2 The nutritional values of finger millet per 100g.

Component	*Amount*
Total Calories (K. Calories)	378
Protein (g)	11
Fat (g)	0.4
Carbohydrate (g)	72.9
Calcium, Ca (mg)	350
Copper, Cu (mg)	0.75
Iron, Fe (mg)	3.9
Magnesium, Mg (mg)	114
Manganese, Mn (mg)	1.63
Phosphorus, P (mg)	285
Potassium, K (mg)	195
Selenium, Se (mcg)	2.7
Sodium, Na (mg)	5
Zinc, Zn (mg)	1.68
Vitamin A (IU)	0
Vitamin C (mg)	0
Vitamin B_6 (mg)	0.38
Vitamin E (mg)	0.05
Vitamin K (mcg)	0.9
Riboflavin (mg)	0.29
Thiamin (mg)	0.42

Contd...

Table 4.2–*Contd...*

Component	*Amount*
Folate, DFE (mcg)	85
Niacin (mg)	4.72
Sugars (g)	–
Fiber (g)	8.5
Cholesterol (mg)	0
Water (g)	8.67
Carotene, alpha (mg)	–
Carotene, beta (mg)	–
Choline (mg)	–
Lycopene (mg)	–

Ragi is the common name of Finger Millet in Southern part of India (and known as Nachni in the northern part of India). Along with oats this is one of the most nutritious food and one of the easy one to digest. It is rich in calcium and protein and also has good amount of iron and other minerals. It is low in fat most of which are unsaturated fats. As any other grains out there it has its shade of good and not so good aspects. Ragi or Nachni is often ground to a fine power and sold and popularly known as Ragi Flour. In this article we will put together all the details Ragi Flour's Nutritional Value and Health Benefits and what ragi has to offer to us. To start with we will look at the per cent share of content of Ragi; second, we will look at the quantity of minerals present and different amino acids present. We will then look at what are the overall health benefits of consuming Ragi or Nachni on a regular basis. Finally we will round it up with comparing Ragi flour to other grains like Wheat, Maize against key staple food ingredients.

Ragi Flour Nutritional Value

In this below section depicts the in a typical content in 100 grams of Ragi Flour (edible portion, 12 per cent moisture). Major portion of Ragi Flour is carbohydrate, followed by protein and fiber. It has the least amount of fat in them. Major portion of Ragi Flour is carbohydrate, around 80 per cent. The Fat percentage is quite less which is good. Protein and fiber contents vary but range from less to moderate. 100 grams of Ragi has roughly on an average of 336 KCal of energy in them.

Minerals

Below table shows typical amount of various minerals present in the Ragi Flour (to be referred as Ragi here after). As indicated in the table that Ragi has high amount of calcium. Calcium is one of the most important ingredients of our body and our daily need is around 350 mg. There are traces of Iron but rests of other minerals are very small.

Amino Acids

Ragi also has some good number of Essential Amino Acids (EAA) which are essential for human body (Table 4.3). Few of the key EAAs which Ragi has are Valine, Methionine, Isoleucine, Threonine and Tryptophan. Below section beiefly explains why these are essential for us.

Table 4.3: Amino Acid Content in Ragi (mg/g of protein)

Amino Acid	*Amount (mg/g of protein)*
Leucine	594
Valine	413
Phenylalanin	325
Isoleucine	275
Threonine	263
Methionine	194
Tryptophan	191
Lysine	181
Cystine	163

Valine is essential for repair of tissues, muscle coordination and metabolism and also helps maintain balance of nitrogen in the body. It also help promote mental calmness and enhance maintain mental vigor.

Isoleucine is essential for ensuring blood formation, keeping a check on blood sugar levels. It also helps heal and repair muscle tissues, bones and skin.

Threonine helps maintain protein levels in the body. It also helps formation of tooth enamel, also prevents formation of fat in the liver.

Tryptophan which act as natural relaxant and help fight anxiety, depression and insomnia. They also help in treating migraine headaches. They reduce excess appetite and help control weight gain; they also release of growth harmones.

Methionine which is sulfur based amino acid which is essential for various activities in human body. It is said to help promote growth of healthy skin and hair. The Sulfur which we get from Methionine helps produce a substance called lecithin which helps reduce cholesterol, reduce fat inside the livers and help protect kidneys. This methionine is generally lacking in staple diet of some of the vegetarians and hence regular Ragi intake will help get enough of these amino acids.

Summary of Health Benefits of Ragi Flour

It is easy to get overwhelmed by so many numbers. What we are need to know is how good Ragi is and what health benefits we get by consuming it regularly. So below is a summary of these benefits.

1. Promote Healthy Eating
2. Promote Natural Weight Loss
3. Help maintain optimum sugar levels in Blood

4. A good Vegan Gluten free source of Protein
5. Acts as a good baby food due to its rich nutrient content
6. Great whole grain source of Calcium

Ragi Vs Other grains (Wheat, Maize, Brown Rice)

Now let us look at How Ragi fairs with other whole grains when it comes to key nutrient content (Table 4.4). Below table shows how Ragi (finger millet) stands against other staple grains such as Wheat, Maize and Brown Rice.

Table 4.4: Comparison of Nutrients of Ragi with other Grains

Content	*Brown Rice*	*Wheat*	*Maize*	*Ragi*
Energy (K Cal)	362	348	358	336
Carbohydrate (per cent)	76	71	73	72.6
Protein (per cent)	7.9	11.6	9.2	7.7
Fiber (per cent)	1.0	2.0	2.8	3.6
Fat (per cent)	2.7	2.0	4.6	1.5
Calcium (mg)	33	30	26	350
Iron (mg)	1.8	3.5	2.7	3.9
Niacin (mg)	4.3	5.1	3.6	1.1
Thiamin (mg)	0.41	0.41	0.38	0.42
Riboflavin (mg)	0.04	0.1	0.2	0.19

We can make out from the chart that Protein content in Ragi is not on the higher side. It is least on Calories, Protein and Fat.

Ragi has very high amount of calcium which are way more than others. The Ragi also have traces of Iron. Other minerals are not present in high numbers.

Verdict

Ragi is a very good to be included as a part of staple diet as it lot of essential nutrients, amino acids and minerals. Add to this is the fact that these have very low amount of fat content and reasonable amount of fiber. Having said that caution should be taken to ensure that you don't take more than your body can take, especially if you have some limitation. Excess eating of Ragi could lead to Kidney stones and hence caution is required to ensure moderate eating. This article is a testament of goodness of ragi and including Ragi into diet plan starting with couple of times a week will help a lot.

Health Benefits of Finger Millet (Ragi)

Finger Millet, also known as Ragi, is cultivated in drier parts of the world - mainly in Asia and Africa. Ragi has a distinct taste and is widely used in Southern Indian and Ethiopian dishes. Ragi is a rich source of Calcium, Iron, Protein, Fiber and other minerals. The cereal has low fat content and contains mainly unsaturated fat.

It is easy to digest and does not contain gluten; people who are sensitive to gluten can easily consume Finger Millet. Ragi is considered one of the most nutritious cereals. It has different names in local languages. It is known as Ragi in Telugu and Kannada, Kelvaragu/aariyam in Tamil and Madua/Mangal in Hindi.

Finger Millet (or Ragi) Nutrition Facts

- ✰ **Finger millet/Ragi for losing weight**: Ragi contains an amino acid called Tryptophan which lowers appetite and helps in keeping weight in control. Ragi gets digested at a slower rate thus keeps one away from in taking excessive calories. Also, fibers present in ragi give a feeling of fullness thus controls excessive food consumption.
- ✰ **Finger millet/Ragi for bone health**: Ragi is rich in Calcium which helps in strengthening bones. It is an excellent source of natural calcium for growing children and aging people. Ragi consumption helps in development of bones in growing children and in maintenance of bone health in adults. Ragi keeps diseases such as osteoporosis at bay and could reduce risk of fracture.
- ✰ **Finger millet/Ragi for diabetes**: Finger Millet's phytochemicals help in slowing digestion process. This helps in controlling blood sugar level in condition of diabetes. In a study conducted in 2000, it was found that Finger Millet based diet helps diabetics as it contains higher fiber than rice and wheat. Also, the study found that diet based on whole finger millet has lower glycemic response *i.e.* lower ability to increase blood sugar level. This is due to presence of factors in ragi flour which lower digestibility and absorption of starch.
- ✰ **Finger millet/Ragi for lowering blood cholesterol**: Finger Millet contains amino acids Lecithin and Methionine which help in bringing down cholesterol level by eliminating excess fat from Liver. Finger Millet also contains Threonine amino acid which hinders fat formation in the liver, which brings cholesterol level of the body down.
- ✰ **Finger millet/Ragi for anemia**: Ragi is a very good source of natural Iron. Ragi consumption helps in condition of Anaemia.
- ✰ **Finger millet/Ragi for relaxation**: Ragi consumption helps in relaxing body naturally. It is beneficial in conditions of anxiety, depression and insomnia. Ragi is also useful for migraines.
- ✰ **Finger millet/Ragi for protein/amino acids**: Ragi is rich in Amino Acids which are vital in normal functioning of body and are essential for repairing body tissues. Finger Millet contains Tryptophan, Threonine, Valine, Isoleucine and Methionoine amino acids. Isoleucine helps in muscle repair, blood formation, contributes to bone formation and improves skin health. Valine is essential amino acid which facilitates metabolism, helps in muscle coordination and repair of body tissues. It helps in balancing nitrogen in the body. Another essential amino acid, not found in most cereals, is Methionine which is useful in various body processes, helps in eliminating fat from the body and is main provider of

sulfur in body. Sulfur is essential for production of Glutathione - body's natural antioxidant.

- ☆ **Finger millet/Ragi for other health conditions**: If consumed regularly, Ragi could help in keeping malnutrition, degenerative diseases and premature aging at bay. Green Ragi is recommended for conditions of blood pressure, liver disorders, asthma and heart weakness. Green Ragi is also recommended to lactating mothers in condition of lack of milk production.

Ragi is an extremely nutritious cereal and is very beneficial for maintaining a good health. However, its high intake could increase quantity oxalic acid in the body. Therefore, it is not advised to patients having kidney stones (Urinary Calculi). Ragi could be enjoyed in different forms and preparations. Ragi Roti, Ragi Dosa,Ragi Porridge, Ragi Upma, Ragi Cakes, Ragi Biscuits are few popular dishes of Ragi.

Nutritional Value of Finger Millet

Finger millet grain has a carbohydrate content of 81.5 per cent, protein9.8 per cent, crude fiber 4.3 per cent and mineral 2.7 per cent which is comparable to other cereals like rice, wheat, maize and millets. Its crude fiber and mineral content is remarkably higher than those of wheat (1.2 per cent fiber, 1.5 per cent minerals) and rice (0.2 per cent fiber, 0.6 per cent minerals); its protein profile is relatively well balanced; as it contains more lysine, threonine and valine than other millets. In addition to this black finger millet contains 8.71 mg/g dry weight fatty acid and 8.47 g/g dry weight protein. Finger millet varieties contain calcium (220–450) and iron (3–20 per cent) respectively. The finger millet contains important amino acids *viz.*, isoleucine (4.4 g), leucine (9.5 g), methionine (3.1 g) and phenyl alanine (5.2 g) which are deficient in other starchy meals. Millets also contains B vitamins, especially niacin, B_6 and folic acid calcium, iron, potassium, magnesium and zinc.

Finger millet is poor in protein content compared with other common cereals. Wide variability in the composition of the grain, including its protein content, was reported (Hulse, Laing and Pearson, 1980). Both genetic and environmental factors appear to have an important role in determining the protein content of finger millet (Pore and Magar, 1977; Virupaksha, Ramachandra and Nagaraju, 1975). Prolamin is the major protein fraction in finger millet. The high protein of white-grain varieties was attributed to the higher prolamin content of the grain, while the lysine content and hence the protein quality of these varieties are low (Virupaksha, Ramachandra and Nagaraju, 1975). Differences in amino acid composition in different varieties of finger millet are large and as in other cereals both the lysine content and the methionine content of the protein are inversely correlated with the protein content of the grain. The protein tractions also showed wide variation in their amino acid composition. While the albumin and globulin fraction was found to contain a good complement of essential amino acids, the prolamin fraction contained higher proportions of glutamic acid, proline, valine, isoleucine, leucine and phenylalanine but low lysine arginine and glycine. The amino acid composition of prolamin was almost the same as that of endosperm protein.

In vitro studies showed that proteins of finger millet and kodo millet were resistant to pepsin digestion unless the millet was first cooked in an autoclave for 15 minutes or boiled for at least two hours in water. Digestibility of the protein was found to be adversely affected by tannin in the grain, which was as high as 3.42 percent in some of the finger millet varieties studied (Ramachandra, Virupaksha and Shadaksharaswamy, 1977). A finger millet diet was found to be adequate to maintain a positive nitrogen balance in adults (Subrahmanyan *et al.,* 1955). The subjects also showed positive calcium and phosphorus balances and the digestibility of finger millet protein was found to be 50 percent. Supplementation of a finger millet protein diet with lysine or with leaf protein in addition to lysine significantly improved nitrogen retention in young boys. They also showed greater increase in height and weight (Doraiswamy, Singh and Daniel, 1969). The use of finger millet in child and infant feeding, however, appeared to be limited because of its poor digestibility and the large quantity required meeting energy requirements. The growth performance of growing rats fed sprouted finger millet was better than that of animals fed raw grain. However, the protein quality as judged by PER was unaltered by sprouting (Hemanalini *et al.,* 1980). Further processing of germinated finger millet grains by drying, roasting and filtering through a cloth gave a product low in fiber. Animals fed this product as a source of protein showed improved calcium retention, probably because of the low fiber content of the flour.

Malted grains of finger millet have a significantly higher saccharifying enzyme activity useful in brewing. This activity is higger than that of malted sorghum, pearl millet or maize (Rao and Mushonga, 1985). A weaning food with low hot paste viscosity and high energy density was developed in which malted finger millet from which the vegetative portion had been removed was combined with green gram. Amix of 70 parts malted finger millet grains and 30 parts green gram, combined with 10 percent skimmed milk powder, had a PER of 2.7 and NPU of 63 percent (Malleshi and Desikachar, 1982). An extruded cooked product prepared from a blend of rice (42.5 parts), finger millet (42.5 parts) and defatted soy flour (15 parts) exhibited a significant improvement in protein quality over the unprocessed blend (Dublish, Chauhan and Bains, 1988). The PER values after extrusion had increased from 1.92 to 2.41, while the trypsin inhibitory activity was reduced by about 70 to 100 percent, the tannin content was below measurable amounts and the phytin phosphorus as percentage of total phosphorus had decreased by about 4 to 13 percent. These changes obviously might have contributed to the improved protein quality. A blend of finger millet and defatted soy flour (85:15) on extrusion had a PER of 1.81 before processing and 2.23 after extrusion.

Polyphenols and Dietary Fiber

Finger millet grain has a dark brown seed coat, rich in polyphenols compared to many other cereals such as barley, rice, maize and wheat. These Phenolics are not equally distributed in the grain, but are mainly concentrated in the outer layers, namely, aleuronelayer, testa and pericarp, which form the main components of the bran fraction. The Phenoliccompounds in grains exist as free, soluble conjugates and insoluble bound forms. Major bound phenolics present in finger millets are ferulic acid and p-coumaric acid and the bound phenolics fraction account for

64–96 and 50–99 per cent of total ferulic acid and p-coumaric acid contents of millet grains, respectively. Varieties of Finger millets are also reported to contain proanthocyanidins, also known as condensed tannin. These are high-molecular weight polyphenols that consist of polymerized flavan-3-ol and/or flavan-3, 4-diol units. Proanthocyanidins are biologically active; these may lower the nutritional value and biological availability of proteins and minerals when present insufficient quantities. So far the millet varieties studied, local finger millet had the highest content(311.28±3.0) mol of catechin equivalent/g of defatted meal) followed by finger (Ravi), foxtail, little, pearl and proso millets. These values for millets were higher than those reported for barley. Finger millet grain genotypes have varied total tannin and Phenolic contents. Grains with light-colored types contain much lower total phenolics and tannins compared to brick red pigmented types. Red colored varieties having pigmented testa are known to contain much tannin content and these are located in the said tissue of the grain. They observed that brown varieties contained (1.2– 2.3 per cent) higher proportions of polyphenols than white (0.3– 0.5 per cent) varieties. Considerable differences (0.19– 3.37 per cent) in the total polyphenol contents (as catechin equivalents) among 85 Indian finger millet varieties have been reported. Tannin content was estimated in hilly region varieties was found to beless compared to base region varieties. These noticeable differences between polyphenols content in white and brown varieties could be due to the presence of the red pigments, such as anthocyanins, which are generally polymerized phenolics present in brown cultivars. Phenolics found in finger millet fall into three classes namely hydroxybenzoic acid derivatives, Hydroxycinnamic acid derivatives and Flavonoids having basic skeleton of C6–C1, C6–C3 and C6-C3–C6. The compounds indentified in this minor millet are as Gallic acid, protocatechuic acid, p-hydroxybenzoic acid, vanillic acid, syringic acid, ferulic acid, trans cinnamic acid- coumaric acid, caffeic acid, sinapic acid, quercetin and proanthocyanidins (condensed tannins). Phenolics present in finger millet grain particularly tannins in the outer layer of grains acts as a physical barrier to fungal invasion and thus imparts resistance of grain to fungal attack. Finger millet seed coat contains high polyphenolic content as compared to its whole flour extracts and these seed coat based polyphenols show high antifungal and antibacterial activity as compared to the flour extracts. The good storage stability of finger millet and its processed foods is attributed due to this phenolic content. The free radicals formed due to the oxidation of microbial membranes and cell components forms irreversible complexation with nucleophillic amino acids leading to inactivation of enzymes are major biochemical benefits of polyphenols towards the antifungal activity. Besides, their functionality loss and also the interaction of polyphenolic compounds, especially tannins with biopolymers and with complex metal ions making them unavailable to micro-organisms are some of the mechanisms involved in the inhibitory effect of phenolic compounds on microorganisms. Flavonoids and tannins present in millet seed coat are multifunctional and they act as reducing agents (free radical terminators), metal chelators and singlet oxygen quenchers. Finger millet being a potent source of antioxidants and this has high radical-scavenging activity higher than that of wheat, rice and other millets. The brown or red variety of finger millet had higher activity (94 per cent) using the DPPH method than the white variety (4 per cent), which had lower activity. Carbohydrates present in finger millet are slowly digested and assimilated than those present in other cereals. Regular consumption of finger millet having high polyphenols and dietary fiber contents are known to reduce the risk of diabetes mellitus and gastrointestinal tract disorders and this property is due to this high polyphenol and dietary fiber content. Finger millet is having high proportion of dietary fiber than many other cereals. Health benefits associated with finger millet are delayed nutrient absorption, increased faecal bulk, lowering of blood lipids, prevention of colon cancer, barrier to digestion, mobility of intestinal contents, increased faecal transit time and fermentability characteristics. Ragi also contains a functional fiber fraction known as RS, this escapes the enzymatic digestion, imparts beneficial effects by preventing several intestinal disorders. As it escapes digestion and provides fermentable carbohydrates for colonic bacteria. It also provide benefits such as the production of desirable metabolites, including short-chain fatty acids in the colon, especially butyrate, which seems to stabilize colonic cell proliferation as a preventive mechanism for colon cancer. Besides its therapeutic effects, resistant starch (RS) provides better appearance, texture and mouth feel than conventional fibers.

Decortications of Finger Millet

Decorticated finger millet is prepared by hardening the endosperm by hydrothermal processing and polishing the processed grains (Figure 4.1). The yield of the decorticated grains is of paramount importance in the millet industry. Milling yield depends on the grain moisture content and incipient moisture conditioning during milling. It was found that steaming conditions such as steaming time and steam pressure significantly influenced the milling yield. Hence, studies were undertaken to determine the influence of moisture and steaming conditions on the yield of decorticated millet. Steaming conditions were optimized through response surface methodology. The responses studied were hardness, milling yield, porosity and water uptake of hydrothermally processed millet. The studies indicated that hydrothermally processed millet with 16±1 per cent moisture content, tempered with 5 per cent added water at I stage and 4 per cent water in the II stage milling, resulted in a yield of 64.6 per cent. The relationship of milling yield, hardness and porosity of the millet was quadratic with the severity of steaming conditions, while water uptake of the steamed millet exhibited a linear relationship. Based on the regression analysis, optimum conditions estimated for steaming time and pressure were 17.5 min and 313.8 kPa, respectively. At this condition, the milling yield, water uptake, porosity and hardness values were also predicted and the values were 68.33 g/100 g, 63.43 g/100 g, 52.23 per cent and 204.01 N, respectively. The results indicate that steaming the millet at elevated pressure and temperature increases the milling yield and steaming beyond the threshold level has a detrimental effect on the yield of head grains.

The food uses of finger millet are confined to flour based products because; it has not been possible to decorticate the millet similar to other cereals (Figure 4.2). This is mainly due to the highly floury endosperm. But, it was observed that, the hydrothermal treatment to the millet hardens the endosperm texture and enables its decortication. Accordingly, the optimal conditions for the hydrothermal treatment

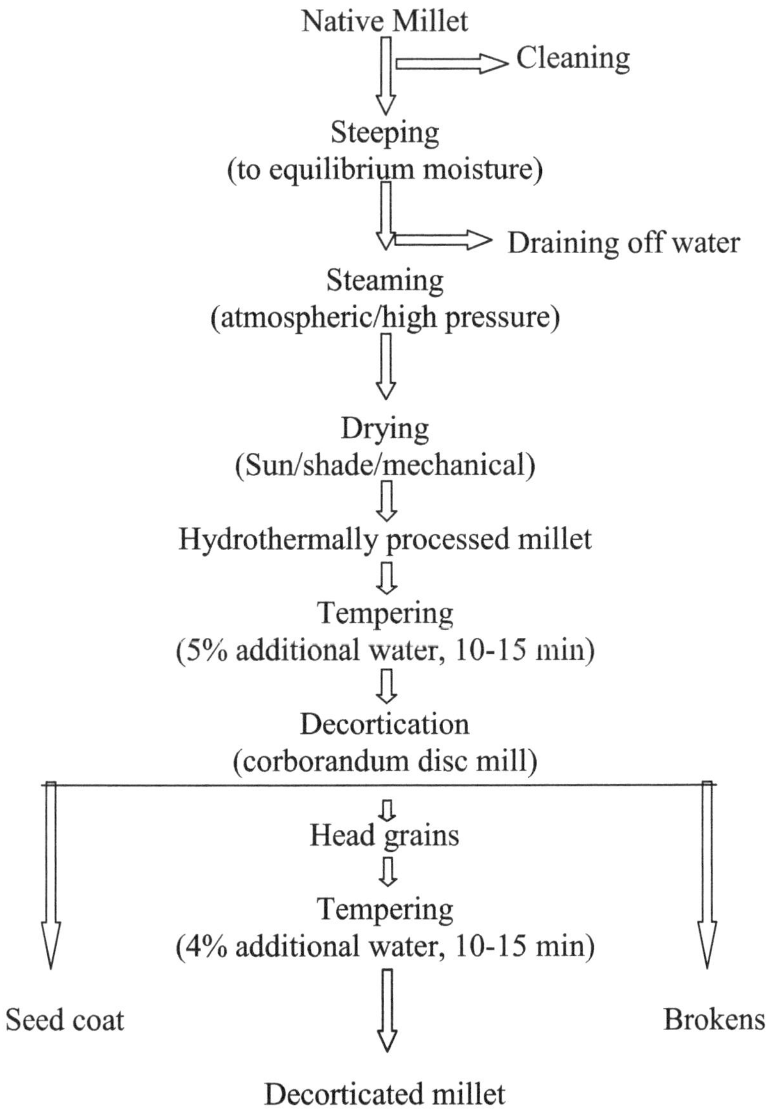

Figure 4.1: Flow Chart for the Preparation of Hydrothermally Treated and Decorticated Finger Millet.

and also for decortication of the millet were determined. Equilibrating the millet to 33 ± 2 per cent moisture content and steaming the same for about 20 min at atmospheric pressure followed by drying to 12 ± 2 per cent moisture content enhanced the hardness of the millet kernel from 1.1 ± 0.2 to 7.1 ± 0.5 kg/cm^2 and enabled its decortication. Among the various cereal pearlers and decorticators, horizontal carborundum disc mill was most suitable and incipient moist conditioning the millet improved the decortication efficiency. The decorticated millet contained 6.3 ± 0.6 g protein, 0.9 ± 0.2 per cent fat, 14.7 ± 1.8 per cent dietary fiber, 0.180 ±

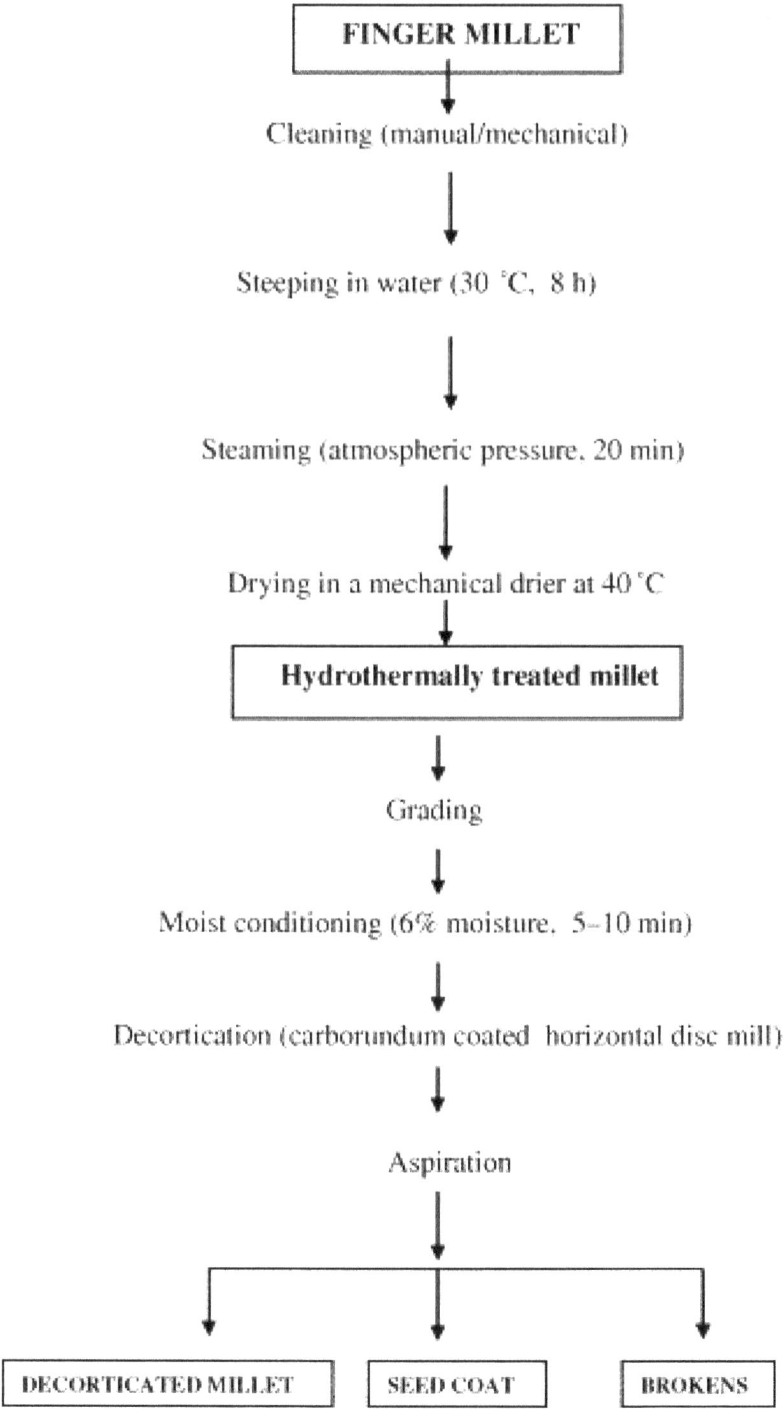

Figure 4.2: Flow Chart for the Preparation of Decorticated Finger Millet.

0.015 per cent calcium and 0.109 ± 0.01 per cent phosphorus. The polyphenols and phytate phosphorus content of the decorticated millets were lower by 74.7 per cent and 39.8 per cent, respectively, as compared to the native millet. The decorticated millet could be cooked as discrete grains similar to rice to soft edible texture within 5 min which was not possible hitherto. The pasting and the dough properties and also some of the functional characteristics of the product indicated its versatility for diversified food uses.

The floury texture of finger millet endosperm could be transformed to corneous texture by hydrothermal treatment followed by low temperature drying. This in vivo bio-transformation of the endosperm texture enables decortication of the millet in horizontal carborundum coated twin disc mill. The decorticated millet could be cooked as discrete grains similar to rice, which was not possible hitherto. Thus, it holds high promise for diversified food uses similar to rice and wheat and also for preparation of value added products.

Value Added Products from Finger Millet

It can be used in a variety of ways and is a great substitute for other grains such as rice and other starchy grains. Some of the examples of value added products and possibilities of utilizing this minor millet as one of the basic ingredients are discussed below. These products are either in practice or have been demonstrated for enhancing consumption of this particular millet.

Chapatti (*Roti*)

Wheat and finger millet in the ratio of 7:3 (wheat: finger millet) is suitable for making chapatti (roti). In this proposed blend, though the gluten content is reduced the making of flattering chapatti is not affected. Moreover, the color of the chapatti turns to slightly dark. Fortification of finger millet in chapattis not only improves the taste but also controls glucose levels in diabetic patients very efficiently. Slower digestion rate and bulkiness of the fibers makes us feel fuller on, fewer calories and therefore may help to prevent from eating excess calories. In addition Finger millet fiber content is helpful to the individuals having the problem of constipation.

Papad

Papad from finger millet flour 15-20 per cent (w/w) along with other essential ingredients such as black or green gram, rice and spices is also prepared in which it is used as base material. In this flour is first cooked in water till it is gelatinized and dough is prepared. From this dough thin sheet are prepared by rolling it and cutting into desired shapes and sizes followed by drying of these papad pieces to desired moisture content of 7- 8 per cent (db). However it gives a little dark color to the papad as the pericarp is not separated out from the starch, which again upon frying or roasting turns to lighter with good consumer acceptability.

Puffing or Popping

Puffing or popping is one of the popular traditional methods done in case of millets. The popped millet and its flour is a RTE product with pleasing texture and appealing flavour. This process improves the nutritional value by inactivating some of the anti- nutritional factors and thereby enhancing the protein and carbohydrate digestibility. The appearance, colour, taste and aroma of the processed raw material it also enhanced by this process. The flour can be used for different types of RTE food preparations depending upon the taste and likings. For puffing, the whole finger millet grain is conditioned by mixing additional water so as to reach its moisture content in the range of 18-20 per cent and tempered for about 4-6 hours under shed. The conditioned grains are puffed by agitation on the hot sand surface maintained at about 250°C for short time following HTST (high temperature and short time) process. During this process, the sugars present in the aleuronelayer react with amino acids of the millet causing Millard reaction and as a result, a pleasant and highly desired aroma is developed. Moreover, during this process, the vapour pressure of the grain increases and the moisture present in the grain turns into steam; gelatinization of the starch takes places and explodes. Since during this process as grains are dehydrated to extremely low level of moisture content, nearly 3-5 per cent, the shelf-life is enhanced. For mass production of puffing millet grain snow day's modern air puffing machines have been developed. These machines have advantage that is no risk of sticking sand particles with the product in machine during popping or puffing. Puffed finger millet grains can be converted into powder by simple grinding which can further be enriched with additional ingredients.

Malting Weaning Food

For infant feeding purpose traditionally the millet malt is utilized and also to prepare beverages either with milk of lukewarm water with the addition of sugar since old times. Finger millet having good malting characteristics is used to prepare weaning foods. Its malting is popular particularly in the areas of Karnataka and some parts of Tamil Nadu. Malting of finger millet improves its digestibility, sensory and nutritional quality as well as has pronounced effect in lowering the antinutrients. The inherent qualities of Finger millet make it superior compare to other cereals and also qualify for malting and preparation of malted foods. As it is resistant to fungal infection and elaboration of alpha and beta amylase during germination and during desirable aroma is developed during roasting/kilning makes it an ideal grain for malt foods. In order to make milk based beverage the malted weaning food is mixed with powdered sugar, milk powder or whole milk along with flavoring agents. This preparation is a good source of nutrition and suitable for all the age groups. Popularly this preparation is known as 'ragi malt' and can be used as health drink or energy drink.

Noodles Vermicelli

Changing food habits of children and teenaged groups have created a good market of noodles in India and abroad. The demand for millet noodles particularly made out of finger millet is growing due to awareness of its nutritional properties. Different combinations of Noodles are prepared, exclusively made of finger millet, finger millet and wheat in the ratio of 1:1 and finger millet blended with wheat and soy flour in the ratio of 5:4:1.Noodles are the pasta like products also known as convenience foods prepared through cold extrusion system which become hard and brittle after drying. The cooking of these noodles is very convenient and requires few minutes.

Extruded Products

Extrusion technology is a novel way of transforming ingredients into value added products. Now days extruded products prepared from different grains are very popular among the all age groups and their demand is growing, one such example is 'Kurkure', very popular among children. Flour or grits of Finger millet exhibits good extrusion characteristics. Finger millet flour like other preparations can be blended with other legume flour ingredients in appropriate proportion with further fortification of minerals and vitamins to design a balanced nutritional extruded food.

Fermented Foods

Dosa and Idli are popular fermented foods in many parts of India. These are also very common breakfast foods and evening meals in southern part of the country. For these kinds of fermented foods finger millet is widely used as one of the ingredient. Addition of this millet not only improves the taste but at the same time enriches the food value in terms of protein, calcium and fiber. Sprouted finger millet grain or the malted grains are also used for fermented foods depending on the taste and choice.

Ragi Soup

Ragi soup is prepared by mixing the ragi flour into water (one part ragi flour and 2.5 parts water). Sufficient mixing or stirring is taken so that lump formation should not occur and the mix should leave a smooth and thick body. Under medium heat this mix is then heated for 15-17 minutes or till it is cooked. Continuous stirring during heating is needed to avoid any further lump formation. After cooking, mix is removed from the heat and mixed with curd (half tea cup) and salt to taste. After leaving it for another 5 minutes it is ready to serve as warm. Further cooling is required in case of cold serving. Besides this Ragi Pakora (finger millet fritters), Ragi Vada are also prepared from this millet. Many other local preparations are in practice in addition to the above preparations making use of finger millet depending upon the local habits and choices. Modern products available in the market incorporated with finger millet are ragi health drink (baby vita), foodless, multi-grain noodle, ragi biscuit, ragi vermicelli etc. Being comparable and even superior to many cereals in terms of mineral and micronutrient contents. Its major use as food has remained only in the area where it is cultivated and to the traditional preparations. This millet has good potential of providing nutritional security to the consumers. With the advancement of post-harvest processing and value addition technologies, its consumption can be increased in urban area.

Finger millet being staple food in different parts of India and abroad is promoted as an extremely healthy food. Its nutritional and functional properties have been reviewed and found best among all cereals grains. Vitamins, minerals, fatty acids and antioxidant properties of this make its strong contribution to human nutrition.

Doughs

Masala Dough

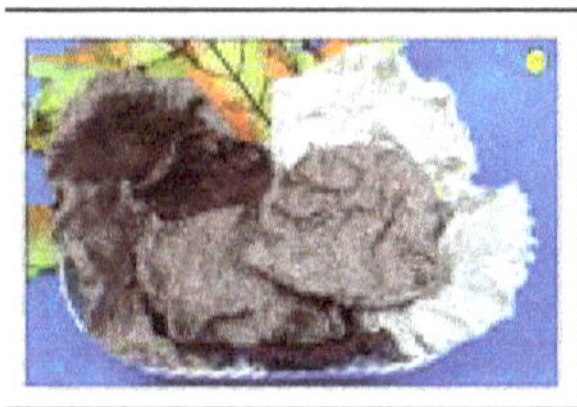

Papads

Extract

Whole Flour

Sweets

Flour

Laddoo

Soup

Pakoda

Masala Pakoda

Salty Soup

Laddoo with Sugar

Laddoo Jaggery

Seeds

Sweets

Meal

Idali **Parot and Soup** **Rice**

Sweets **Biscuits** **Cookies**

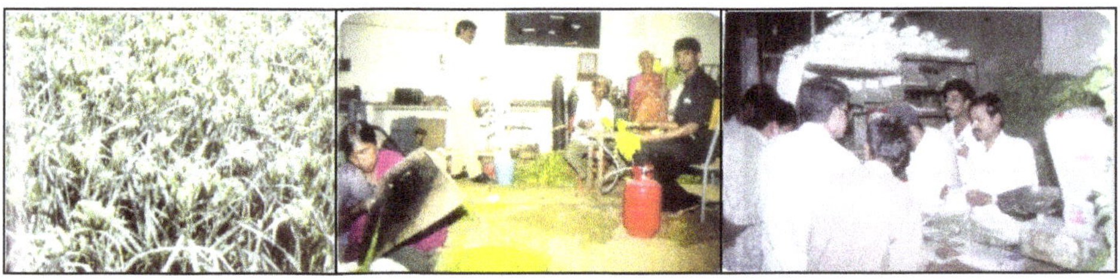

Earheads **Processing Unit** **Marketing Products**

This could be a strategic grain used to complement the diet in rural/marginal regions where energy-protein malnutrition affects most of the population. Finger millet can be used in different food formulations for making value added products due to its well-balanced protein profile and gluten free properties. Although the consumption pattern of this millet is specific and continue to remain as such, therefore its popularization in the broader range is essential and specific design of foods acceptable to the population can help in promoting the consumption of this millet.

References

AACC, 2000. Approved methods of the AACC. Moisture (method 44-19), protein (method 46-12), fat (method 30-25) ash (method 8-01), rapid visco-analyser (method 76-21), Brabender farinograph characteristics (method 54-21)

Alina Paul Bossuet, *Nourishing Communities Through Holistic Farming* http://exploreit.icrisat. org/page/sorghum/882/249. ICRISAT. Downloaded 26 January 2014.

AOAC, 1999. Official methods of the AOAC international. Calcium (method 944.03)

Asp, N. G., C.G. Johansson, H. Hallmer, M. Siljestrom. 1983. Rapid enzymatic assay of insoluble and soluble dietary fiber. *Journal of Agricultural Food Chemistry*, 31 (1983), pp. 476–482.

Association of Official Analytical Chemists, Gaithersburg, USA (1999).

Bhattacharya, K. R. S.Z. Ali. 1985. Changes in rice during parboiling and properties of parboiled rice. Y. Pomeranz (Ed.), *Advances in cereal science and technology*, Vol. 7. American Association of Cereal Chemists, St Paul, MN (1985), pp. 105–167

BSBI List 2007 (xls). Botanical Society of Britain and Ireland. Retrieved 2014-10-17.

D'Andrea, A. C., D.E. Lyons, Mitiku Haile, E.A. Butler, "Ethnoarchaeological Approaches to the Study of Prehistoric Agriculture in the Ethiopian Highlands" in Van der Veen, ed., *The Exploitation of Plant Resources in Ancient Africa*. Kluwer Academic: Plenum Publishers, New York, 1999.

Doraiswamy, T.R., Singh, N. and Daniel, V.A. 1969. Effects of supplementing ragi (*Eleusine coracana*) diets with lysine or leaf protein on the growth and nitrogen metabolism of children. *Br. J. Nutr*. 23: 737-743.

Dublish, R.K., Chauhan, G.S. and Bains, G.S. 1988. Nutritional quality of extruded rice, ragi and defatted soy flour blends. *J. Food Sci. Technol*. 25: 35-8.

Eleusine coracana - (L.) Gaertn. Plants For A Future. Retrieved 19 November 2012.

Fiske, C. H., Y. Subba Row. 1983. Phosphorus estimation. A manual of laboratory techniques, National Institute of Nutrition, Indian Council of Medical Research, Hyderabad (1983), pp. 140–142

Hemanalini, G., Umapathy, K.P., Rao, J.R. and Saraswathi, G.1980. Nutritional evaluation of sprouted ragi. *Nun: Rep. Int.*, 22: 271 -277.

Hulse, J.H., Laing, E.M. and Pearson, O.E. 1980. *Sorghum and the millets: their composition and nutritive value*. New York, Academic Press. 997 p.

Kurien, P. P., K. Joseph, M. Swaminathan, V. Subrahmanyan, V.A. Daniel. 1959. The distribution of nitrogen, calcium and phosphorus between the husk and endosperm of Ragi (*Eleusine coracana*). *Food Science* (Mysore), 8 (1959), pp. 353–355

Malleshi, N. G. and Ushakumari, S. R. (2003). A process for preparation of expanded finger millet. Indian Patent Application No. 360/NF/03.

Malleshi, N. G. and N.A. Hadimani. 1991. Nutritional and technological characteristics of small millets and preparation of value added products from them. K.W. Riley, S.C. Gupta, A. Seetharam, J.N. Mushonga (Eds.), *Advances in small millets*, Oxford and IBH Publishing Co, New Delhi, pp. 271–288.

Malleshi, N.G. and Desikachar, H.S.R. 1982. Formulation of a weaning food with low hot paste viscosity based on malted ragi (Eleusine coracana) and green gram (*Phaseolus radiatus*). *J. Food Sci. Technol.* 19: 193- 197.

Malleshi, N.G. (2002). A process for preparation of decorticated finger millet (*Eleusine coracana*). Indian Patent Application No. 242/DEL/02.

Malleshi, N.G. (2003). Decorticated finger millet (*Eleusine coracana*) and process for its preparation. US Patent No. 2003/0185951 A1.

McDonough, C. M, L.W. Rooney, C.F. Earp. 1986. Structural characteristics of *Eleusine coracana* (finger millet) using scanning electron and fluorescence microscopy. *Food Microstructure*, 5: 247–256.

Pore, M. S. and N.G. Magar. 1979. Nutrient composition of hybrid varieties of finger millet. *Indian Journal of Agricultural Sciences*, 49: 526–531

Pore, M.S. and Magar, N.G. 1977. Nutritive value of hybrid varieties of finger millet. *Indian J. Agric. Sci.* 47: 220-228.

Ragi is one of the important crops in the Indian state of Goa

Ragi. Tamilnadu.com. 8 April 2013.

Rajasthan: by Gopal K. Bhargava, Shankarlal C. Bhatt, p. 319

Ramachandra, G., Virupaksha, T.K. and Shadaksharaswamy, M. 1977. Relationship between tannin levels and *in vitro* protein digestibility in finger millet (*Eleusine caracana* Gaertn.). *J. Agric. Food Chem.* 25: 1101 - 1104.

Rao, S.A. and Mushonga, J.N. 1985. Traditional food crops in Zimbabwe. l. Finger millet. *Zimbabuwe Agric.* J., 82: 101-104.

Ravindran. G. 1991. Studies on millets: Proximate composition, mineral composition and phytate and oxalate contents. *Food Chemistry*, 39: 99–107

Serna-Saldivar, S. O., C. Clegg and L.W. Rooney. 1994. Effects of parboiling and decortication on the nutritional value of sorghum (*Sorghum bicolor* L. Moench) and pearl millet (*Pennisetum glaucum* L.). *Journal of Cereal Science*, 19: 83–89.

Subrahmanyan, V., Narayana Rao, M., Rama Rao, G. and Swaminathan, M. 1955. The metabolism of nitrogen. calcium and phosphorus in human adults on a poor vegetarian diet containing ragi (*Eleusine coracana*). *Br. J. Nutr.*, 9: 350-357.

The Plant List: A Working List of All Plant Species". Retrieved 8 January 2015.

Virupaksha, T.K., Ramachandra, G. and Nagaraju, D. 1975. Seed proteins of finger millet and their amino acid composition. *J. Sci. Food Agric.* 26: 1237-1246.

Chapter 5

Foxtail Millet (*Setaria italica*)

Foxtail millet (*Setaria italica*, formerly *Panicum italicum* L.) is the second-most widely planted species of millet and the most important in East Asia. It has the longest history of cultivation among the millets, having been grown in China since sometime in the sixth millennium BC. Other names for the species include dwarf Setaria, foxtail bristle grass, giant setaria, green foxtail, Italian millet, German millet, Chinese millet and Hungarian millet.

Foxtail millet is an annual grass with slim, vertical, leafy stems which can reach a height of 120–200 cm (3.9–6.6 ft). The seed head is a dense, hairy panicle 5–30 cm (2.0–11.8 in) long. The small seeds, around 2 mm (less than 1/8 in.) in diameter, are encased in a thin, papery hull which is easily removed in threshing. Seed colour varies greatly between varieties.

Threshing is the process of loosening the edible part of cereal grain (or other crop) from the scaly, inedible chaff that surrounds it. It is the step in grain preparation

after harvesting and before winnowing, which separates the loosened chaff from the grain. Threshing does not remove the bran from the grain. Threshing may be done by beating the grain using a flail on a threshing floor. Another traditional method of threshing is to make donkeys or oxen walk in circles on the grain on a hard surface. A modern version of this in some areas is to spread the grain on the surface of a country road so the grain may be threshed by the wheels of passing vehicles.

Hand threshing was laborious, with a bushel of wheat taking about an hour. In the late 18th century, before threshing was mechanized, about one-quarter of agricultural labor was devoted to it. Industrialization of threshing began in 1786 with the invention of the threshing machine by Scotsman Andrew Meikle. Today, in developed areas, it is now mostly done by machine, usually by a combine harvester, which harvests, threshes and winnows the grain while it is still in the field.

Cultivation

In South India, it has been a staple diet among people for a long time from the sangam period. It is popularly quoted in the old Tamil texts and is commonly associated with Lord Muruga and his consort Valli. In China, foxtail millet is the most common millet and one of the main food crops, especially among the poor in the dry northern part of that country. In Europe and North America it is planted at a moderate scale for hay and silage and to a more limited extent for birdseed. It is a warm season crop, typically planted in late spring. Harvest for hay or silage can be made in 65–70 days (typical yield is 15,000-20,000 kg/ha of green matter or 3,000-4,000 kg/ha of hay) and for grain in 75–90 days (typical yield is 800–900 kg/ha of grain). Its early maturity and efficient use of available water make it suitable for rising in dry areas.

Diseases of foxtail millet include leaf and head blast disease caused by *Magnaporthe grisea*, smut disease caused by *Ustilago crameri* and green ear caused by *Sclerospora graminicola*. The unharvested crop is also susceptible to attack by birds and rodents.

History and Domestication

The wild antecedent of foxtail millet has been securely identified as *Setaria viridis*, which is interfertile with foxtail millet; wild or weedy forms of foxtail millet also exist. Zohary and Hopf note that the primary difference between the wild and cultivated forms is "their seed dispersal biology. Wild and weedy forms shatter their seed while the cultivars retain them. The earliest evidence of the cultivation of this grain comes from the Peiligang culture of China, which also cultivated the common millet, but foxtail millet became the predominant grain only with the Yangshao culture. Foxtail millet arrived in Europe later; carbonized seeds first appear in the second millennium BC in central Europe. The earliest definite evidence for its cultivation in the Near East is at the Iron Age levels at Tille Hoyuk in Turkey, with an uncorrected radiocarbon date of about 600 BC.

Setaria viridis is a species of grass known by many common names, including green foxtail green bristle grass and wild foxtail millet. It is sometimes considered a subspecies of *Setaria italica*. It is native to Eurasia, but it is known on most continents as an introduced species and is closely related to *Setaria faberi*, a noxious weed. It is a hardy grass which grows in many types of urban, cultivated and disturbed habitat, including vacant lots, sidewalks, railroads, lawns and at the margins of fields. It is the wild antecedent of the crop foxtail millet. This is an annual grass with decumbent or erect stems growing up to a meter long and known to reach two meters or more at times. The leaf blades are up to 40 centimeters long and 2.5 wide and glabrous. The inflorescence is a dense, compact, spike like panicle up to 20 centimeters long, growing erect or sometimes nodding at the tip only. Spikelets are 1.8 - 2.2 mm long. Each is subtended by up to three stiff bristles. Its fertile lemmas are finely cross-wrinkled.

Setaria viridis is often confused with *S. faberi*, (Chinese or Giant Foxtail), which has sparse, soft hairs on the leaves and a nodding inflorescence. *Setaria viridis* is closely related to *S. italica* (Foxtail Millet), which has larger spikelets about 3 mm long and usually smooth, shiny upper lemmas. Foxtail Millet was cultivated in China by 2700 B.C. and during the Stone Age in Europe. *S. viridis* has been proposed as a model to study C4 photosynthesis and related bioenergy grasses. It has a short life cycle (6-8 weeks), is transformable and is currently being sequenced. Genetic resources are currently being developed by a number of groups. A method to break the prolonged seed dormancy has been discovered recently and all these could contribute towards making *S. viridis* a choice monocot genetic model system.

The protein in foxtail millet is also deficient in lysine its amino acid score is comparable to that of maize (Baghel, Netke and Bajpai, 1985). Monteiro, Virupaksha and Rajagopol Rao (1982) observed large variation in the grain protein content and in the distribution pattern of different solubility fractions. Prolamin constituted the major storage protein and showed a positive correlation with total protein in the grain. Evaluations of the amino acid composition of the protein fractions and of total protein in different varieties have confirmed that lysine is the first limiting amino acid, followed by tryptophan and sulphur amino acids.

With increase in grain protein, the lysine content of the protein decreased. The protein was found to be high in leucine. Naren and Virupaksha (1990) observed that the prolamin was relatively rich in the sulphur amino acid methionine and that the sulphur status of the soil affected the synthesis of prolamin in the grain. In studies on in vitro protein digestibility, 90.5 to 96.9 percent of the protein in foxtail millet was digestible by pepsin and 89.7 to 95.6 percent by papain (Monteiro *et al.*, 1988). Poor digestibility with trypsin (21.6 to 36.9 percent) was improved by prior treatment with acid. The protein quality of dehusked grain was the lowest among the minor millets tested (Geervani and Eggum, 1989). Heat treatment or lysine supplementation improved the protein quality (Geervani and Eggum, 1989). In growing rats fed 10 percent protein, the nitrogen balance improved from 19 to 31 percent when the foxtail millet diet was fortified with lysine protein digestibility and biological value were also enhanced (Ganapathy, Chitra and Gokhale, 1957). Supplementation of dehusked millet with chickpea rose the PER from 0.5 to 2.2.

Foxtail Millet Good for Diabetes Patients

The area under which minor millets is grown is shrinking year after year in Kurnool district as has been the trend elsewhere in Andhra Pradesh State. Foxtail millet (*Sataria italica*), which used to be cultivated in an area of two lakh hectare in the district two decades ago has now shrunk to 20,000 hectare. The millet, known as Korra in local parlance, remains to be the staple diet of rural people. However, after the introduction of the subsidized rice scheme in the early eighties, people switched over to rice and the demand started declining for millet. However, of late, the importance of foxtail millet was recognized as diabetic food. The millet is rich in dietary fiber (6.7 per cent), protein (11 per cent) and low in fat (four per cent).

Unlike rice, foxtail millet releases glucose steadily without affecting the metabolism of the body. The incidence of diabetes is rare among the population which consumes foxtail millet diet. According to G. Narasimha Rao, former scientist of Agriculture University, foxtail millet is a versatile crop which can be grown in any season of the year with the shortest crop cycle of 60 to 90 days. No other food crop can be grown in such a short period. Considering the importance of the millet in Kurnool district, the government has sanctioned a research centre for the district, which is located at Nandyal. So far, the centre has released new varieties of the millet such as Nallama Korra, Krishnadevaraya, Narasimharaya, Srilakshi and Suryanandi. Srilakshmi, which gave a yield of 14 quintal per acre this year, has been popular among farmers. The duration of Suryanandi was reduced to 60 days which could be accommodated in any season.

The foxtail millet grain is available only in select urban shops at present in ready-to-cook form and is priced at Rs. 28 a kg, less than the price of rice. Other millets of the group such as Arika, Sama, Ooda and ragi are vanishing gradually even from the homes of farmers.

1. Foxtail Millet Khichdi (Kichadi Healthy Lunch Ideas)

Millets are highly nutritious and non-glutinous. They are considered to be the least allergenic and most easily digestible grains available. Since millet does not contain gluten, it is a wonderful grain alternative for people who are gluten-sensitive. Millets are particularly high in minerals like iron, magnesium, phosphorous and potassium. Foxtail millet is rich in dietary fiber, protein and low in fat unlike rice, foxtail millet releases glucose steadily without affecting the metabolism of the body. The incidence of diabetes is rare among the population which consumes foxtail millet diet. Millet is a grain that should also be included on your list of heart-healthy choices because of its status as a good source of magnesium. A cup of cooked millet provides 19 per cent of the daily value for magnesium (source-whfood, Hindu). Foxtail millet is known as kangni in Hindi, tinai in Tamil, Korra in Telugu and Navane in Kannada and thina in Malayalam. Many people think that healthy food is boring and not delicious, but that is not true. Today we are going to learn how to make nutritious and delicious khichdi with foxtail millet following our easy recipe.

Foxtail Khichidi

Foxtail Upma

How to make Foxtail Millet Khichdi

Soaking time - under pressure 10 minutes and cook time - under pressure 20 minutes.

Ingredients Needed

Onion – 1; Green chilli -1-2 slit; Ginger - 1 inch piece finely chopped; Foxtail millet/tinai - 1/2 cup (50g); Moong dal split yellow variety - 1 tbsp (optional); Turmeric powder - a pinch; Ghee - 1 tsp; Salt as needed.

Vegetables and Greens

Carrot -1 medium size; Beans -3-4; Potato - small 1 or 1/2 a medium sized potato; Tomato -1/2 a tomato; Drumstick leaves – fistful (you can add peas also).

For the Seasoning

Oil - 1 tbsp; Mustard seeds - 1/2 tsp; Urad dal - 1/2 tsp; Cumin seeds -1/2 tsp; Whole black pepper - 2-3; Red chilli – 1; Hing - a pinch; Curry leaves few.

Preparation

Chop all the vegetable into medium sized pieces. Wash and soak millets along with moong dal in 1 1/2 cup of hot water for 2 -2 1/2 hours.

Method

Heat oil in a pressure pan, add mustard seeds, when it splutters, add urad dal, cumin seeds, red chillies, hing black pepper and curry leaves. Add chopped onions, ginger, green chillies and saute until onions turn pink. Then add all the vegetables, drumstick leaves, turmeric powder, salt needed and saute for 2-3 minutes. Add the soaked millet along with the water. Check for salt by tasting the water. Close the cooker and cook for 2 whistles on medium flame, then simmer for another 5 minutes and switch it off. Once the pressure subsides, open the cooker, add a tsp of ghee and mix the kichadi well. Serve hot with any raita of your choice.

2. Foxtail Millet Upma

Thinai Arisi (Foxtail Millet) - 1/2 cup; Big Onion - 1 chopped finely; Water - 3 cups; Salt - to taste.

To Temper

Oil - 1 tsp; Coconut Oil - 1 tsp; Mustard seeds - 1 tsp; Urad Dal - 3/4 tsp; Curry leaves - a sprig; Green Chilli - 1 chopped roughly.

Method:

1. Rinse it well if you feel there is any husk. Measure 1/2 cup of thinai arisi. Dry roast it for 3 min, don't let it change the color. Heat oil - add the items listed under 'to temper' let it splutter.
2. Add onions and fry till transparent, then add water and let it come to a rolling boil then add roasted thinai arisi little by little and keep stirring.
3. Once the water is absorbed fully, check if the grains have become soft then cook covered for 2-3mins until its fluffy.
4. Serve hot with sugar or any side dish of your choice.

3. Foxtail Dosa

Urad dal – 1 cup; Foxtail – 3 cups; Methi seeds – 1 tbsp; Salt. Soak the urad dal, Foxtail millet and methi seeds over night or for 8 hours. Grind it to a soft paste, add salt and let it ferment for 8-10 hours. Once the batter has been fermented, pour out the dosas.

4. Foxtail Idli

Foxtail Millet – 3 cups; Whole Urad Dal – 1 cup; Methi (Fenugreek) seeds – 1 tbsp; Salt to taste. Soak the millet and dal separately for 8-10 hours/overnight. Soad the methi seeds with the dal. Grind them separately – the dal to a fine batter and the millet batter should be slightly coarse. Add salt and let the batter ferment for 8-10 hours. Grease idli plates, pour out the batter and let it steam for 10 minutes.

Foxtail Millet Thalpit

Foxtail Millet Pancake

Foxtail Millet Sweets

Foxtail Millet Upma

Foxtail Millet Laddoo

Foxtail Millet Masala Rice

References

Baghel, R.P.S., Netke, S.P. and Bajpai, L.D. 1985. Nutritive value of kangni. *Poult. Guide* 22: 28-29.

Brutnell, T. (2010). "Setaria viridis: a model for C4 photosynthesis". *Plant Cell* 22: 2537–44. doi:10.1105/tpc.110.075309.

Daniel Zohary and Maria Hopf, *Domestication of plants in the Old World*, third edition (Oxford: University Press, 2000), p. 86-88.

Dundoo, Sangeetha Devi (August 18, 2013). "Magic of millets". *The Hindu*.

Ganapathy, N.S., Chitra, R.G. and Gokhale, S.K. 1957. The effect of the protein of Italian millet (*Setaia italica*) on nitrogen retention in albino rats. *Indian J. Med. Res*. 45: 395-399.

Geervani, P. and Eggum, B.O. 1989. Nutrient composition and protein quality of minor millets. *Plant Foods Hum Nutr*. 39: 201-208.

Jiang, Hui; Barbier, Hugues; Brutnell, Thomas (2013). "Methods for Performing Crosses in *Setaria viridis*, a New Model System for the Grasses".*Journal of Visualized Experiments* (80). doi:10.3791/50527.

Lillian M. Li (2010). *Fighting Famine in North China: State, Market and Environmental Decline, 1690s-1990s*. Stanford University Press. p. 93-94.

Monteiro, P.V., Gopal, D.N., Virupaksha, T.K. and Ramachandra, G. 1988. Chemical composition and in vitro protein digestibility of Italian millet (*Setaria italica*). *Food Chem.*. 29: 19-26.

Monteiro, P.V., Virupaksha, T.K. and Rajagopol Rao, D. 1982. Proteins of Italian millet: amino acid composition, solubility fractionation and electrophoresis of protein fractions. *J. Sci. Food Agric.*, 33: 1072-1079.

Naren, A.P. and Virupaksha, T.K. 1990. Effect of sulfur deficiency on the synthesis of ocsetarin, a methionine-rich protein of Italian millet. *Cereal Chem.* 67: 136138.

Sebastian, J *et al.* (2014). "Methods to Promote Germination of Dormant Setaria viridis Seeds". *Plos One* 9 (4): 2537 –44.doi:10.1371/journal.pone.0095109.

Chapter 6

Kodo Millet (*Paspalum scrobiculatum*)

Kodo millet (*Paspalum scrobiculatum*), is an annual grain that is grown primarily in India, but also in the Philippines, Indonesia, Vietnam, Thailand and in West Africa where it originates. It is grown as a minor crop in most of these areas, with the exception of the Deccan plateau in India where it is grown as a major food source. It is a very hardy crop that is drought tolerant and can survive on marginal soils where other crops may not survive and can supply 450–900 kg of grain per hectare Kodo millet has large potential to provide nourishing food to subsistence farmers in Africa and elsewhere.

Kodo millet is a monocot and an annual grass that grows to heights of approximately four feet. It has an inflorescence that produces 4-6 racemes that are 4 - 9 cm long. Its slender, light green leaves grow to be 20 to 40 centimeters in length. The seeds it produces are very small and ellipsoidal, being approximately 1.5 mm in width and 2 mm in length; they vary in colour from being light brown to a dark grey. Kodo millet has a shallow root system which may be ideal for intercropping.

Ethnography

Paspalum scrobiculatum var. scrobiculatum is grown in India as an important crop, while *Paspalum scrobiculatum var. commersonii* is the wild variety indigenous to Africa. The kodo millet, also known as cow grass, rice grass, ditch millet, Native Paspalum, or Indian Crown Grass originates in tropical Africa and it is estimated to have been domesticated in India 3000 years ago. The domestication process is still ongoing. In southern India, it is called kodo or kodra and it is grown as an annual. It is a minor food crop eaten in many Asian countries, primarily in India where in some regions it is extremely important. It grows wild as a perennial in the west of Africa, where it is eaten as a famine food. Often it grows as a weed in rice fields. Many farmers do not mind it, as it can be harvested as an alternative crop if their primary crop fails. In the Southern United States and Hawaii, it is considered to be a noxious weed.

Growing Conditions

Kodo millet is propagated from seed, ideally in row planting instead of broadcast sowing. Its preferred soil type is a very fertile, clay-based soil. Var. *scrobiculatum* is better suited to dried conditions than its wild counterpart, which requires approximately 800–1200 mm of water annually and is well suited to sub-humid aridity conditions. With very low competition from other plants or weeds for nutrients, it can grow well in poor-nutrient soils. However, it does best in soils supplemented with a general fertilizer. The recommended dose for optimal growth is 40 kg of nitrogen plus 20 kg of phosphorus per hectare. A case study in India's Rewa district in 1997 showed a 72 per cent increase in kodo millet grain yields as opposed to no fertilizer. It should be noted that lodging issues may accompany this. (Please see section "Other farming issues"). Kodo millet prefers full light for optimal growth, but can tolerate some partial shading. Its ideal temperature for growth is 25-27°C. It requires four months until maturity and harvesting.

Other Farming Issues

The kodo millet is prone to lodging at maturity, causing loss of grain. To prevent this, limited fertilization is recommended. While plenty of fertilizer dramatically improves yields, there is the risk of lodging accompanying vigorous growth. A good balance is applying 14–22 kg of nitrogen. Lodging also occurs due to heavy rains. Kodo millet is harvested by cutting the stalk of the grass and allowing it to dry in the sun for a day or two. It is then ground to remove the husk. Weather dependency is a major issue related to proper harvesting and storage. Additionally, threshing on roads damages the grains and husking is a very time consuming process. Kodo millets are believed by farmers to be the toughest grain to de-husk.

Stress Tolerance

The kodo millet can survive well on marginal soils; var. *scrobiculatum* requires very little water in order to grow and thus has very good drought tolerance.[18] It can be cultivated without an irrigation system. Farmyard manures provide adequate nutrients in terms of adding fertilizer, but kodo millets can still survive on low-nutrient soils. The wild variety is better suited to wetter conditions and can tolerate flooded areas and swampy ground.

Major Weeds, Pests and Diseases

Paspalum ergot is a fungal disease to which kodo millet is susceptible. Hardened masses of this fungus, called sclerotia, will grow in place of the millet grain.[20] These compact fungi growths contain a chemical compound that is poisonous to humans and livestock if consumed and potentially fatal. It causes damage to the central nervous system, causing excitability in animals and eventually loss of muscle control. If the symptoms are caught early and the animals are removed from the infected food, they have a good chance of recovery. Cleaning the seeds by winnowing them before storage may remove the fungal spores.

Consumption and Uses

In India, kodo millet is ground into flour and used to make pudding. In Africa it is cooked like rice. It is also a good choice for animal fodder for cattle, goats, pigs, sheep and poultry. In Hawaii, var. scrobiculatum is found to grow well on hillside slopes where other grasses do not flourish. It has the potential to be grown as a food source on hillside farms. It may also have potential to be used as grass ties on hillside plots to prevent soil erosion, while also providing a famine food as a secondary purpose. It has been noted that it makes a good cover crop.

Nutritional Information

Kodo millet is a nutritious grain and a good substitute to rice or wheat. The grain is composed of 11 per cent of protein, providing 9 grams/100 g consumed. It is an excellent source of fibre at 10 grams (37-38 per cent), as opposed to rice, which provides 0.2/100 g and wheat, which provides 1.2/100 g. An adequate fibre source helps combat the feeling of hunger. Kodo millet contains 66.6 g of carbohydrates and 353 kcal per 100 g of grain, comparable to other millets. It also contains 3.6 g of fat per 100 g. It provides minimal amounts of iron, at 0.5/100 mg and minimal amounts of calcium and 27/100 mg. Kodo millets also contain high amounts of polyphenols, an antioxidant compound.

Practical Information

If fertilizer is available, supplemented nitrogen and phosphorus in limited amounts can increase yields dramatically. Before applying fertilizer, soil pH tests should be used to make sure it ideal for proper nutrient uptake. While pH levels can vary across fields, a few may be able to appropriate a rough guess. If pH levels are not suitable, fertilizer cannot be taken up by plants and will be wasted. Planting kodo millet in rows, instead of broadcasting the seeds, will increase yields and make weeding easier. Kodo millet will grow on marginal soils, but only if it has little competition from weeds. Finally, proper cleaning of the seeds by wind winnowing will help prevent the sclerotia of fungal diseases being consumed accidentally. Kodo millet seeds can be obtained from the International Crop Research Institute for the Semi-Arid Tropics.

Inhibition of Collagen Glycation and Crosslinking

This is the first report to show the potential usefulness of Finger millet and Kodo millet in preventing glycation and crosslinking of collagen. The above millets could have potent therapeutic role as dietary supplements for the prevention of glycation induced complications, as in diabetes or aging after subjecting to further *in vivo* studies.

Kodo Millet Products

Kodo Millet Coconut Rice

Kodo millet rice – 1 cup; Water – 2 cups; 1/2 grated coconut; Carrot – 1, chopped and steamed (optional); Onion – 1, chopped.

Seasoning – Curry leaves, Channa Dal, Mustard Seeds, Oil.

Cook the kodo millet rice in water. Let it cool down completely, so it does not become mushy when you prepare the coconut rice. Add seasoning to oil. Then saute onions and add the steamed carrots. Add grated coconut and let it cook for a minute. Then add the rice and mix everything well.

Kodo Millet Mango Rice

Kodo millet rice – 1 cup (or 1/2 cup kodo millet, 1/2 cup paddy rice); Water – 2 cups; Raw mango, grated – 1 or per taste; Groundnuts – 2 tbsp.

Seasoning – curry leaves, chillies. mustard seeds, urad dal, turmeric, asafoetida, oil.

Salt to taste. Cool the kodo millet rice in water and let it cool before mixing the rest of the ingredients. Fry groundnuts in oil, keep aside. Prepare the seasoning. Add grated mango and saute for a minute. Add the rice and groundnuts. Tangy mango rice is ready.

Kodo Millet Carrot Rice

Millet - 1/2 cup; Grated carrot -1 cup; Onion - 1 cup finely chopped; Turmeric powder -1/4 tsp; Roasted Peanuts -2 tbsp (optional). Fry and grind: Coriander seeds - 2 tsp; Cumin seeds - 2 tsp; Red Chillies -3-4. For the Seasoning: Oil -2 tbsp; Mustard seeds -1 tsp; Urad dal -3/4 tsp; Curry leaves - a sprig. For Garnishing: Coriander leaves - 2 tbsp finely chopped.

Preparation

Cook millet and keep it aside. Chop onions and grate carrots using the blade with bigger hole. Heat 1/4 tsp of oil and fry cumin seeds, coriander seeds and red chillies on medium flame until you get a nice aroma. Leave it to cool. Then powder it slightly coarse and keep it ready. This gives a nice flavor to the dish. This is the spice powder we are going to use in this dish.

Method

Heat oil, add mustard seeds, when it splutters, add urad dal and curry leaves. When dal turns golden brown, add chopped onions and saute until it turns transparent. When onions turn transparent, add grated carrots and mix well. Add the spice powder, salt needed, turmeric powder, sprinkle water and cook covered. As we have grated the carrots, it needs very less time to get cooked. Once the carrot is soft, fluff the millets with a fork and add to the carrot mixture. Mix well. Heat 1/2 tsp of oil, fry the peanuts and add to the rice. Since I did not have peanuts at home, I have used cashew nuts. This is optional. You can also add this at the time of seasoning. Garnish with coriander leaves and serve hot with any vegetable curry or pachadi/raitha.

Kodo Millet *Idali*

Kodo Millet *Dosa*

Kodo Millet *Dosa*

Kodo Millet *Rabadi*

Kodo Rice

Kodo Millet *Upama*

References

Agroclimatic Zones. Production Estimates and Crop Assessment Division Foreign Agricultural Service. (2013). http://www.fas.usda.gov/pecad2/highlights/2002/10/ethiopia/baseline/Eth_ Agroeco_Zones.htm

Board on Science and Technology for International Development, Office of International Affairs, National Research Council. "Kodo Millet". Lost Crops of Africa; Volume 1: Grains. (1996).http://books.nap.edu/openbook.php?record_id=2305 and page=249

de Wet, J. M. J., K. E. Prasada Rao, M. H. Mengesha and D. E. Brink. "Diversity in Kodo Millet". New York Botanical Garden Press. (1983). http://www.jstor.org.subzero. lib.uoguelph.ca/stable/pdfplus/4254476.pdf? acceptTC =true and acceptTC =true and jpd Confirm=true

Hedge, P.S., Chandra, T.S. "ESR spectroscopic study reveals higher free radical quenching potential in kodo millet (Paspalum scrobiculatum) compared to other millets". Food Chemistry. (2005).http://www.sciencedirect.com/science/article/pii/S0308814604005886

Heuzé V., Tran G., Giger-Reverdin S. "Scrobic (Paspalum scrobiculatum) forage and grain". Feedipedia.org. (2012). http://www.feedipedia.org/node/401.

http://www.icrisat.org/gene-bank-home.htm

Johns, M. "Millet for Forage Use: Frequently asked Questions". Alberta Agriculture and Rural Development. (2007). http://www1. agric.gov.ab.ca/$department/deptdocs.nsf/all/faq8355

Kodo millet. International Crop Research Institute for the Semi-Arid Tropics. (December 4, 2013). http://www.icrisat.org/crop-kodomillet.htm

Kodomillet. United States Department of Agriculture. (No date given, accessed November 11, 2013). http://plants.usda.gov/core/profile?symbol=Pasc6

Millets. Earth360. (2010-13). http://earth360.in/web/Millets.html

Millets: Future of Food and Farming. Millet Network of India. (No date given, accessed November 13th 2013.) http://www.swaraj.org/shikshantar/millets.pdf

Paspalum scrobiculatum (grass). Global Invasive Species Database. (2010). http://www. issg.org/database/species/ecology.asp?si=1423 and lang=EN

Report on Survey of Post-Harvest Technology and Constraints Faced by Women Farmers related to Small Millets and Associated Crops". DHAN Foundation. (2011). http://www. dhan.org/smallmillets/docs/report/PHT_final_report.pdf

Chapter 7

Little Millet (*Panicum sumatrense*)

Little Millet (*Panicum sumatrense*, Syn.: *Panicum miliare* auct. non Lam.) is a species of millet in the family *Poaceae*. In the temperate zones of Asia: the Caucasus, China, East Asia and also in the tropics of the continent: India, Indochina and Malaysia. The plant can be cultivated in India up to 2000 m above sea level. It can withstand both drought and waterlogging. It can be cultivated up to 2000 m above sea level. This species of cereal is similar in habit to the proso millet except that it is smaller. It is an annual herbaceous plant, which grows straight or with folded blades to a height of 30 cm to 1 m. The leaves are linear, with the sometimes hairy laminae and membranous hairy ligules. Thepanicles are from 4 to 15 cm in length with 2 to 3.5 mm long awn. The grain is round and smooth, 1.8 to 1.9 mm long.

Common Names for Little Millet

In India Moraiyo, Kutki, Shavan are some of the names in Hindi for little millet. In Bengali it is called as Sama. In Tamil it is called Samai. In Gujarati Gajro, kuri. In Telugu, it is called as Samalu. In Marathi it is called as Sava, Halvi, vari. In oriya it is called as Suan.

Subspecies

There have been two subspecies described:

- *Panicum sumatrense* Roth ex Roem. and Schult. subsp. *psilopodium* (Trin.) Wet.
- *Panicum sumatrense* Roth ex Roem. and Schult. subsp. *sumatrense*

Cultivation

The largest cultivation is in central India. Usually it is planted using a seed drill. It can also if necessary be planted spoiled. The green plant can also be used in part as cattle feed. The straw can mixed with clay or cement be used in construction.

The harvest yield is from 230 to 900 kg/ha.

Preparation

Little Millet is cooked like rice. Sometimes the millet is also milled and baked. The protein content of the grain is 7.7 per cent.

Health Benefits of Millet

Foxtail Millet may Help Control Blood Sugar and Cholesterol

Foxtail millet (*Setaria italica*) is a common food in parts of India. Scientists at Sri Venkateswara University in that country studied its health benefits in diabetic rats and concluded that the millet produced a "significant fall (70 per cent) in blood glucose" while having no such effect in normal rats. Diabetic rats fed millet also showed significantly lower levels of triglycerides and total/LDL/VLDLcholesterol, while exhibiting an increase in HDL cholesterol.

Sprouting (Malting) Millet makes some Minerals more Bioavailable

In India and some other countries, sprouted (malted) grains are commonly used as weaning foods for infants and as easily-digested foods for the elderly and infirm. A study at the Central Food Technological Research Institute in Mysore, India, measured the changes caused by malting finger millet, wheat and barley. They found that malting millet increased the bioaccessibility of iron (> 300 per cent) and manganese (17 per cent) and calcium ("marginally"), while reducing bioaccessibility of zinc and making no difference in copper. The effects of malting on different minerals varied widely by grain.

All Millet Varieties Show High Antioxidant Activity

At the Memorial University of Newfoundland in Canada, a team of biochemists analyzed the antioxidant activity and phenolic content of several varieties of millet: kodo, finger, foxtail, proso, pearl and little millets. Kodo millet showed the highest phenolic content and proso millet the least. All varieties showed high antioxidant activity, in both soluble and bound fractions.

Naturally Gluten-Free Grains may be Cross-Contaminated

A Polish team from the Instytut Zywnosci in Warsaw analyzed 22 gluten-free products and 19 naturally gluten-free grains and flours, for gluten content. Gluten content in the products ranged from 5.19 to 57.16 mg/kg. In the inherently gluten-free grains and flours, no gluten was detected in rice and buckwheat samples, but was detected in rice flakes (7.05 mg/kg) in pearl millet (27.51 mg/kg) and in oats (>100 mg/kg).

Meanwhile, in the U.S., Tricia Thompson, MS, RD, a nutrition consultant specializing in gluten-free diets, arranged for gluten-testing of 22 retail samples of inherently gluten-free grains, seeds and flours. She found contamination of 20 to 2925 ppm in seven of 22 samples, putting them over the proposed FDA limit of 20 ppm, with lower levels in some others. Both articles point to the importance of gluten-free certification even on foods that are naturally gluten-free, such as millet.

Millet Consumption Decreases Triglycerides and C-Ractive Protein

Scientists in Seoul, South Korea, fed a high-fat diet to rats for 8 weeks to induce hyperlipidemia, then randomly divided into four diet groups: white rice, sorghum, foxtail millet and proso millet for the next 4 weeks. At the end of the study, triglycerides were significantly lower in the two groups consuming foxtail or proso millet and levels of C-reactive protein were lowest in the foxtail millet group. The researchers concluded that millet may be useful in preventing cardiovascular disease.

Indian Diabetics Turn to Ragi (Finger Millet) and other Millets

Diabetes is rising rapidly in India, as it is in many nations. Researchers at Sri Devaraj Urs Medical College in Tamaka, Kola, India decided to study the prevalence and awareness of diabetes in rural areas, in order to inform health policy. While there was widespread lack of awareness of the longterm effects of diabetes and diabetic care, common perception favored consumption of ragi, millet and whole wheat chapatis instead of rice, sweets and fruit.

Finger Millet (Ragi) Tops in Antioxidant Activity among Common Indian Foods

The National Institute of Nutrition in Hyderabad, India, carried out a study of the total phenolic content and antioxidant activity of various pulses, legumes and cereals, including millets. Finger millet and Rajmah (a type of bean) were highest in antioxidant activity, while finger millet and black gram dhal (a type of lentil) had the highest total phenolic content.

Increase your Heart Health

Millet is one of the best possible grains to add to your diet if you want to protect your heart, which is something that everyone can relate to. Millet is a rich source of magnesium, which is an important mineral for reducing blood pressure and the risk of heart attacks of strokes, particularly in the case of atherosclerosis. Millet is also a great source of potassium, which further keeps blood pressure low by acting as a vasodilator. Reducing your blood pressure and optimizing your circulatory system is one of the best ways to protect your cardiovascular health. Furthermore, the plant lignans found in millet can be converted to animal lignans by the microflora in our digestive system and those animal lignans have been shown to protect against certain chronic diseases, like cancer and heart disease.

Cholesterol Levels in Check

Cholesterol levels go hand-in-hand with heart health, so the high fiber levels in millet make for an ideal cholesterol-lowering approach. Dietary fiber actually eliminates dangerous "bad cholesterol" (LDL) from the system, while promoting the effects of "good cholesterol" (HDL).

Protection against Diabetes

Diabetes is a disease found in millions of people around the world. Millet is a beneficial food staple in many developing countries (where diabetes is less frequently found), perhaps because one of the effects of millet is a reduced chance of Type 2 diabetes, thanks to the significant levels of magnesium found in this particular grain. Magnesium is considered one of the most important minerals for increasing the efficiency of insulin and glucose receptors in the body, thereby preventing this disease. A 30 per cent reduction in diabetes has been seen in populations divided between diets with or without magnesium.

Digestive Health

As most fiber-rich foods boast, millet can help move your gastrointestinal system along and eliminate problems like constipation, excess gas, bloating and cramping. By regulating your digestive process, you also improve your nutrient retention and reduce your chance of more serious gastrointestinal conditions like gastric ulcers or colon cancer. Regular digestion and elimination of waste also helps to optimize your kidney, liver and immune system health, as those organ systems are closely related to the body's metabolic activities.

Cancer Risk

Recent research has revealed fiber to be one of the best and easiest ways to prevent the onset of breast cancer in women. In fact, women can reduce their chances of breast cancer by more than 50 per cent by eating more than 30 grams of fiber every day. Given that breast cancer is one of the most common and deadliest forms of cancer, this is advice that every woman should hear!

Detoxify the Body

Many of the antioxidants found in millet, in addition to their beneficial impact on neutralizing free radicals, which can cause cancer, they can also clean up other toxins from your body, such as those in your kidney and liver. Quercetin, curcumin, ellagic acid and various other beneficial catechins can help to rid your system of any foreign agents and toxins by promoting proper excretion and neutralizing enzymatic activity in those organs.

Respiratory System

Research has come to light showing that millet can significantly improve the quality of life for people suffering from childhood asthma and can also prevent it from developing in the first place. Although some of the evidence is controversial, it is shown that significantly less wheezing and asthma attacks (by more than 15 per cent) was seen in children who had large intakes of grains like millet. However, as wheat is a common allergen that is associated with asthma and wheezing, millet does not have the same components and does not have this negative effect.

A Final Word of Caution

Given the modern stresses on our body, particularly to our glandular system, the excessive work needed to properly digest and process millet may be damaging. Consider speaking to your doctor about your glandular and thyroid health before making a major shift to a millet-heavy diet.

Little Millet Products

Little Millet Bread

Use 50:50 (little millet flour and wheat flour) and other all ingredients and procedure remain same as wheat bread.

Little Millet Dosa

Little millet rice – 2 cups or 3 cups. Follow your usual *dosa* ratio; *Urad Dal* – 1 cup; *Methi* Seeds – 1 tsp; Salt to taste. Soak the little millet rice, *urad dal* and *methi* seeds for 8-10 hours. Grind to a fine paste and allow it to ferment overnight or however long it takes to ferment. Add salt and pour out *dosas* on a hot pan. Serve with your favorite *chutney* and *sambar*.

Little Millet Khichdi

2 cups sama (little millet); 1 cup sprouted mung dal; 6 cups water; carrots, beans (can add peas also); 13 ginger chopped small; 4-6 cloves; 2-3 13 cinnamon

Little Millet Bread

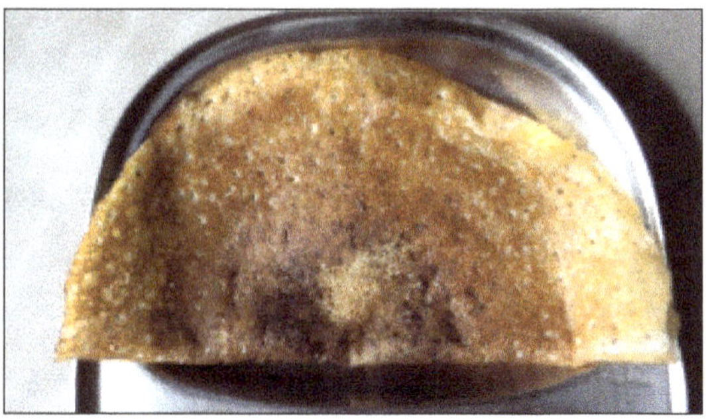

Little Millet Dosa

Little Millet *Chakali*

Little Millet *Roti*

sticks; turmeric; salt. Seasoning: jeera, 1 onion chopped, oil. Chop the carrots and beans. Pressure cook the veggies with sama, spouted mung, ginger pieces, cloves, cinnamon, turmeric, salt and water. Saute jeera and onion in oil. Add the seasoning to the khichdi. It's ready to eat.

References

Anona C. and F. Shahidi. 2010. Content of insoluble bound phenolics in millets and their contribution to antioxidant capacity. *Journal of Agricultural and Food Chemistry*. 58(11): 6706-14.

Daniewski, W., Wojtasik, A. and Kunachowicz, H. 2010. Gluten content in special dietary use gluten-free products and other food products. *Rocz Panstw Zaki Hig* (Poland). 61(1): 51-5.

Franke, W. 1985. *Nutzpflanzenkunde*. Stuttgart.

Genaust, H. 1996. *Etymologisches Wörterbuch der botanischen Pflanzennamen*. 3. Aufl., 701 S. Basel-Boston-Stuttgart.

Heywood, V. H. 1978. *Blütenpflanzen der Welt*. Basel-Boston-Stuttgart. This article is based on a translation of the corresponding article in the German Wikipedia.

Horwath_Winter, J. 2010. Pathophysiology of dry eye syndrome. *Pathophysiology*. Sept 23, 2010. DOI 10.111/j.1755-3768.2010.3221x.

Muninarayana, C., G. Balachandra, S. G. Hiremath, Krishna Iyengar and N. S. Anil 2010. Prevalence and awareness regarding diabetes mellitus in rural Tamaka, Kolar. *International Journal of Diabetes in Developing Countries*. 30(1): 18.

Platel, K., Sushma, W. E. and Srinivasan, K. 2010. Bioaccessible mineral content of malted finger millet (*Eleusine coracana*), wheat (*Triticum aestivium*) and Barley (*Hordeum vulgare*). *Journal of Agricultural and Food Chemistry*. 58(13): 8100-3.

Sreeramulu, D; Reddy, C Vijaya Kumar; Raghunath, M 2009. Antioxidant activity of commonly consumed cereals, millets, pulses and legumes in India. *Indian Journal of Biochemistry and Biophysics*. 46(1): 112-5.

Sun Hee Lee, III-Minchung, Youn-Soo Cha and Park, Y. 2010. Millet consumption decreased serum concentration of triglyceride and C-reactive protein but not oxidative status in hyperlipidemic rats. *Nutrition Research*. 30(4): 290-6.

Thompson, T., Lee, A. R. and Thomas Grace 2010. Gluten contamination of grains, seeds and flours in the United States: A Pilot Study. *Journal of the American Dietetic Association*. 110(6): 937-40.

Chapter 8

Maize Coarse Millet (*Zea mays*)

Introduction

Corn is one of the world's most versatile seed crops. Its botanical name is *Zea mays*. Corn is used as food and feed. Com can be processed into various foods and feed ingredients, industrial products and alcoholic beverages. But the modem com milling technology developed for the above products is mostly confined to some of the developed countries only. However, modem com milling technology is to be suitably adopted for producing the types of products required for other countries. At present, there are two modern methods of milling of corns, dry milling and wet milling. Besides germ for com oil extraction and husk and deoiled germ, etc., for feed, grits (mainly used for the breakfast cereals) are the main products of corn dry milling whereas pure starch, germ and feed are the major products of wet milling.

Composition and Structure

The mature com kernel is composed of four major parts: (a) endosperm (82 per cent), (b) germ (12 per cent), (c) pericarp (5 per cent) and (d) tip cap (1 per cent).

Pericarp: The pericarp is mainly composed of four successive layers, namely; outermost thick layer of tough cells, spongy layer of cells, seed coat or testa layer and aleurone layer. The spongy layer is continuous with spongy cells of tip cap.

Germ: The germ is mainly composed of scutellum and embryonic axis. The major parts of lipids and proteins are reserved in the scutellum.

Endosperm: The endosperm of corn is composed of floury and horny parts. The proportion of horny to floury parts varies widely from dent corn to floury corn variety. The ratio of horny to floury endosperm is about 2: 1 in dent whereas the floury corn contains little horny part. The horny parts being hard and floury parts being soft, the latter is milled into flour easily during rolling. The horny regions have 1.5 to 2 per cent higher protein content than the floury regions.

Com Dry Milling

Com dry milling system can be divided into two groups: the traditional non degerrning system and modem degerming system. In the non degerming system, the whole com is ground into meal of high fiber as well as high protein contents by a stone grinder without removing germ. After grinding certain amount of germ and hull can be removed from the meal by sifting. In the degerrning system the com is moistened with a little amount of water and tempered for moisture equilibration. After degerming the stock is dried, milled and classified into different products. The purpose of all dry degerming com milling methods is to remove hull, germ and tip cap from the com kernel as far as practicable and primarily produce com grits with some meals and flours. The germ is then used for oil extraction and deoiled germ, hull, etc., are used as feed which is known as hominy feed. The yield of endosperm products and hominy feed are about 70 per cent and 30 per cent respectively.

Tempering-Degerming (T.D.) Method of Dry Milling

The major objectives of this method are: (a) to remove essentially all germ and hull so that endosperm contains as low fat and fiber as possible, (b) to recover a maximum amount of the endosperm as large clean grits without any dark speck and (c) to recover a maximum amount of germ as large and pure articles. *Description of the T.D. system* The basic operations/processes involved 10 the T.D. methods are as follows:

1. Cleaning of the corn.
2. Conditioning of the corn by addition of control amount of moisture either at ordinary temperature or at an elevated temperature to toughen the germ and husk and facilitate their removal from the endosperm.
3. Releasing hull, germ and tip cap from the endosperm in a degermer.
4. Drying and cooling the degermer products obtained from the degermer.
5. Fractionating degermer stock by multistep milling through a series of machines namely roller mills, sifters, aspirators, gravity table separators and purifiers to separate and recover the various products.
6. Further drying of the products is done as and when necessary.
7. Blending and packaging of products.

Cleaning of Corn

Thorough cleaning of com is essential for the subsequent milling operations. Pieces of iron, etc., are removed by magnetic separators. Dry cleaners consisting of sieves and aspirators and sometimes a wet cleaner consisting of a washing destoning unit and a mechanical type dewatering unit, known as whizzer, are used for cleaning of com.

Hydrothermal Treatment/Conditioning

Predetermined amount of moisture is added to the com in the form of cold or hot water or steam in one, two or three stages with appropriate tempering times after each stage; the tempering times (rest periods) vary according to the hydration methods. So also tempering temperatures vary from room temperature to about 50°C accordingly. The optimum moisture content for degerming in the Beall. Degermer is 21-25 per cent. Either cold or hot water is used for the addition of moisture. A little heat in the form of open steam is added as and when necessary.

Degerming

The purpose of degerming is to remove hull, tip cap and germ as far as practicable and leave toe endosperm into large grits. However, the products from degermer consists of a mixture of kernel components, freed from each other to varying degrees, with the endosperm particles varying in sizes from grits to flour.

The Beall Degermer consists of a rotating cast iron conical roller mounted on a horizontal shaft in a conical cage. Part of the cage is fitted with perforated screens and the remainder with plates having conical projections on its inner surface the rotating cone has similar projections over most of its surface. The feed end of the cone has spiral corrugations to move the com forward whereas the large end has corrugations in an opposite direction to retard the flow. The product leaves the unit in two streams. The major portions of the released germ, husk and fines as well as some of the grits are discharged through the perforated screens. Tail stock containing large amount of grits, escapes through an opening fitted with the large end of the cone. A hinged gate with an adjustable weight adjusts pressure inside the chamber and controls the flow of the stock.

Drying and Cooling of Degermer Stock

The degenner products are to be dried to 15 to 18 per cent moisture content for proper grinding and sifting. Generally rotary steam tube dryers are used for drying the product. Rotary Louver type dryer 'can also be employed. The stock is heated to about 50° C. Counter-flow or cross-flow rotary, vertical gravity or fluidized bed types of cooler can be used for cooling the dried products.

Rolling and Grading

Recovery of various primary products is the next step. Further release of germ and husk from the endosperm products occurs during their gradual size reduction roller mills. The germ, husk and endosperm fragments are then separated by means of sifters, aspirators, specific gravity table separators or purifiers. Sifting is an important operation and is variously referred to as scalping, grading, classifyirrg, or bolting depending upon the means used and purpose. Sifting is actually a size separation operation on sieves. Scalping is the coarse separation made on the product leaving a roller mill or degermer. Grading or classifying is the separation of a single stock (usually endosperm particles) into two or more groups according to particle size. Bolting is the removal of hull fragments from a com meal or flour.

Another modem Com dry milling method, known as Miag process. Only the flow diagram of the Miag process is given in the figure but detailed process has not been' described here.

Corn Wet Milling

It has been discussed earlier that pure starch, pure germ and feed are the basic products of com wet milling. But a few hundreds of byproducts can be produced from these three main products. A list of these byproducts with their uses is given in Table 8. l. The raw com for wet milling should contain 15-16 per cent moisture and it should be physically sound. Insect and pest infested, cracked and heat damaged corns (treated at temperature around 75° C during drying) are unsuitable for wet milling. The heat damaged com affects the quality of oil extracted from its germ. Sufficient amount of moisture is added to the com during steeping in the wet milling process in order to prepare the com for subsequent degerming, grinding and separation operations. The wet milling process consists of the following steps: (a) cleaning, (b) soaking, (c) germ separation and recovery, (d) grinding and hull recovery and (e) separation of starch and gluten.

Cleaning

Impurities such as dust, chaff, cobs, stones, insect-infested grain and broken grain and other foreign materials are removed from corn by screening and aspirating. The clean grains are conveyed to the storage bins.

Table 8.1: Corn Wet Milling Products

Product	*Feed/Food Uses*	*Industrial uses*
Germ oil and	Livestock meal foots	Soap, glycerine, leather dressing, etc.
Refined oil	Salad and table oils, cooking oils, margarine	Pharmaceutical,
Steep water	Yeast-food	Phytic acid, inositol
Gluten and hulls	Livestock and Poultry feed	–
Starch	Corn starch, chewing gum, bakeries, baking powder, brewing confectionery	Textiles, laundry, paper and paper boxes, explosives, cosmetics, adhesives.
Syrup	Bakery products, canned fruits, ice cream, confectionery, soft drinks, chewing gums, mixed syrups and jellies.	Textiles, leather tanning, pharmaceuticals, tobacco.
Sugar	Bakery products, pharmaceuticals, jams and jellies, ice cream, canned foods, confectionery.	Rayon, tanning, fermentation, brewing, vinegar, caramel colour, fermentation products, tobacco.

Steeping

The major objectives of steeping are (1) to soften the kernel for grinding, (2) to facilitate separation of germ, (3) to facilitate separation of gluten from the starch granules and (4) to remove solubles, mainly from the germ. Water impregnated with S02 (*i.e.*, acidulated: water with H_2SO_4) is used for steeping. It helps in arresting certain fermentation during long steeping process. The steeping is carried out at about 50° C for a period varying from 28 to 48 hours in different plants. The steeped corn attains a moisture content of about 45 per cent. The flow diagram of corn wet milling process is shown in Figure 8.1.

Germ Recovery

The wet and softened corn kernels containing about 45 per cent moisture are conveyed to the degerminating unit. This machine consisting of a metallic stationary plate and a rotating plate with projecting teeth is employed only for tearing the soft kernels apart and freeing the germs without grinding them. The pulpy mixture containing germs, husk, starch and gluten is passed through hydroclones, where the germ being lighter is separated from other heavier ingredients, by centrifugal force. Only modern starch plants employ hydro clones for germ separation. Otherwise the floatation method of germ separation is still in use in. old types of mills.

Milling and Fiber Recovery

After separation of germ and screening of the coarse particles, the mixture contains starch, gluten and hulls. Mainly horny endosperm and hull are then generally ground by either traditional Burstone mill or modern entoletor impact mills to release the rest of the starch. Material to be ground enters the machine through a spinning rotor and is thrown out with great force against the impactors at the periphery of the rotor and also against a stationary impactor resulting in considerable reduction in particle size. Here only the starch is readily released, with a very little size reduction of hulls. The milled slurry, containing the ground starch, gluten and hulls, is passed through a series of hexagonal reels where the coarser hulls and fibers are removed.

Starch-Gluten Separation

In the modern process, the slurry containing starch and gluten is concentrated and then the lighter gluten particles are separated from the relatively heavier starch particles by the centrifugal force in high speed centrifuges. The centrifuging of starch is carried out in two stages. In many modern plants, the second stage of centrifugation is performed by a number of hydroclone types of equiprnents. The starch obtained from the second stage of separation is filtered and then dried to produce dry starches.

Health Benefits from Maize/Corn

Braving and bearing the brunt of summers can be quite challenging. From health to skin and hair, maintaining all can certainly be daunting. We often underestimate the goodness and power of natural resources to combat most of our everyday health concerns. Surprisingly, corn is one such wonder crop that can get us rid of most of our probable nightmares this summer.

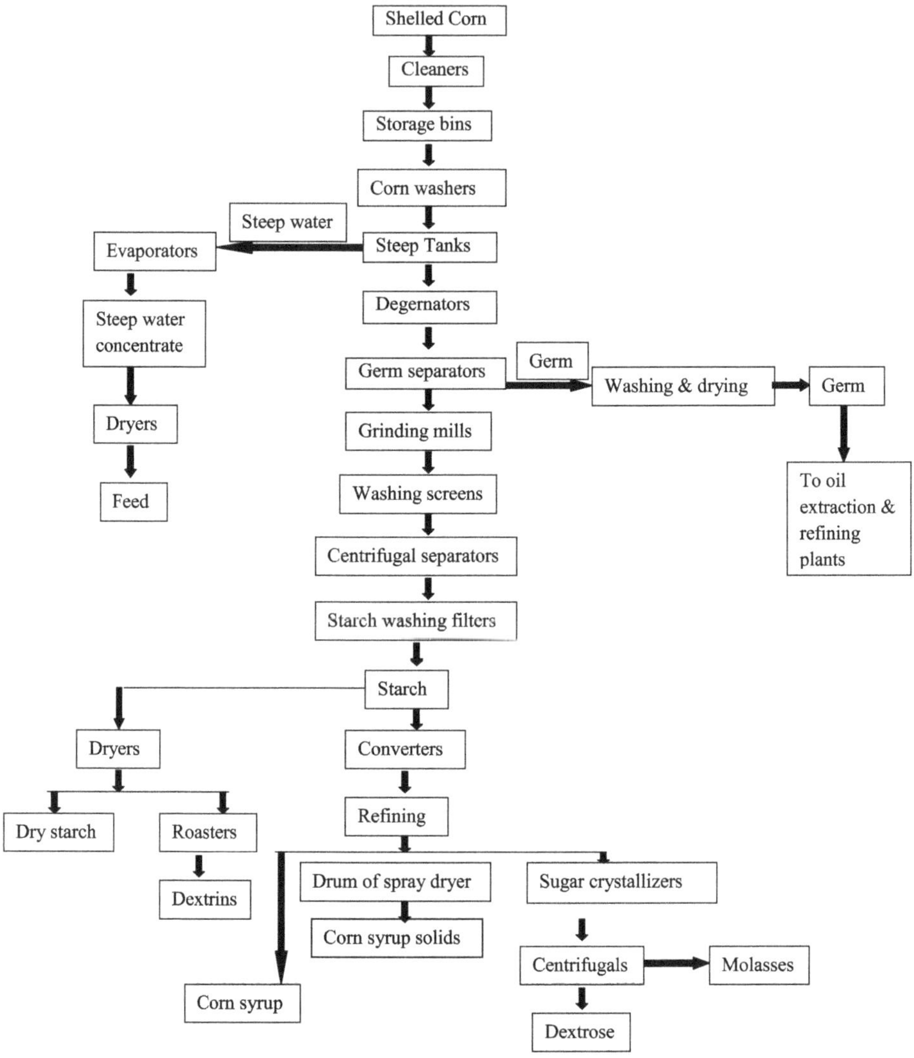

Figure 8.1: Flow Diagram of Corn Wet Milling and Refining Processes.

Good for Digestion

During summers it is ideal to consume foods that are light on stomach. Corn is high on fibber which facilitates proper digestion warding off numerous digestive problems including constipation.

Power Packed with Energy

One tends to lose a lot of essential bodily fluids during summers and in that situation the energy levels take a dip as well. Corn is full of starch, water content and is high in carbohydrates which provides for long term energy. This is particularly beneficial for athletes as they require more carbs to optimize their performance.

For Glowing Skin

The scorching heat and the play of dry spells of weather can cause immense damage to skin and hair. Take a sigh of relief as corn is a one stop solution for all your skin and hair woes. It is a good source of vitamin C, thiamine, niacin, vitamin E and other minerals and antioxidants that are essential for skin care. It contains lycopene which fights against UV rays and assists in the increased production of collagen that ensures radiantly smooth skin.

Damage Control for Hair

Summer is supposedly the worst for hair. Hair care and maintenance becomes quite challenging with excess heat, dust and UV rays blazing hot on you every day. No amount of cosmetics or external application will help unless we eat right. Consumption of corn strengthens hair follicles and speeds up collagen production, thus maintaining smooth and silky hair. Topical use of hot corn oil can help prevent dryness of the scalp and hair loss.

Long term Benefits

Including corn in your diet can result in long term health benefits as well. The essential minerals, vitamins and antioxidants can help combating anemia, diabetes, hypertension, increased cholesterol levels, heart risks and even Alzheimer's disease. Recent research has established that corn contains ferulic acid which helps in fighting cancerous tissues and tumours in breasts and liver.

Corn Flakes

Corn Pops

Corn Cracker

Corn Crackers

Corn Starch

Corn Bakery Products

Corn Oil

Corn Sweets

Bean and Cheese Burgers

Open Bean Burger

Broccoli Burger

Hara Tava Burger

Healthy Burger

Veggie Burger

Masala Toast

Chunky Cheese and Vegetable

Mexican

Rajma Burger

Spinach and Corn Open Burger

Mini Pav Bhaji Burgers

Three Bean Burger **Veggie Burger**

Masala Toast

Cheese and Spring Onion Sandwich

Bread Pakoda

Marinated Bell Pepper Open Burger

How to eat Corn, Differently

The best thing about cooking is the freedom that you get to try things out, to experiment and to create an appetizing masterpiece of your own by a calculated yet bold mix and match of ingredients and flavours. Yet, foods like corn have mostly and widely remained underrated on the Indian plate. Whereas western cuisines have experimented and incorporated corn effortlessly, it's only in the recent past that we are finding that it is now being liberally used with Indian flavours. If the taste isn't enough to make you bite into it, its health quotient could.

References

Anderson, R. A. (1962). A note on wet-milling of high amylose corn containing 75 per cent-amylose starch, *Cereal Chem.*, 39, 406-408.

Anderson, R. A. (1963). Wet-milling Properties of grains: Bench- Scale Study, *Cereal Sci. Today*, 8, No.6, 190-92, 195, 221.

Anon, 1964, Joint United States-Canadian tables of feed composition, *Natl. A cad. Sci.*, Nath. Res. Council Publication 1232.

Anonymous, (1982). Prevention of Food Adulteration Act, Government of India Publication, New Delhi.

Atkins, T. A. and Geddes, W. F. (1939). The relationship between protein content and strength of flours in gluten enriched flours, *Cereal Chem.*, 16, 223-31.

Autrey, H. S. (1953). Effect of variables upon milling yields, *Rice J. Annual.*

Baird, P. D., Mac Masters, M. M. and Rist, C. E. (1950). Studies on a rapid test for the viability of corn for industrial use, *Cereal Chem.;* 27, 508-13.

Banwart, G.J. (1979). Basic Food Microbiology. The AVI Publishing Company, Inc., Westport, Connecticut.

Bender, A. E. (1973). Composition of infant foods, In Nutritional and Dietetic Foods, International Textbook Co. Ltd., London, U. K., p. 64.

Bhat, C.M., Sharma, R.N. and Sehgal, S. (1982). A Manual on Food Preservation at Home. Directorate of Publication, Haryana Agricultural University, Hissar.

Bishop, W. and Dustin, J. (1951). Topnotch wet com mill house, *Food Ind.*, 23, No.3, 121-24, 211-12.

Boundy, J. A., Woychik, J. H., Dimler, R. J. and Wall, J. S. (1967)., Protein composition of dent, waxy and high-amylose corns, *Cereal Chem.*, 44, 160-169.

Brautlecht, CIA, (1953). *Starch, Its sources, Production and Uses*, Reinhold Publishing Corp., New York.

Brekke, 0 L. and Kwolek, W. F. (1969). Com dry milling: Cold tempering and degermination of com of various initial moisture contents, *Cereal Chern.* 46, 545-49.

Brekke, O. L. (1967). Com dry milling: Pretempering low-moisture com, *Cereal Chem.*, 44, 521-31.

Brekke, O. L. (1968). Com dry-milling: Stress crack formation in tempering low-moisture com and effect on degerminator performance, *Cereal Chem.*, 45, 291-303.

Brekke, O. L. (1970). Dry milling artificially dried com: Roller milling of degerminator stock at various moistures, *Cereal Science Today*, 15, 37-42.

Brekke, O. L. and Weinecke, L. A. (1964). Com dry-milling, A comparative evaluation of commercial degermer samples, *Cereal Chern.*, 41, 321-28.

Brekke, O. L., Weinecke, L. A., Boyd, J. N. and Griffin, E. L., Jr. (1963). Com dry milling: Effects of first-temper moisture, Screen perforations and rotor speed on Beall egerminator throughput and products, *Cereal Chem., 40*, 423-29.

Chandra, P. K. (1972). Methodology and Procedures for evaluating performance of commercial rice mills, Unpublished M. Tech. Thesis, L I. T., Kharagpur.

Chavan, J. K. and Kadam S. S. (1989). Nutritional improvement of cereals by fermentation. *CRC in Food Sci. and Nut.* 28: 349-399.

CSIR, (1980). The Wealth of India: Raw Materials, Councill of Scientific and Industrial Research, New Delhi.

Dandekar Machinery Works, Catalogue for rice and pulse milling machinery, *Mis.* G. G. Dandekar Machine Works Ltd., Bhiwandi, Thana, Maharashtra, India.

Desikachar, H. S. R. (1977). Processing of sorghum and millets for versatile food uses in India. In Sorghum and millets for human food. pp. 41-45. *Symp. Proc. IACC*, Vienna, 1976. Tropical Products Institute, London.

Desrosier, N.W. and Desrosier, J.N. (1987). The Technology of Food Preservation. CBS Publishers and Distributors, New Delhi.

Desrosier, N.W. and Tressler, D.K. (1977). Fundamentals of Food Freezing. The AVI Publishing Company, Inc., Westport, Connecticut.

Earle, F.,R., Curtis, J. J. and Hubbard, J. E. (1946). Composition of the component parts U1 the com kernel, *Cereal Chem.*, 23, 504-11.

Elias, D. G. and Scott, R. A. (1957). British flour milling technology, *Cereal Sci. TorJay*, 7, 180-84.

Fennema, O.R. Ed.(1976). Principles of Food Science, Part-1. Food Chemistry, Marcel Dekker Inc, New York.

Frazier, W.e. and Westhoff, D.C. (1983). Food Microbiology. Tata McGraw-Hill Publishing Company Limited, New Delhi.

Goel, A. K., Kumar, R. and Mann, S. S. (2007). Postharvest management and value addition. Daya Publishing House, Delhi.

Gopalan, C, Rama Sastri, B.V. and Bala Subramanian, S.C. (1982). Nutritive Value of Indian Foods. National Institute of Nutrition, R.C.M.R., Hyderabad.

Hardenburg, R. E., A.E. Watada and C. Y. Wang, (1986). The Commercial Storage of Fruits, Vegetables and Florist and Nursery Stocks, Agricultural Handbook 66, US Dept. Agric. Washington D.C.

Harris, RS. and Karmas, E. (1975). Nutritional Evaluation of Food Processing. 2nd Edition. AVI Publishing Co. Westport, Connecticut.

Houston, D. F. (1973). Western Marketing and Nutrition Research Division, Agricultural Research Service, U. S. Department of Agriculture, Berkeley, California.

Ilyas, S.M. and Coyal, R.K. (2002). Quality assurance in food processing-impact of WTO, In: Proc. National Seminar on Development of Food Processing Industries in U.P., held at Lucknow during January 16-17.

Jacob, T. (1987). Poisons in our Food. Publications Division, Ministry of Information and Broadcasting, Government of India, New Delhi.

Javaraman. J. (1981). Laboratory Manual in Biochemistry. Wiley Eastern Limited, New Delhi.

Jones, CR. (1940). The production of mechanically damaged starch in milling as a governing factor in the diastatic activity of flour, *Cereal Chem.*, 17, 133-69.

Karel, M., O.R. Fennema and D.S. Lund, Eds. (1975). Principles of Food Science Part II. Physical Principles of Food Preservation. Marcel Dekker, Inc. New York.

Kent-jones, D. W. and Amos, A. J. (1957). Modern Cereal Chemistry, Fifth Ed., Northern Publishing Co. Ltd., Liverpool.

Kerr, R. W. (1950). Chemistry and Industry of starch, 2nd Edition, Academic Press, New York.

Koga, Y. (1969). Drying, husking and milling in Japan IV and V., *Farming,* Japan.

Kordylas, J.M. (1991). Processing and Preservation of Tropical and Subtropical Foods. Macmillan Education Ltd., Houndmills, Basingstoke, Hampshire.

Lockwood, J. F. (1952). *Flour Milling,* Third ed., Northern Publishing Co. Ltd., Liverpool.

Lockwood, J. F. (1960). *Flour Milling,* The Northern Publishing Co., Liverpool.

Macrae, R., R. K. Robinson and M. J. Sadler, Eds, (1993). Encyclopaedia of Food Science, Food Technology and Nutrition, Academic Press, London.

Matz, S. A. (1959). The Chemistry and Technology of Cereals as Food and Feed, Avi Publishing Co., Westport, Conn.

Miller, O. H. and Burns, E. E. (1970). Starch characteristics of selected grain sorghum as related to human food. *J. Food Sci.* 35: 666-668.

Mudambi, S.R and Rao, S. (1986). Food Science. Wiley Eastern Limited, New Delhi.

Narsirnha, H. V, Ramakrishnan. N., Pratape, V M., Sashikala, V B. and Narsirnhan, K. S. (2000). Status of pulse milling industries in India. Presented in Workshop on Recent Development in Pulse Milling, CFTRI, Mysore on 14 February.

Odunfa, S. A., Nordsrom, J. and Adeniran, S. A. (1994). Development of starter cultures for nutrient enrichment of ogia West African fermented cereal gruel. Report submitted to HBVC research grants program. USAID, Washington, U.S.A.

Paulus, K. O. (1984). Modelling in Industrial Cooking. In: *Thermal Processing and Quality of Foods.* Zeuthen *et al.* (eds.), Elsevier Applied Science Publishers, London.

Potter, N. N. (1984). Food Science. 3rd Edition. The AVI Publishing Company, Inc., Westport, Connecticut.

Potter, N.N. (1973). Food Science. The AVI Publishing Company, Inc., Westport Connecticut.

Potter, N.N. and Hotchkiss, J.H. (1996). Food Science. 5th Edition. CBS Publishers and Distributors, New Delhi.

Quick, Graeme R. (1980). An In-depth Look at Farm Fuel Alternatives. Power Farming Magazine, Feb. 10-17 pp.

Raghavandra Rao, S. N., Malleshi, N. G., Sreedharamurty, S., Viraktamath, C. S. and Desikachar, H. S. R. (1979). Characteristics of roti, dosa and vermicelli from maize, sorghum and bajra. *J. Food Sci. Technol.* 16: 21-24.

Rai, Mangla and Mauria S. (1999). Coarse cereal: Not for the poor alone. The Hindu Survey of Indian Agriculture. pp. 55-58.

Rai, Mangla. 2002. Amazing nutrient composition. The Hindu Survey of Indian Agriculture. pp. 59-62.

Reichert, R. D. and Young, C. G. (1976). Dehulling cereal grains and grain legumes for developing countries. I. Quantities comparison between attrition and abrasive type mills. *Cereal Chem.* 53: 829-839.

Salunkhe, D.K., S.S. Kadam and S. J. Jadhav, (1991). Potato: Production, Processing and Products, CRC Press, Soca Raton.

Satake Owner Manual, Type-03, 91973). Satake Enzz, Co., Ltd., Japan.

Schoch, T. J. (1941). Physical aspects of starch behavior, *Cereal Chem.*, 18, 121-28.

Schrimshaw, N. S. (1974). Nutritional requirements for high protein foods, Proc. Int. Symp. Protein Foods and Concentrates, Central Food Technol. Res. Institute, Mysore, India, pp. 13-18.

Scott, J. H. (1951). *Flour Milling Processes,* Second ed., Chapman and Hall, London.

Seetharam A. (1997). Coarse cereals: Waiting for incentives. The Hindu Survey of Indian Agriculture. pp. 55-59.

Seetharama N. and Rao B D (2004). Sustaining nutritional security. *The Hindu Survey of Indian Agriculture.* pp. 37-38.

Shibano, M. (1973). Construction and Function of Abrasive Roll Type Rice Whitening Machine, Technical note, JRMA, Japan.

Shibano, M. (1973). Construction and Function of Friction Type Rice Whitening Machine Technical note, JRMA, Japan.

Shreve, N. R. (1956). The Chemical Precess Industries, McGraw-Hili Book Co. Inc., Tokyo.

Singh, G. (1997). Data book on Mechanization and Agro-Processing since Independence, CIAE, Bhopal-462038.

Smith, L. (1944), *Flour Milling Technology,* Third ed., Northern Publishing Co., Liverpool.

Srivastasva, M (1985). Essentials of Food and Nutrition Vol. II (Applied Aspects). The Bangalore Printing and Publishing Company Limited, Bangalore.

Srivastava, R.K and Singh, S. (1987). Utilization of Mushrooms for Preparation of Different Types of Preserved Products. Government Fruit Preservation and Canning Institute,.U.P., Lucknow.

Stermer, R A, (1968). Environmental conditions and stress cracks in milled rice, *Cereal Chem.*, 45, 365.

Stivers, T. *E.*, Jr. (1955), American corn milling systems for degermed products, *Assoc. Operators Millers Bull.*, 2168-79.

Sugden, G. H. (1956). Various aspects of wheat conditioning, *Cereal Science Today,* I, 136-42.

Thompson, R. A. and r-oster, G. H. (1963). Stress cracks and breakage in artificially dried corn, U. S. Dept. Agr., AMS, 631.

Tuite, J. and Foster, G. H. (1993). Effect of artificial drying on the hygroscopic properties of corn, *Cereal Chem., 40,* 630-37.

Watson, S. A. and Yahl, K. R. (1976b). Comparison of wet-milling properties of opaque-2 high-lysine corn and normal corn, *Cereal Chem.*, 44, 488-98.

Watson, S. A., Hirata, Y. (1962). Some wet-milling properties of artificially dried corn, *Cereal Chem.*, 39, 35-43.

Wolf, M. J., Buzan, C. L., Mac Masters, M. M. and Rist, C. E. (1952a). Structure of the mature Corn Kernel, Gross anatomy and structural relationship, *Cereal Chem.*, 29, 321-33.

Wolf, M. J., Buzan, C. L., Mac Masters, M. M. and Rist, C. E. (1952b). Structure of the mature Corn Kernel, II, Microscopic structure of pericarp, seed coat and hilar layer of dent corn, *Cereal Chem., 29,* 334-48.

Chapter 9

Pearl Millet (*Pennisetum glaucum*)

Pearl millet is a dual-purpose crop with food, feed and fodder value. Its grain is a staple food of people living in arid regions of India and also has a high feed value for livestock, poultry and fish. It also provides high quality green forage in seasons of fodder scarcity. Both grain and stover of pearl millet have a better mineral profile than many other cereals. The bioavailability of these minerals is low because of presence of some inherent anti-nutritional factors *e.g.* phytate and polyphenols in grain; and oxalic acid in fodder and forage. Its flour acquires a rancid odour within a few days of milling because of high concentration of lipids that contribute to the development of fat acidity, lipolytic activity and accumulation of peroxides of lipids in the meal during storage. The typical grey colour of pearl millet grain and its products due to polyphenolic pigments present in peripheral area of the endosperm further restricts efficient utilization of pearl millet. However, several processing techniques have been developed to enhance food value and shelf-life of pearl millet products and to improve the availability of starch, protein and minerals.

Malting reduces anti-nutritional factors and imparts desirable flavour and taste. Shelf life of pearl millet flour is also increased by malting as this lowers the levels of lipids that are responsible for off-flavors. Blanching and heat treatment improve the storability and stability of flour. Acid treatment of grain may bleach grey colour, remove anti-nutritional factors and improve digestibility and shelf life of pearl millet flour. A wide range of value-added products may be prepared from pearl millet processed flour.

Formulation of poultry feed using pearl millet may help to reduce costs and also ease competition for energy sources between man and monogastrics. Pearl millet-supplemented poultry feeds are generally superior to sorghum and equivalent to maize in broiler diets. Pearl millet is also suitable for use in cattle feeding if diets are adjusted to accommodate higher protein levels in the grain. In general, pearl millet is at least equivalent to maize and often superior to sorghum in cattle, pig and sheep rations because of its high energy and grain protein levels. Its grain, as a high-energy alternative source of inexpensive feed, could be exploited and utilized in production of high quality fish.

Pearl millet (*Pennisetum glaucum*) is the most widely grown type of millet. It has been grown in Africa and the Indian subcontinent since prehistoric times. The center of diversity and suggested area of domestication, for the crop is in the Sahel zone of West Africa. Recent archaeobotanical research has confirmed the presence of domesticated pearl millet on the Sahel zone of northern Mali between 2500 and 2000 BC. Cultivation subsequently spread and moved overseas to India. The earliest archaeological records in India date to around 2000 BC and it spread rapidly through India reaching South India by 1500 BC, based on evidence from the site of Hallur. Cultivation also spread throughout eastern and southern Africa. Records exist for cultivation of pearl millet in the United States in the 1850s and the crop was introduced into Brazil in the 1960s. Pearl millet is well adapted to growing areas characterized by drought, low soil fertility and high temperature. It performs well in soils with high salinity or low pH. Because of its tolerance to difficult growing conditions, it can be grown in areas where other cereal crops, such as maize or wheat, would not survive. Today pearl millet is grown on over 260,000 km^2 of land worldwide. It accounts for approximately 50 per cent of the total world production of millets.

There's great news for all you gluten intolerant people. If you are looking for alternatives, then don't forget to incorporate the pearl millet in regular diet. The most widely grown millet, the pearl millet, is known for its culinary uses as well as health benefits. It is cultivated in countries of Africa and the Indian subcontinent since prehistoric times. Pearl millets, earlier known as 'bird food', come in several delicious flavours. What is unique about this cereal is that it may be as creamy as mashed potatoes or as fluffy as rice. In India, pearl millets are regarded as one of the major sources of dietary energy and nutritional security for poor farmers and consumers. Apart from offering excellent taste, these millets contain essential mineral and nutrients (Table 9.1), which provide the body with a variety of advantages.

Table 9.1: Nutrient Content in Pearl Millet Grains

Nutrients	*Amount*
Basic Components (per cent)	
Proteins	11
Water	8.7
Ash	3.3
Energy in Calories (K. Calories)	
Total Calories	378
Calories From Carbohydrates	300
Calories From Fats	36
Calories From Proteins	43

Contd...

Table 9.1–*Contd...*

Nutrients	*Amount*
Carbohydrates (per cent)	
Total Carbohydrates	73
Dietary Fiber	8.5
Fats and Fatty Acids (per cent)	
Total Fat	4.2
Saturated Fat	0.7
Monounsaturated Fat	0.75
Polyunsaturated Fat	2.2
Omega-3 Fatty Acids	0.12
Omega-6 Fatty Acids	2
Vitamins (μg)	
Vitamin E	50
Vitamin K	0.9
Thiamin	421
Riboflavin	290
Niacin	4.7
Vitamin B_6	384
Folate	85
Pantothenic Acid	0.85
Minerals (mg)	
Calcium	8
Iron	3
Magnesium	114
Phosphorus	285
Potassium	195
Sodium	5
Zinc	1.7
Copper	0.75
Manganese	1.65
Selenium	2.7

History

Since prehistoric times, pearl millet has been grown prominently in Africa and Indian Subcontinent. It is believed that pearl millet originated in Africa and was later introduced to India. The earliest archaeological evidences show that millet was cultivated in India around 2000 BC; hence, domestication in Africa must have taken place before that. Its origin can be trailed back to North Africa, specifically in Ethiopia but the center of diversity for the crop was in the Sahel zone of West

Africa. From there, its cultivation subsequently spread to east and southern Africa and southern Asia. Millet is an important food staple in Africa where it is used to make a traditional flatbread known as injera. The ground millet seeds are used for making the Indian flatbread, called *roti.* In the middle Ages, millet was a staple grain in Europe, especially in countries in Eastern Europe. It was introduced in the United States in the 19th century. Millet is popular as birdseed and livestock fodder in Western Europe and North America, but it has recently gained popularity as a delicious and nutritious grain due to its nutritious benefits and gluten-free status. Currently, India is the leading commercial producer of pearl millet, followed by China and Nigeria.

Common Names for Pearl Millet

- ✰ In Africa: gero (Hausa),Uwele (Kiswahili), Oka (Yoruba), mahangu (Namibia), sar (Bambara), gawri (Fula), babala, nyoloti, dukkin, souna, petit mil (French), heyni (Zarma), masago (Somali), mexoeira (Mozambique), biltug (Tigrinya), biltug (Blin), mhunga (Shona, Zimbabwe), inyawuthi (Northern Ndebele, Zimbabwe), lebelebele (Setswana, Botswana), zembwe (Ikalanga, Botswana), (Tunisian Arabic), dokhn (Yemeni Arabic).
- ✰ In Australia: bulrush millet.
- ✰ In Brazil: milheto.
- ✰ In Europe: candle millet, dark millet.
- ✰ In India: *Bajri* in Rajasthani, Gujarati and Marathi), *Sajje/kambu* in Kannada); *Kambu* in Tamil; *Bajra* inHindi, Urdu and Punjabi and *sajjalu* in Telugu and Kambam in Malayalam and bajra in Bengali.
- ✰ In Pakistan: *Ba'ajra*, in Urdu, Sindhi, Kashmiri, Balochi, Pashto, Punjabi, Saraiki.
- ✰ In USA: cattail millet (*Pennisetum americanum*).

Pearl Millet Around the World

India

India is the largest producer of pearl millet; rajasthan is highest producing state in India (Table 9.2).

Sahel

Pearl millet is an important food across the Sahel. It is a main staple (along with sorghum) in a large region of northern Nigeria, Niger, Maliand Burkina Faso. In Nigeria it is usually grown as an intercrop with sorghum and cowpea, the different growth habits, growth period and drought vulnerability of the three crops maximising total productivity and minimising the risk of total crop failure. It is often ground into flour, rolled into large balls, parboiled, liquefied into a watery paste using fermented milk and then consumed as a beverage. This beverage called "fura" in Hausa is popular drink in northern Nigeria and southern Niger.

Table 9.2: State-wise Average Area, Production and Yield of Pearl Millet

State	*Area*		*Production*		*Yield (t/ha)*	*Percent Irrigated Area (2008-09)*
	Million ha	*Per cent to All India*	*Million tones*	*per cent to All India*		
Rajasthan	5.14	56.55	3.51	41.54	0.68	5.2
Uttar Pradesh	0.85	9.31	I1.34	15.90	1.59	7.8
Haryana	0.61	6.71	1.06	12.50	1.73	37.0
Gujarat	0.76	8.40	1.03	12.23	1.35	22.1
Maharashtra	1.06	11.66	0.85	10.10	0.81	5.5
Madhya Pradesh	0.18	1.94	0.25	2.92	1.40	0.2
Karnataka	0.34	3.70	0.23	2.76	0.69	15.1
Tamil Nadu	0.06	0.62	0.08	0.99	1.47	11.1
Andhra Pradesh	0.06	0.66	0.07	0.79	1.11	26.3
J and K	0.02	0.22	0.01	0.12	0.50	0.3
Others	0.02	0.18	0.01	0.16	–	–
All India	9.09	100	8.45	100.00	0.93	9.4

Source: Agricultural Statistics at a Glance. 2011, DAC, Min. of Agriculture, Government of India.

Namibia

In Namibia, pearl millet is locally known as "mahangu" and is grown mainly in the north of that country, where it is the staple food. In the dry, unpredictable climate of this area it grows better than alternatives such as maize. Mahangu is usually made into a porridge called "oshifima" (or "oshithima"), or fermented to make a drink called "ontaku" or "oshikundu". Traditionally the mahangu is pounded with heavy pieces of wood in a 'pounding area'. The floor of the pounding area is covered with a concrete-like coating made from the material of termite mounds. As a result, some sand and grit gets into the pounded mahangu, so products like oshifima are usually swallowed without chewing. After pounding, winnowing may be used to remove the chaff. Some industrial grain processing facilities now exist, such as those operated by Namib Mills. Efforts are also being made to develop smaller scale processing using food extrusion and other methods. In a food extruder, the mahangu is milled into a paste before being forced through metal die. Products made this way include breakfast cereals, including puffed grains and porridge, pasta shapes and "rice". Recently more productive varieties of pearl millet have been introduced, enabling farmers to increase production considerably.

Millet Production and Consumption

Several kinds of millets are grown in the world, but FAO data on area, yield and production of all millets are given together under the general heading of millet. Pearl millet, finger millet and prove millet account for a large proportion of the world production. Millet production increased from 26 million tonnes in 197981 to 31 million tonnes in 1988 and was similar in 1989 and 1990. Asia, Africa and the former Soviet Union produce almost all the world's millets, as shown in Table 9.3. The major producers of millets in 1990 were India (39 percent), China (15 percent), Nigeria (13 percent) and the Soviet Union (12 percent) (Table 9.4). The millet production and consumption are shown in various graphs (see graphs).

Table 9.3: Area, Yield and Production of Millet, by Region, 1990

Region	*Area*		*Yield (kg/ha)*	*Production*	
	(10³ ha)	*(Per cent of total)*		*(10³ t)*	*(Per cent of total)*
Asia	20 853	55.5	804	16 767	56.2
Africa	13 548	36.1	669	9 066	30.4
USSR	2903	7.7	1 256	3647	12.2
North and Central America	150	0.4	1 200	180	0.6
South America	55	0.2	1 655	91	0.3
Oceania	34	0.1	882	30	0.1
World	37565	100	794	29817	100

Source: FAO, 1991.

The area under millet production decreased marginally from 38.1 million hectares in 1979-81 to 37.6 million hectares in 1990. However, production increased by 17 percent, from 25.6 million tonnes in 1979-89 to 29.8 million tonnes in 1990, largely because of production increases in Nigeria (65 percent), India (25 percent) and the Soviet Union (207 percent). However, there was a 24 percent decrease in production in China during the same period.

Table 9.4: Leading Millet Producers, 1990

Country	*Area*		*Production*	
	(10^3 ha)	*(Per cent of total)*	*(10^3 t)*	*(Per cent of total)*
India	17 000	45.3	11 500	38.6
China	2 601	6.9	4 401	14.8
Nigeria	4 000	10.7	4 000	13.4
USSR	2 903	7.7	3 647	12.2
Niger	3 100	8.3	1 133	3.8
Mali	900	2.4	695	2.3
Uganda	400	1.1	620	2.1
Burkina Faso	1 150	3.1	597	2.0
Senegal	865	2.3	514	1.7
Nepal	200	0.5	240	0.8
Total	33 119	88.2	27 347	91.7
World(1990)	37 565		29 817	
World (1989)	37 409		29 962	

Source: FAO. 1991.

Table 9.5: Sources of Energy and Protein in the Food Supply of the World's Ten Leading Millet Producers, 1987-89

Country	*Energy per Caput per Day (kcal)*				*Protein per Caput per Day (g)*			
	Total	*Vegetable Products*	*Percentage of Total*	*Animal Products*	*Total*	*Vegetable Products*	*Percentage of Total*	*Animal Products*
India	2196	2 048	93.3	148	53.2	45.6	85.7	7.6
China	2634	2 365	89.8	269	62.8	50.7	80.7	12.1
Nigeria	2306	2 248	97.5	58	49.5	43.6	88.1	5.9
USSR	3380	2 444	72.3	936	106.2	50.1	47.2	56.1
Niger	2297	2 152	93.7	145	64.0	53.2	83.1	10.8
Mali	2234	2 090	93.6	144	62.5	50.1	80.2	12.4
Uganda	2136	2 010	94.1	126	48.1	38.7	80.5	9.4
Burkina Faso	2286	2 186	95.6	100	69.8	62.6	89.7	7.2
Senegal	2374	2 160	91.0	214	68.2	49.9	73.0	18.3
Nepal	2074	1 937	93.4	137	52.5	44.8	85.3	7.7

In all of the ten leading millet-producing countries except the Soviet Union, vegetable products supplied 90 percent or more of total dietary energy in 198789

(Table 9.5). In India, China, Nigeria, the Niger, Mali, Uganda, Burkina Faso and Nepal, vegetable products supplied more than 80 percent of protein. Thus in many sorghum- and millet-producing countries, vegetable products, especially cereals, provide the bulk of energy and protein.

Health Benefits of Pearl Millet

- Millets are a great source of starch, making it a high-energy food. It is also an excellent source of protein and fiber. It is said that the amino acids in the pearl millet are more easily digestible than the ones found in wheat.

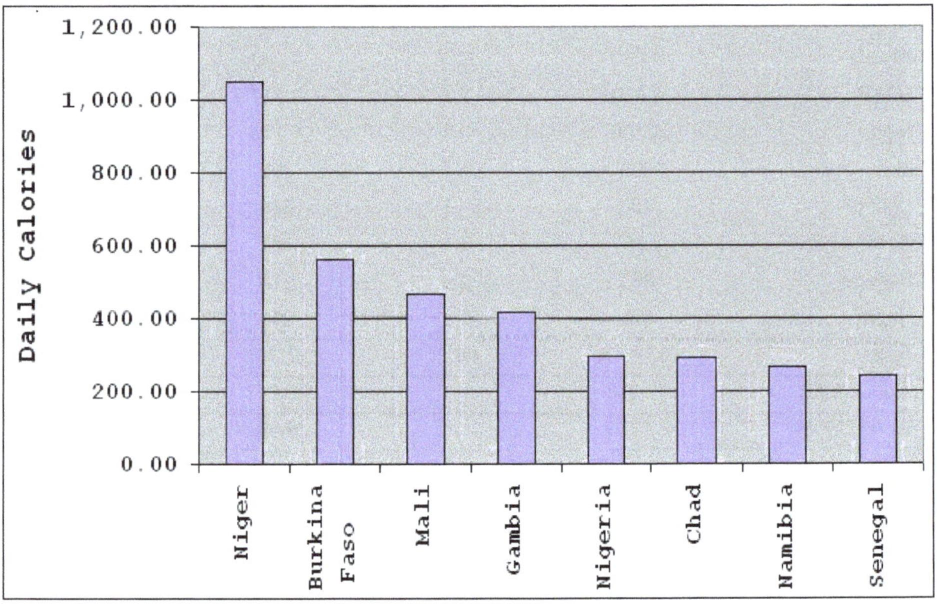

Figure 9.1: Millet Consumption (5 year average).

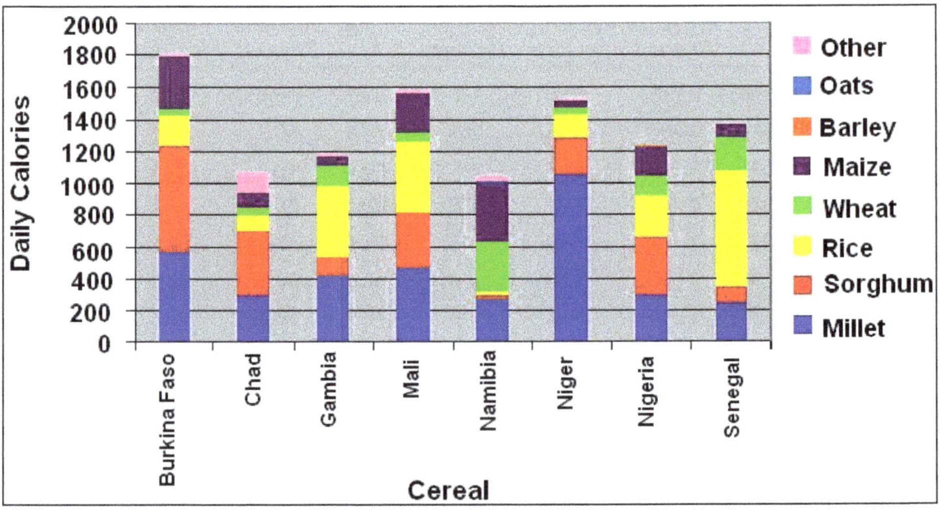

Figure 9.2: Cereal Consumption.

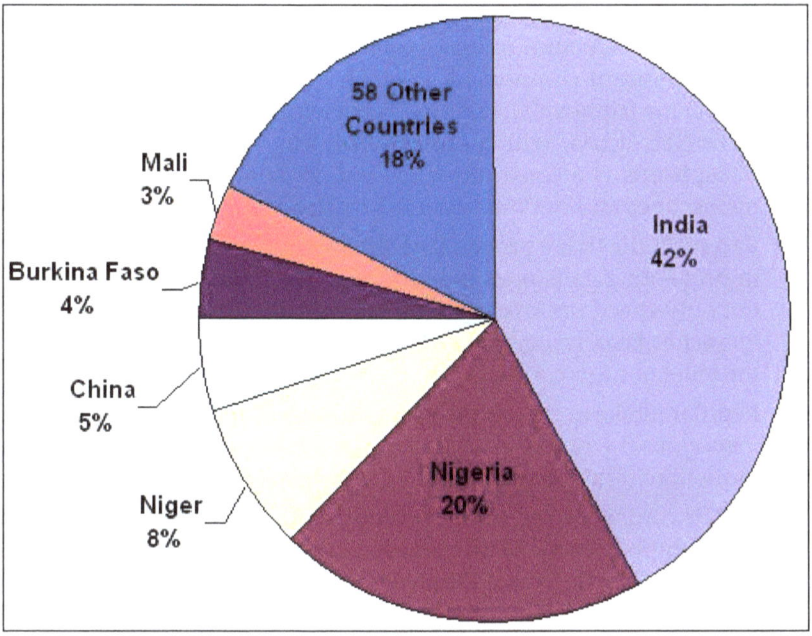

Figure 9.3: Average Millet Consumption (1999-2003).

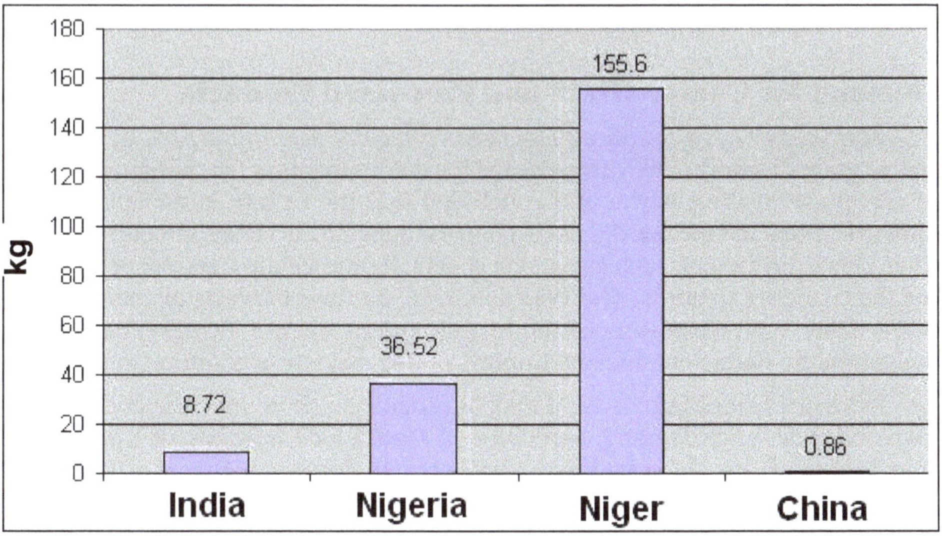

Figure 9.4: Millet Consumption (kg/Cap/Yr).

☆ Due to essential nutrients such as methionine (an amino acid), B complex vitamins (niacin, thiamin and riboflavin), folic acid, lecithin, potassium, magnesium, manganese and zinc, millets are very effective in several roles. Niacin reduces cholesterol while magnesium is essential for maintaining good heart health, as it lowers blood pressure and reduces the risk of heart attacks.

- Pearl millet is a rich source of phosphorus, which plays an important part in the structure of body cells. Phosphorus, found in pearl millets, is a significant component of several necessary compounds including adenosine triphosphate (ATP). This element is also a crucial component of nucleic acids, which are the building blocks of the genetic code. Phosphorus is a constituent of lipid-containing structures such as cell membranes and nervous system structures.
- Recent studies have proven that regular consumption of pearl millets help in preventing gallstones in women. They contain insoluble fibers which not only speed up intestinal transit time but also reduce the secretion of bile acids. Pearl millets are known to increase insulin sensitivity and lower the level of triglycerides.
- Regular intake of millets provides protection against breast cancer in pre-menopausal women. Apart from that, it has also shown a considerable reduction in the occurrence of wheezing and asthma in children.
- Millets contain an essential phytonutrient, lignin, which is very beneficial for the human body. With the help of natural flora, lignans get converted to mammalian lignans and they fight against hormone-dependent cancers and reduce the risk of cardiac arrests.
- Consumption of pearl millets helps in minimizing the risk of type 2 diabetes. Being a good source of magnesium, millets act as a cofactor in a number of enzymatic reactions.

Demand for Coarse Grain and Processed Products

Few studies have specifically focused on factors determining the demand for coarse grains. Debrah (1993) investigated the determinants of the changing patterns of cereal use in West Africa and concluded that the pattern is demand driven. However, price and income did not explain the demand but rather, structural factors. Even short-run factors, such as harvest short fall or price dips, are not responsible for the changing patterns. The 1985 and 1986, bumper harvests of coarse grains in the Sahel, with the associated fall in grain prices, did not increase coarse grain consumption. Rather commercial imports of rice and wheat continue to rise.

Evidence from Mali, Senegal and Burkina Faso show inelastic coarse grain own and cross-price demand elasticities. In Mali, a long-term rise of 1 per cent in coarse grain price is associated with a 0.07 per cent decrease in quantity demanded of coarse grains. Similarly, in Burkina Faso, a long-term rise of 1 per cent in coarse grain price is associated with a 0.5 per cent decrease in quantity demanded of coarse grains. Cross-price elasticities show virtually no impact of wheat and rice price on coarse grain consumption. For example, a long-term rise of 1 per cent in wheat price in Mali is associated with a 0.01 per cent increase in quantity demanded of coarse grains. Demand for coarse grains is also not responsive to income. A rise of 1 per cent in income leads to a decrease of only 0.24 per cent in coarse grain demand in Mali. The situation is, however, different for processed pearl millet and sorghum products. Micro-level evidence from Mali supports the contention that household structural and non-price factors are major determinants for the changing patterns of cereal consumption. Sorghum and pearl millet grain is still cheaper than maize and rice. Following currency devaluation in 1994, millet and sorghum became more competitive than maize. In Niger, maize is mainly imported from Nigeria. With currency devaluation, it became more expensive to import maize. However, this situation did not last long due to an exchange rate re-alignment between the Nigerian's Naira and the CFAF. In contrast to grains, processed products should be more responsive to income. In Urban Mali, Boughton and Reardon (1997) examined the determinants of household purchases of already processed millet, sorghum and maize and found that the purchases of these processed products rise with the opportunity cost of women's time and household income. Processed millet and processed sorghum are similarly affected by the opportunity cost of women's time and household income. As with grain, processed sorghum and millet are cheaper than rice, despite their high processing costs. Dibley *et al.* (1995) assessed the processing and preparation costs for rice and coarse grains in urban Mali to show that coarse grain dishes are still cheaper than rice based dishes. The main contributing factors to the higher cost of rice based dishes were the sauce, cereal and preparation costs. This result held through a wide range of rice and coarse grain prices and opportunity costs of time. The study should be applicable to urban Sahel, not just Bamako. Coarse grains need not take longer to process and prepare than rice dishes if the dehulling stage is mechanized. Pre-processed coarse grains are not competitive with household processing given present technology and the opportunity cost of women's time. Coarse grains will not be able to retain the extent of their cost advantage over rice as income and opportunity cost of labor rise without improvement in the efficiency of processing services. This review suggests that sorghum and millet are still competitive compared to maize. Sorghum and millet grains are still cheaper than maize in the markets. Processed sorghum and millet products are still cheaper than maize and rice products. However, to remain competitive the demand for processed coarse grains must increase. In Mali, Boughton *et al.* (1997) concluded that to increase demand for processed coarse grains, it is necessary to promote access to and use of abrasive-disk dehullers at the semi-whole sale level; promote improvements in the quality and cleanliness of the grains available to dehullers; institute grading by grain quality in the marketing system; and finally, increase consumer information about possible savings realized on purchases of dehulled cereals, via, for example, the Malian Market Information System (SIM). A demand enhancing strategy requires improving product attributes, such as enhanced nutritional qualities and greater convenience in the form of decreased processing time. Demand expansion also requires that questions of consumer acceptance and product characteristics that affect storage, processing, marketing and consumption be addressed. The end-markets of sorghum and pearl millet may be diversified by introducing use as stock feed, industrial utilization for clear beer, opaque beer, composite flour and weaning foods. Sorghum and pearl millet demand expansion will be conditional on the availability of a well functioning marketing system and credit facilities, as well as government policies that affect imports and the purchasing power of consumers.

Utilization of Pearl Millet

The detail procedure is given for processing of pearl millet into ben-saalga (Figure 9.5).

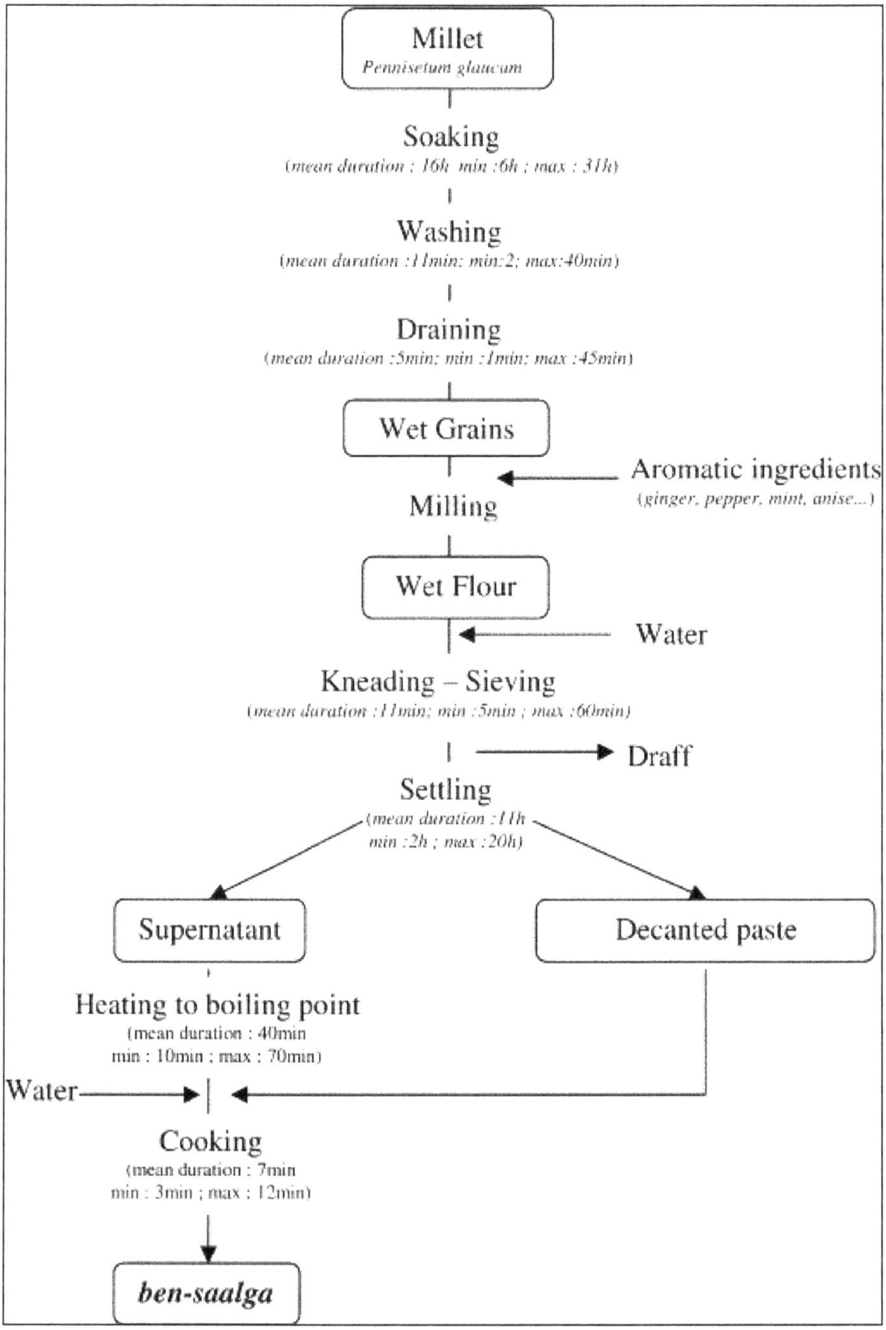

Figure 9.5: Description of the Traditional Processing of Pearl Millet into *ben-saalga*.

Pearl millet is a multipurpose crop, which is grown for food, feed, green and dried (Karvi) forages. The presence of all required nutrients in millets make them suitable for large-scale utilization in the manufacture of various food products such as traditional meals, baby foods, snack foods, dietary foods in both grain and flour form.

Nutritional Composition of Pearl Millet

Both grain and stover of pearl millet have a better mineral profile than many other cereals. The protein contents in grain ranges from 9 to 15 percent, which is much higher than the protein contents in other millets of sorghum and maize. The green fodder is of higher quality with crude protein content ranging from 5 to 15 per cent, total mineral content 3 to 16 per cent, ether extract 0.5 to 2.3 per cent, nitrogen free extract 38 to 56 per cent and crude fiber 25 to 38 per cent at various stages of the crop growth.

Table 9.6: Nutritional Compositions of Millets, Coarse Cereals and Fine Cereals (Per 100 g)

Commodity	*Protein (g)*	*Carbo-hydrates (g)*	*Fat (g)*	*Crude Fibre (g)*	*Mineral Matter (g)*	*Calcium (mg)*	*Phosphorus (mg)*
Sorghum	10. 4	72. 6	1. 9	1. 6	1.6	25	222
Pearl millet	11. 6	67. 5	5. 0	1. 2	2. 3	42	296
Finger millet	7. 3	72. 0	1. 3	3.6	2. 7	344	283
Proso millet	12. 5	70. 4	1. 1	2. 2	1. 9	14	206
Foxtail millet	12. 3	60. 9	4. 3	8. 0	3. 3	31	290
Kodo millet	8. 3	65. 9	1. 4	9. 0	2. 6	27	188
Little millet	8. 7	75. 7	5. 3	8. 6	1. 7	17	220
Barnyard millet	11. 6	74. 3	5. 8	14. 7	4. 7	14	121
Barley	11.5	69.6	1.3	3.9	1.2	26	215
Maize	11.5	66.2	3.6	2.7	1.5	20	348
Wheat	11. 8	71. 2	1. 5	1. 2	1. 5	41	306
Rice	6. 8	78. 2	0. 5	0. 2	0. 6	10	160

Source: National Institute of Nutrition (NIN), Hyderabad (Adapted from DMD 2010).

Pearl millet contains 1.6-2.3 per cent ash and is very rich in major mineral especially Phosphorus. Minerals are also high specially Fe content. HC1-extractability of minerals (as an index of their bioavailability) varies between the cultivars and type of mineral as well as contents of antinutritional factors like phytic acid and polyphenoles. In a study Phytic acid content ranged from 969.3 to 1101.0 mg/100 g and polyphenol ranged from 306.7 to 669.4 mg/100 g. Cooking was found to reduce the these antinutritional factors and caused a slight increase in mineral content while bioavailability was significantly improved for all minerals. Germination for 2, 4 and 6-days significantly decreased major and trace minerals (except Cu and Co) contents. Bioavailability of all minerals was significantly improved throughout the germination time and fermented/germinated foods are very promising. Composition of Millets, Coarse cereals and fine cereals is given in Table 9.6 which indicates that Pearl Millet is better than rice in terms of total protein, fat, crude fibers, mineral matters, calcium and phosphorus. Except protein and phosphorus, same is true when compared with another staple grain of wheat.

Anti Nutritional Factors in Pearl Millet

Beside its nutritional qualities, the bioavailability of various minerals like Ca and P and trace elements such as Zn, Fe, Cu and Mn is low because of presence of some inherent anti-nutritional factors *e.g.* phytate and polyphenols in grain; and oxalic acid in fodder and forage, which are undesirable. Its flour acquires a rancid odour within a few days of milling because of high concentration of lipids that contribute to the development of fat acidity, lipolytic activity and accumulation of peroxides of lipids in the meal during storage. The typical grey colour of pearl millet grain and its products due to polyphenolic pigments present in peripheral area of the endosperm further restricts efficient utilization of pearl millet.

However, several processing techniques have been developed to enhance food value and shelf-life of pearl millet products and to improve the availability of starch, protein and minerals. Decreasing or degrading of phytic acid is very advantageous, due to its influence on nutrition.

Processing of Pearl Millet into Various Products

Technologies for various processing treatments, such as milling, malting, blanching, acid treatment, dry heating and fermentation, which reduce antinutritional factors and increase the digestibility and shelf life of various alternative food products made from pearl millets such as unleavened flat bread (roti/chapati), porridges, noodles, bakery products and extruded and weaning food products have been developed and tested at the laboratory scale. Some of the techniques are Dehulling, Milling, Malting, Blanching, Acid treatment, Dry heat Treatment and parboiling.

Dehulling

Both whole grains and dehulled (decorticated) grains of pearl millet are used for preparing various types of food products. Decortication is generally to the extent of removing 12-30 per cent of the outer grain surface. Increased decortication leads to greater loss of fiber, ash and fat. It also reduces protein, lysine, histidine and arginine. Decorticated grains improve the nutritional quality and sensory properties of various food products, but these also have cost considerations in terms of the time and investments and grain weight losses. Further, these also lead to micronutrient losses, which are more concentrated in the outer layers of the grain. Pearl millet grains can be decorticated in rice mills or other modified mills. In some villages and urban areas, pearl millet grains are decorticated with abrasive disks in mechanical dehullers.

Milling

Grains can be milled either by using a hammer mill or a roller mill. The flour produced using a hammer mill has large particle size and is not uniform; hence it is suitable for preparing thin and stiff porridge of rough texture and not suitable for preparing baked and steamed food products of smooth texture. Shelf life of grains is almost 18 months whereas that of flour is about 2 weeks due to its high fat contents and ranicificaton after grinding. This is a limitation as well as opportunity for preparing fermented products. A new method for improving the shelf life of pearl millet has been developed at the Central Food Technological Research Institute (CFTRI), Mysore. It involves moist heating of the grains followed by drying to about 10-12 per cent moisture and decortication to the desired degree or pulverization. This process improves the milling characteristics of pearl millet varieties which have high proportions of floury endosperm. Flour from treated and decorticated pearl millet could be stored for about 3-4 months, during which the free fatty acid (FFA) content remained below 10 per cent, which is the limit of perceptible deteriorative condition. The oxidative rancidity also remains low, as the flours are refined. Another advantage of this process is that the microbial load on the grain surface is drastically reduced.

Malting

This process involves limited germination of cereal in moist air under controlled conditions. Malting reduces protein, but improves the quality of protein compared to that in the bran, so a small loss in protein in milling of the malted pearl millet is compensated for by protein quality. The process results in a higher protein efficiency ratio and bioavailability of malting of pearl millet grains with 48-h germination reduced polyphenols and phytic acid by more than 40 per cent. Malting also increases vitamins such as riboflavin, thiamin, ascorbic acid and vitamin A. There was little effect of malting on increasing the shelf-life of flour.

In Africa, millet beer branded as Opaque Beer, Bantu, Kaffir Beer, Boza, etc. are quite popular. In India, pearl millet is procured by breweries especially in glut years and actual figures and its uses are not known. Published information about malting is available in the journals and articles. It will be advisable to come out with a cheaper brand of pearl millet bear to ensure proper price to the farmers.

Blanching

It is one of the effective processing technologies to increase the shelf life of pearl millet. Blanching is usually done by boiling water at 98°C in a container then submerging the grains in the boiling water (1:5 ratios of seeds to boiling water) for 30 seconds and drying at 50°C for 60 minutes. Blanching has been observed to be effective in the retardation of enzymatic activity and thus improve the shelf life of pearl millet flour without much altering the nutrient content (Chavan and Kachare, 1994). Blanching of seeds at 98°C for 10 seconds in boiling water before milling has been reported to effectively retard the development of fat acidity in meal and enhance shelf life by 25 days (Kadlag *et al.*, 1995). Fat acidity increased about 6-fold in untreated pearl millet flour, whereas it remained almost unchanged in flour obtained from boiled water-blanched grains (98°C for 30 seconds) (Chavan and Kachare, 1994). As compared to the high levels of polyphenols (755 mg/100 g grains) and phytic acid (858 mg/100 g grains) in the untreated controls, blanching of pearl millet seeds reduced the polyphenol and phytic acid anti-nutritional contents considerably. Also, fat acidity was reduced significantly in the case of blanched pearl millet flour as compared to raw flour after 28 days of storage (Rekha, 1997).

Acid Treatment

The dark-grey grain in pearl millet is highly preferred in Maharashtra state of India. Elsewhere in India and most of the world, this grain color is not preferred for food purposes. Treating the decorticated seed with mild organic acids, such as acetic, fumaric, or tartaric and also with the extracts of natural acidic material such as tamarind (Hadimani and Malleshi, 1993) has been found to improve the product quality by reducing polyphenols and other anti-nutritional factors, thereby also increasing consumer acceptability. Various studies have reported that soaking of pearl millet in acidic solutions, like sour milk or tamarind, markedly reduced the colour of the grain. Dehulled grains decolorized faster than whole grains because the acidic solution penetrates the grain at a faster rate (Reichert and Youngs, 1979). Among the various acidic solutions tried, dilute hydrochloric acid was more effective and suitable chemical treatment to remove pigments from whole grain before milling as compared to citric acid and acetic acid (Naikare *et al.*, 1986). Soaking grains in dilute HCl for 15 to 24 h reduces a major portion of these pigments and thus helps in the production of creamy white grains. Soaking of pearl millet in 0.2 N HCl for 24 h reduced polyphenols by 76 per cent and phytic acid by 82 per cent as compared to 755 mg/100g polyphenol and 858 mg/100g grains of phytic acid in the untreated control. While fat acidity of the flour during 28 days of storage increased 4-fold in the untreated control, there was very marginal increase in the flour produced from the acid-treated grains. Similar patterns of changes were observed in the acid-treated and control treatments with respect to free fatty acids and lipase activity. In another study, pearl millet grain samples given acid treatments for 6, 12, 18 and 24 h had in vitro protein digestibility increased by 29, 44, 56 and 59 per cent, respectively and the in vitro starch digestibility increased by 40, 57, 76 and 85 per cent, respectively.

Dry Heat Treatment

Lipase activity is the major cause of spoilage of pearl millet meal, so its inactivation before milling improves the meal quality. The application of dry heat effectively retards lipase activity and minimizes lipid decomposition during storage. It has been observed that when pearl millet grains were given a dry heat treatment in a hot air oven at 100±2°C for different time periods ranging between 30 and 120 min and then cooled to room temperature, there was about 50 per cent increase in fat acidity, free fatty acids and lipase activity during the 28 days of the storage of flour produced from the acid-treated grains, while there was a 4-fold increase in these parameters in the flour produced from untreated grains. Heating grains for 120 min has been found to be most effective for maximum retardation of the lipolytic decomposition of lipids during storage (Kadlag *et al.*, 1995). Fat acidity, free fatty acid presence and lipase activity decrease significantly during storage of

28 days in pearl millet flour given an 18-h acid treatment and a 120 minutes heat treatment. Results also showed that heat treatment increased the shelf life of pearl millet flour as compared to raw flour (Poonam, 2002).

Parboiling

Parboiling is especially beneficial for soft-textured grains. Parboiled grains decorticate more efficiently in removing the germ and the pericarp. Parboiled decorticated grains have slightly lower protein digestibility than the raw grains decorticated to the same extent. In practical terms, however, this detrimental effect is negligible since most traditional food processes involve cooking of flour or decorticated grains. The parboiled grains can be used for various snack food items, especially for diabetics (Sehgal *et al.*, 2004). Parboiled grains can also be cooked to produce rice-like products. In pearl millet, parboiling can prolong the shelf life of the products such as milri.

Alternate Uses of Pearl Millet

During the last 2 decades the new sources of demand for sorghum and pearl millet have been emerging from different sectors. As evident from the Table 21, the share of sorghum and millet alternative uses to availability increased phenomenally, especially after 1990's. The demand for sorghum and pearl millet has potential in the poultry feed (especially as layer feed) and animal feed sector. Additionally, there is a growing demand for the grain of these two crops from alcohol industry for manufacture of potable alcohol. Thus, the demand for sorghum and pearl millet is derived from the demand for feeds both for animals and poultry and other industrial uses. This requires efforts from breeders for development of varieties based on industrial preferences traits.

Pearl Millet for Brewing and Production of Alcohol

Seagram is the only company in India which uses grain and not molasses, in the production of its blended whiskies. This company has already invested Rs. 35 million (0.83 million dollars) on a research and development centre in the Western State of Maharashtra to develop methods for converting pearl millet, grown by farmers in Maharashtra and ragi cultivated in Rajasthan State into whisky. This company has plans to invest Rs. 100 million ($2.38 million) over the next five years in grain alcohol research using locally grown raw material. The Department of Scientific and Industrial Research (DSIR) under the Ministry of Science, has also lent its support to the grain alcohol project. It recently awarded full recognition to the Seagram R and D centre, making it eligible for coveted government incentives such as tax benefits and import of duty-free equipment.

Use in Poultry and Animal Feed

Pearl millet-supplemented poultry feeds are generally superior to sorghum and equivalent to maize in broiler diets. Pearl millet is also suitable for use in cattle feeding if diets are adjusted to accommodate higher protein levels in the grain. In general, pearl millet is at least equivalent to maize and often superior to sorghum in cattle, pig and sheep rations because of its high energy and grain protein levels.

Its grain, as a high-energy alternative source of inexpensive feed, could be exploited and utilized in production of high quality fish.

Use of Pearl Millet as Fodder

Pearl Millet is also cultivated for forage mostly in North India. Pearl millet varieties and hybrids are available exclusively for forage purpose. Green forage yield, dry forage yield and crude protein of pearl millet is more compared to sorghum and maize. Yield of green and dry forage and crude protein contents of sorghum, pearl millet, maize and pigeon pea are given in Table 9.7.

Table 9.7: Yield of Green and Dry Forage and Crude Protein Contents of Sorghum, Pearl Millet, Maize and Pigeon Pea

Crop	*Green Forage Yield (t/ha)*	*Dry Forage Yield (t/ha)*	*Crude Protein (per cent)*
Sorghum	32.7	7.7	6
Pearl millet	37.6	8.5	8.7
Maize	30.9	6.5	5.5
Pigeon pea	40.6	12.6	23.7

Use of Pearl Millet as Roof Material

Dried pearl millet stalk makes a very good roofing material for constructing traditional houses at a very cheap cost.

Value Added Food Products

Processed pearl millet grains and meals from them are used to prepare various types of traditional and nontraditional food products. Murty and Kumar (1995) summarized and classified these into 9 major food categories (thick porridge, thin porridge, steam-cooked products, fermented breads, unfermented breads, boiled rice-like products, alcoholic beverages, nonalcoholic beverages and snacks); and they provided the details of their preparations and the various common names in many countries.

Traditional Food Products

The simplest and the most common traditional food made from pearl millet are thin porridge (gruel); thick porridge (fermented and unfermented) and flat and unfermented bread such as chapati. Flat, unleavened bread prepared from pearl millet flour enriched with soy flour has been reported to have high protein efficiency ratio, minimal thickness, puffing and uniform color and texture. Chapati prepared from pearl millet flour produced after the grains had been bleached or acid-treated for heat-treated has been reported to have enhanced overall acceptability as compared to the chapati prepared from the raw untreated grains (Poonam, 2002). Use of processed flour, in comparison to raw flour, in the product development has been found to reduce anti-nutritional factors and increase the digestibility (Singh, 2003). Various types of snacks are also made from pearl millet in India. Products like laddoo, namkeen sev and matari have been made using blanched and malted pearl millet flour. These products were highly acceptable and have shown to have longer shelf life and stored well up to 3 months. Rekha (1997) incorporated blanched and malted pearl millet flour in various products like bhakri, suhali, khichri, churma, shakkarpala, mathari and the products were found to be organoleptically acceptable. An earlier study (Chaudhary, 1993) also indicated that the traditional food products including chapati, khichri, bhakrii, popped grain, dalia and shakkarpala prepared from pearl millet were not only acceptable but their protein and starch digestibilities were also better.

Baked Products

Pearl millet flour is not a good raw material for the baking industry, since it does not contain gluten and this forms dough of poor consistency. For instance, cookies made from pearl millet flour do not spread during baking, have a poor top grain character and are dense and compact (Badi *et al.*, 1976). However, pearl millet flour hydrated with water, dried and supplemented with 0.6 per cent unrefined soy lectin can produce cookies with spread characteristics equal to those made from soft wheat flour. Various types of biscuits and cakes produced using blanched pearl millet have been found to be organoleptically acceptable. Various types of biscuits developed by incorporating different levels of blanched as well as malted pearl millet flour have been found to be acceptable and store well up to 3 months (Singh, 2003).

Extruded Products

Extrusion is being used increasingly for making ready-to-eat foods. In extrusion processes, cereals are cooked at high temperature for a short time. Starch is gelatinized and protein is denatured, which improves their digestibility. Anti-nutritional factors that are present may be inactivated. Pearl millet grit and flour can be used to prepare ready-to-eat (RTE) products. Such products have crunchy texture and can be coated with traditional ingredients to prepare sweet or savory snacks. Alternatively, the grits could be mixed with spices and condiments prior to extrusion to obtain RTE snacks of desirable taste. The acid-treated pearl millet yields products of better acceptability as compared to that from just decorticated pearl millet. Pearl millet, blended with soy or protein-rich ingredients, such as legumes or groundnut (peanut) cake, on extrusion gives nutritionally balanced supplementary foods (Malleshi *et al.*, 1996). Sumathi *et al.* (2007) showed that extruded pearl millet products prepared from a blend of 30 per cent grain legume flour or 15 per cent defatted soybean had, respectively, 14.7 per cent and 16.0 per cent protein and 2.0 and 2.1 protein efficiency ratio. The shelf life of the extrudates was about 6 months in different flexible pouches under ambient storage conditions. Noodles, macaroni and pasta-like extruded products could be prepared from pearl millet flour. (Desikachar, 1975). Extruded snacks prepared with mixed millet flour containing rice flour and/or corn flour and/or tapioca starch in various proportions have been shown to have acceptable appearance, color, texture and flavor (Siwawij and Trangwacharakul, 1995). Extrusion-cooking also enhances the in vitro protein digestibility of foods (Malleshi *et al.*, 1996). Utilization of pearl millet for producing soft-c ooked products such as vermicelli noodles is very rare, although these grains are unique with respect to taste and aroma and provide dietary fiber. Research at the CFTRI, has led to a process to prepare noodles (Sowbhaghya and Ali, 2001a). The noodles on cooking in water retained the texture of their strands and firmness without disintegration and the solid loss is less than 6 per cent (Sowbhaghya and Ali, 2001b). The noodles from pearl millet were readily acceptable in the savory and sweet formulations.

Flakes and Pops

Extensive work has been carried out on sorghum flaking at CFTRI, Mysore and various process parameters, such as soaking time, temperature, wet-heat or dry-heat treatment conditions, have been standardized. The grain soaked to its equilibrium moisture content is steamed or roasted to fully gelatinize the starch, dried to about 18 per cent moisture content, conditioned, decorticated and then flaked immediately by passing through a pair of heavy-duty rollers. The flakes can also be used for the preparation of traditional snacks like 'uppitu' after boiling and seasoning. The thicker flakes could be deep-fried or dry-roasted to prepare expanded and crunchy snack products. Results of exploratory studies on flaking of pearl millet following the method adopted for sorghum have been promising. Pearl millet flaking would be a new avenue for its widespread utilization. Since stabilization of the oil occurs during flaking, pearl millet flakes will have longer shelf life.

Since popping involves formation of steam and development of pressure inside the grain, the optimum moisture level and popping temperature play important roles in the quality of the popped cereal. Varietal differences exist largely with respect to popping characteristics. The optimum conditions for grain popping, according to the CFTRI process, are equilibrating pearl millet to about 16 per cent moisture and subjecting the grains to a high-temperature, short-time treatment (about 230°C for a fraction of a minute) in an air popper developed at the Institute (CFTRI, 1985). The machine is highly suitable for value addition to pearl millet by popping. Popping of pearl millet is not very popular, but the popped pearl millet is a good source of energy, fiber and carbohydrates. Varieties with hard endosperm and medium-thick pericarp exhibit superior popping quality (Hadimani *et al.*, 2001). The lipolytic enzymes are denatured during the process of popping. The nutritional advantage of the popped millet is utilized in developing formulations for supplementary foods or weaning foods for children and mothers (Bhaskaran *et al.*, 1999). Since pearl millet is a rich source of micronutrients and phytochemicals, such products may score over similar products made from rice and wheat.

Weaning Foods

Pearl millet can be successfully utilized for the development of weaning foods, as it can satisfy the nutritional requirement of infants during the crucial transitional phase of life from breast milk to other type of food, at reasonable cost. Keeping in view the delicate digestive system and nutritional requirement of the infants, malting seems to be an effective process as it provides an opportunity to develop easily digestible and nutritious weaning foods of low viscosity, low dietary bulk and of high calorie density. In addition malting also improves the availability of protein, minerals, free sugars, vitamin B and ascorbic acid by reducing the level of anti-nutritional factors. It also imparts desirable flavour and taste to the product. Blanched pearl millet can also be used for weaning foods. Blanching improves the storage stability by retarding the lipolytic spoilage of meal without much altering its nutrients. Nutritive value of pearl millet-based weaning foods can further be enhanced when mixed with legumes like cowpea or green gram because these pulses complement the profile of essential amino acids which is beneficial for infants' optimum growth.

Health Foods

Pearl millet can find uses in preparing various types of health foods and food ingredients as it contains a relatively higher proportion of insoluble dietary fiber. This causes slow release of sugar, thus making the food products based on them especially suitable for those suffering from or prone to diabetes. For instance, various pearl millet-based food products were found to have a lower glycemic index (GI) than those based on wheat, with the extent of reduction in the GI trait ranging from 20 per cent for biscuits to 45 per cent for dhokla (Table 9.8).

Table 9.8: Health Value of Pearl Millet-Based Diabetic Food Products

Product	*Glycemic Index*	
	Control (Wheat flour)	*Pearl Millet Based Products*
Biscuit	72.7	58.1
Chapati	69.4	48.0
Dhokla	68.4	38.0
Instant Idli	69.8	52.1
Pasta	71.3	54.1

Source: Mani *et al.* (1993).

Gluten intolerance, leading to protein allergy (specifically gliadin allergy), is a physiological disorder from which about 500,000 people suffer in the USA alone (Dahlberg *et al.*, 2004). Pearl millet is gluten-free and, hence, has a good chance of being commercialized for the food-based management of this problem. Pearl millet is rich in oil and linoleic acid accounts for 4 per cent of the total fatty acids in this oil, giving it a higher percentage of n-3 fatty acids as compared to maize in which linoleic acid accounts for only 0.9 per cent of the total fatty acids and, hence, is highly deficient in n-3 fatty acids. The n-3 fatty acids play an important role in many physiological functions, including platelet aggregation, cholesterol accumulation and the immune system. Pearl millet in poultry feed can have a significant effect on the fatty acid composition of eggs and, consequently, on human health. In a poultry feeding trial, it was observed that eggs produced from layers fed a pearl millet-based diet had lower n-6 fatty acids and higher n-3 fatty acids and, thus, led to lower n-6:n-3 fatty acid ratios than those fed corn-based diets. These eggs are of special health value, especially for those prone to high levels of LDL in the cholesterol.

The bran separated as a by-product during grain processing could serve as a source of the edible oil similar to that of rice bran oil. Deoiled bran from pearl millet has lower ash and silica contents as compared to that of deoiled rice bran. Thus, it could be efficiently used as a source of dietary fiber. Pearl millet bran contains a high proportion of soluble dietary fiber and could be tapped for hypocholesterolemic and hypoglycemic effects. In view of this, fiber regulated pearl millet flakes could be an ideal snack for the obese and for calorie-conscious people (Hadimani and Malleshi, 1993).

Drinks

Pearl millet flour is used in making different types of drinks. A fermented drink known as rab/rabari is consumed widely during summer months in Rajasthan. Similarly, traditional drink called as 'Cumbu Cool' is consumed in Tamil Nadu. The National Dairy Research Institute, Karnal has recently developed and launched pearl millet 'lassi' made from pearl millet flour.

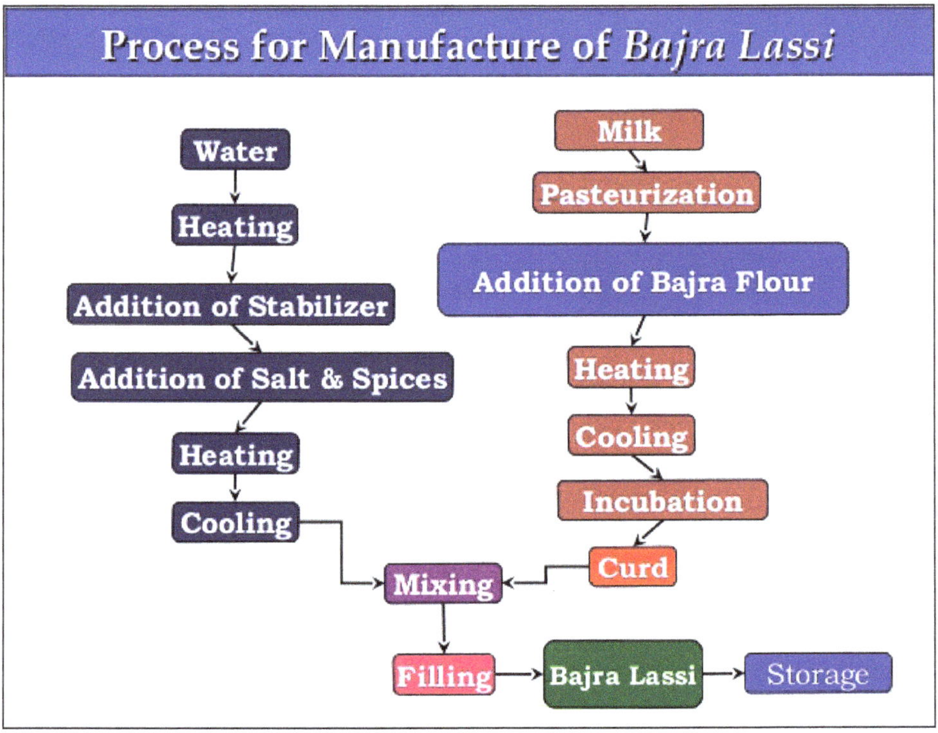

Figure 9.6: Process for Manufacture of *Bajra Lassi*.

Opportunities for Commercialization

One of the greatest constraints in the commercialization of pearl millet grain for food purposes has been a misplaced social stigma dubbing this as poor man's crop primarily because it is grown in marginal environments, where poverty is common. Thus, pearl millet could not make it to the food basket of the urban elite whose consumption choices play a dominant role in the commercialization of any food product. Grain quality and nutritional studies now show that pearl millet grains

Bajara Laddoo **Bajara Chapati** **Bajara Suhali**

Bajara Bakali **Bajara Khichari** **Bajara Mathi**

Chocolate Cake **Nan-khatai** **Bajara Cookies**

Kaju Biscuit **Bajara Eggless Cake** **Melting Movement**

are more appropriate choice for the nutritional security of the rural and urban poor who have limited access to other sources of dietary components. In addition, pearl millet grains could also be more appropriate choices than the fine cereals such as wheat and rice for the elite who will benefit from their high nutraceutical properties. This will require different approaches to commercialize pearl millet to serve the widely different consumer classes. Policy support from the governments plays a significant role in product and process commercialization, at least in the initial stages when the food products from grains of new crop species have to compete with those from the established crop species. For instance, subsidy on wheat and

Bajara Sev

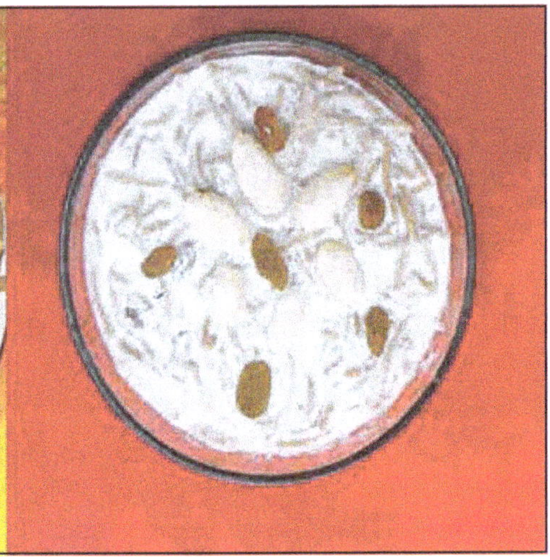

Bajara Sweet Vermicelli

Bajara Thalpit

Bajara Utaapav

Bajara Sweet Cakes

Bajara Khichadi

rice production almost all over the world plays a big role in their production and marketing. On top of this, is the subsidized procurement and supply of wheat and rice through the public distribution system in India. Similar support is not available to pearl millet. This leaves farmers with little incentive for investment in production as the returns are not economical when increased production leads to a drop in

Bajara Pan Cakes

Bajara Salty Cake

Bajara Idali

Bajara Masala Puri

Bajara Chapati

Bajara Cakes

Bajara Muffins

Bajara Chocolate Biscuits

grain prices. The low-resource agriculture, characterized by rainfed cultivation of pearl millet crops with negligible external inputs, leads to low productivity with large variation in production and grain surplus across the years. The low volume and inconsistency in grain supplies reduce the dependability of producers for grain supplies, which is so essential for commercialization. Opportunities exist to drastically reduce or even eliminate these uncertainties through governmental policy support for increased and stable production and marketing of pearl millet

Bajara Bread

Bajara Decorated Cake

Bajara Sweet Laddoo

Bajara Sweet Laddoo

Bajara Idli

Bajara Processing Machines

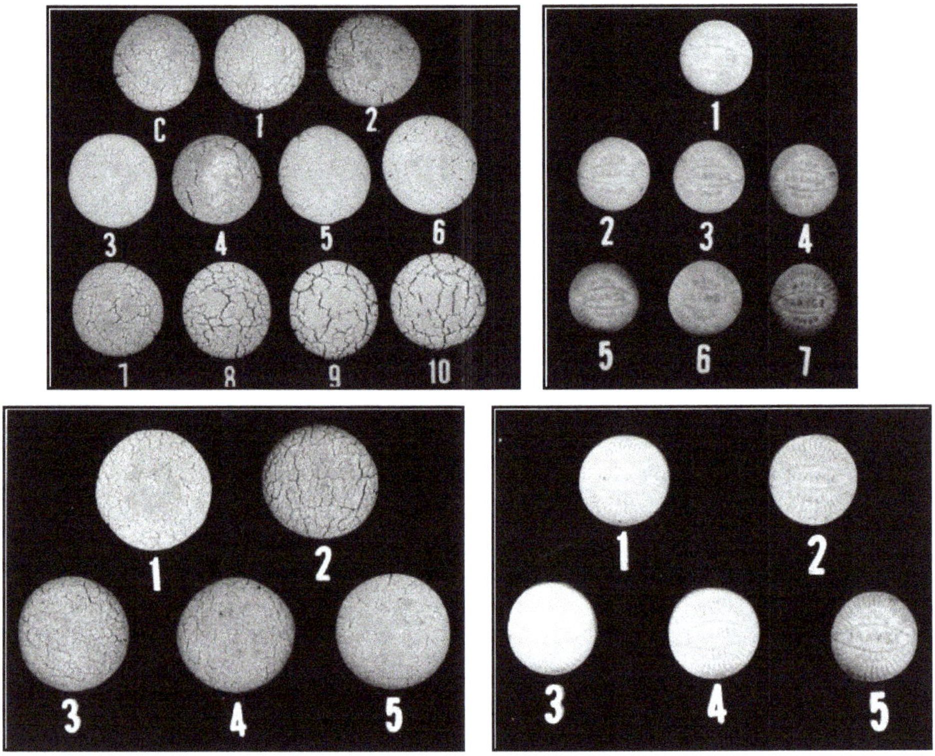

Bajara Various Types of Biscuits

Bajara Discolored Grains

Bajara Beer

Bajara Masala Puri

Bajara Palak Puri

Bajara Roti

Bajara Roti

Bajara Sweets

Polished Bajara Roti

Bajara Mehta Puri

Bajara Parotta

Bajara Sweets

Bajara Chakali

Bajara Peanut Cookies

Bajara Beer

Bajara Masala Sev

Bajara Pan Cake

Bajara Sweet Porgi

Bajara Khichadi

Bajara Masala Flakes

Bajara Flakes

Bajara Pops

grain surpluses. Most of those involved in commercial grain processing and food manufacturing are not familiar with the possible alternative food uses and health value of pearl millet. Commercialization of pearl millet grains for alternative and health food uses needs to be viewed in a broader context from production to utilization and emerging challenges and opportunities.

Research and Development

To combat the problem of micronutrient malnutrition in Africa and Asia, a study of serving iron-biofortified pearl millets which is bred conventionally without genetic modification to a control group is proved to have higher level of iron absorbance by the group. The most widely grown millet is pearl millet, which is an important sized crop in India and parts of Africa. Finger millet, proso millet and foxtail milletare also important crop species. In the developed world, millets p9-yhn n97 are less important. For example, in the United States the only significant crop is proso millet, which is mostly grown for bird seed.

How to Buy Pearl Millet

- ✰ Pearl millets are available in hulled and whole-grain forms. It is commonly found in pre-packaged and bulk containers in the grain section of supermarkets.
- ✰ While purchasing them in bulk, check that the bins are covered and the store has a good product turnover. This is done to ensure maximum freshness.
- ✰ Check for any signs of moisture while purchasing millets.

Pearl Millet Storage Tips

- ✰ When stored in an airtight container in a cool, dry and dark place, pearl millets can be kept for several months.

References

Badi, S.M. and Hoseney, R.C. 1976. Use of sorghum and pearl millet flours in cookies. *Cereal Chemistry* 53 (5): 733-738.

Basavaraj, G., Parthasarathy, R., Bhagavatula, Shraavya and Ahmed, Wasim. 2010. Availability and Utilization of Pearl Millet in India. *Ejournal.icrisat.org (An open Access Journal of ICRISAT)*; 8: 1-6.

Bhaskaran, V., Mahadevamma, Malleshi, N.G., Shankara, R. and Lokesh, B.R. 1999. Acceptability of supplementary food based on popped cereals and legumes suitable for rural mothers and children. *Plant Foods and Human Nutrition* 53: 237-247.

Board on Science and Technology for International Development; Office of International Affairs; National Research Council (1996-02-14). "Pearl Millet: Subsistence Types". *Lost Crops of Africa: Volume I: Grains*. Lost Crops of Africa **1**. National Academies Press. p. 108. ISBN 978-0-309-04990-0. Retrieved 2007-11-07.

Boughton, D. and T. Reardon. 1997. Will Promotion of Coarse Grain Processing Turn the Tide for Traditional Cereals in The Sahel? Recent Empirical Evidence from Mali. *Food Policy*, 224:307-316.

CFTRI. 1985. Annual Report 1984-85. Mysore, India: CFTRI. 11 pp.

Chaturvedi, A. and Sarojini, G. 1996. Malting of pearl millet (*Pennisetum typhoidum*): its effects on starch and protein digestibilities. *Journal of Food Science and Technology* 33 (4): 343 - 344.

Chaudhary S. 1993. Preparation of nutritional evaluation of some traditional foods from pearl millet. *M.Sc thesis*, CCS Haryana Agricultural University, Hisar, Haryana, India. 98 pp.

Chavan, J.K and Kachare, D.P. 1994. Effect of seed treatment on lipolytic deterioration of pearl millet flour during storage. *Journal of Food Science and Technology* 31: 81-82.

Collins, V.P., Cantor, A.H., Pescatore. A.J., Straw, M.L. and Ford, M.J. 1997: Pearl Millet in Layer Diets Enhances Egg Yolk n-3 Fatty Acids. *Poultry Science* 76: 326–330.

Debrah, S.K. 1993. The competitiveness of sorghum and millet in the food system of West Africa in Sorghum and Millet Marketing and Utilization in Tanzania. SADC/ICRISAT. *National Workshop on Sorghum and Millet Marketing and Utilization for Food Security and Development*. Sponsored by the International Development Research Center. Arusha International Conference Center, 3-5 May 1993.

Desikachar, H.S.R. 1975. Processing of maize, sorghum and millets for food uses. *Journal of Scientific and Industrial Research* 34: 231-236.

Dibley, D., D., Boughton and T. Reardon. 1995. Processing and Preparation Costs for Rice and Coarse Grains in Urban Mali: Subjecting an Ipse Dixit to Empirical Scrutiny. *Food Policy*, 20(1): 41-50.

Directorate of Millets Development 2010. *Status Paper on Millets*. Directorate of Millet Development, Ministry of Agriculture, Government of India, Jaipur

Enhancing food security in Namibia through value-added products. Council for Scientific and Industrial Research. March 2003. Archived from the original on 6 December 2005. Retrieved4 March 2012.

Fuller, D. Q. (2003). African crops in prehistoric South Asia: a critical review. in Neumann,K., Butler,A., Kahlheber,S. (ed.) Food, *Fuel and Fields. Progress in Africa Archaeobotany*. Africa Praehistorica 15 series. Cologne: Heinrich-Barth-Institut, 239-271.

Hadimani, N. A. and Malleshi, N. G. 1993. Studies on milling, physico-chemical properties, nutrient composition and dietary fibre content of millets. *Journal of Food Science and Technology* 30(1): 193-198.

Hadimani, N.A., Murali Krishna, G., Tharanathan, R.N. and Malleshi, N.G. 2001. Nature of carbohydrates and proteins in three pearl millet varieties varying in processing Characteristics and kernel texture. *Journal of Cereal Science* 32:17-25.

Haryana, India. 106 pp. Rai, K.N., Gowda, C.L.L., Reddy, B.V.S. and Sehgal, S. 2008. Adaptation and Potential Uses of Sorghum and Pearl Millet in Alternative and Health Foods. *Comprehensive Reviews in Food Science and Food Safety*. 7(4):340-352

http://www.holidaytravel.com.na/index.php?fArticleId=222

Kadlag, R.V., Chavan, J.K. and Kachare, D.P. 1995. Effect of seed treatment s and storage on the changes in lipids of pearl millet meal. *Plant Foods for Human Nutrition* 45: 279-285.

Malleshi, N.G. and Klopfenstein, C.F. 1998. Nutrient composition and amino acid contents of malted sorghum, pearl millet and finger millet and their milling fractions. *Journal of Food Science and Technology* 35(3): 247-249.

Malleshi, N.G., Hadimani, N.A., Chinnaswamy, R. and Klopfenstein, C.F. 1996. Physical and nutritional qualities of extruded weaning foods containing sorghum, pearl millet, or finger millet blended with mung beans and non fat dried milk. *Plant Foods and Human Nutrition* 49: 181-189.

Mani, U.V., Prabhu, B.M., Damle, S.S. and Mani, I. 1993. Glycemic index of some commonly consumed foods in western India. *Asia Pacific Journal of Clinical Nutrition* 2: 111-114.

Manning, Katie, Ruth Pelling, Tom Higham, Jean-Luc Schwenniger and Dorian Q Fuller (2010) 4500-year old domesticated pearl millet (Pennisetum glaucum) from the Tilemsi Valley, Mali: new insights into an alternative cereal domestication pathway. *Journal of Archaeological Science* 38 (2): 312-322

Meera, M.S., Ali, SZ., Narasimha, H.V., Bhashyam, M.K., Srinivas, A. and Sathyendra, Rao B.V. 2002. A process for the preparation of sorghum and pearl millet grain based product having enhanced shelf life. Patent filed in India and Egypt. Indian Patent 306/DEL/02.

Millet. Consultative Group on International Agricultural Research.

Munyaradzi, Makoni (29 August 2013). "Biofortified pearl millet 'can combat iron deficiency'". *SciDev Net*. Retrieved 29 August 2013.

Murty, D.S. and Kumar, K.A. 1995. Traditional uses of sorghum and millets. Pages 185—221 in *Sorghum and millets: Chemistry and technology* (Dendy DAY ed.). St. Paul, Minnesota, USA: American Association of Cereal Chemists.

Naikare, S.M., Chavan, J.K. and Kadam, S.S. 1986. Depigmentation and utilization of pearl millet in the preparation of cookies and biscuits. *Journal of Maharashtra Agricultural Universities* 11:90.

Poonam. 2002. Effect of acid and heat treatment on nutrient composition and shelf life of pearl millet (*Pennisetum glaucum*) flour. *M.Sc. Thesis,* CCS Haryana Agricultural University, Hisar.

Rao, N.B.S. 1987. Nutritional implications of millets and pseudo millet. *Nutrition News* 8 (5): 1- 3.

Reichert, R.D. and Youngs, C.G. 1979. Bleaching effect of acid in pearl millet. *Cereal Chemistry* 56: 287.

Rekha. 1997. Efficacy of processing techniques in the utilization of pearl millet for value added products. *M.Sc. Thesis*, CCS Haryana Agricultural University, Hisar, Haryana, India. 125 pp.

Singh, G. 2003. Development and nutritional evaluation of value added products from pearl millet (*Pennisetum glaucum*). *Ph.D. Thesis*, CCS Haryana Agricultural University, Hisar, Haryana, India.

Siwawij, S. and Trangwacharakul. 1995. Study on sorghum snack production by extrusion process. *Thailand Journal of Agricultural Sciences* 28 (3): 253 - 261.

Sowbhagya, C.M. and Ali, S.Z. 2001a. Vermicelli noodles and their quality characteristics. *Journal of Food Science and Technolology* 38: 423-432.

Sowbhagya, C.M. and Ali, S.Z. 2001b. A process for manufacturing high fibre low fat vermicelli noodles from millets. Indian Patent 1128/DEIV99.

Sumathi A, Ushakumari S.R., Malleshi N.G. 2007. Physico-chemical characteristics, nutritional quality and shelf-life of pearl millet based extrusion cooked supplementary foods. *Int J Food Sci Nutr* 58:350–62.

Yadav, O.P. Project-Coordinator, Pearl Millet, ICAR, 2012: Personal Communication.

Chapter 10

Proso Millet (*Panicum miliaceum*)

Proso millet (*Panicum miliaceum*) with many common names including proso millet, common millet, broomtail millet, hog millet, red millet and white millet, is a grass species used as a crop. Both the wild ancestor and the location of domestication of proso millet are unknown, but it first appears as a crop in both Transcaucasia and China about 7,000 years ago, suggesting it may have been domesticated independently in each area. It is still extensively cultivated in India, Russia, Ukraine, the Middle East, Turkey and Romania. In the United States, proso is mainly grown for birdseed. It is sold as health food and due to its lack of gluten, it can be included in the diets of people who cannot tolerate wheat. The name comes from the pan-Slavic general and generic name for millet (Russian, Serbian, Macedonian, Bulgarian: ioini and Polish, Czech, Slovakian, Slovenian, Croatian: *proso*).

Proso is well adapted to many soil and climatic conditions; it has a short growing season and needs little water. The water requirement of proso is probably the lowest of any major cereal. It is an excellent crop for dryland and no-till farming. Proso millet is an annual grass whose plants reach an average height of 100 cm (4 feet.). Like corn, it has C4 photosynthesis. The seed heads grow in bunches. The seeds are small (2–3 mm or 0.1 inch) and can be cream, yellow, orange-red, or brown in colour. Proso is an annual grass like all other millets, but it is not closely related to pearl millet, foxtail millet, finger millet, or the barnyard millets.

History and Domestication

Unlike the foxtail millet, the wild ancestor of the proso millet has not yet been satisfactorily identified. Weedy forms of this grain are found in central Asia, covering a widespread area from the Caspian Sea east to Xinjiang and Mongolia and it may be that these semiarid areas may harbor "genuinely wild *P. miliaceum* forms." This millet has been reportedly found in Neolithic sites in Georgia (dated to the fifth and fourth millennia BC), as well as excavated Yangshao culture farming villages east in China. Proso millet appears to have reached Europe not long after its appearance in Georgia, first appearing in east and central Europe; however, the grain needed a few thousand more years to cross into Italy, Greece and Iran and the earliest evidence for its cultivation in the Near East is a find in the ruins of Nimrud, Iraq dated to about 700 BC. While proso millet is not a member of the Neolithic Near East crop assemblage, it arrived in Europe no later than the time these introductions did and proso millet as an independent domestication could predate the arrival of the Near East grain crops.

Cultivation

Proso millet is a relatively low demanding crop and diseases aren't known. That's why Proso millet is often used in organic farming systems in Europe. In the United States it is often used as an intercrop. Thereby, proso millet can help to avoid a summer fallow and continuous crop rotation can be achieved. Its superficial root system and its resistance to atrazine residue make proso millet a good intercrop between two water and pesticide demanding crops. The stubbles of the last crop, by allowing more heat into the soil, result in a faster and earlier millet growth. While millet occupies the ground, because of its superficial root system, the soil can replenish in water for the next crop. The later, for example a winter wheat, can in turn benefits from the millet stubbles, which can notably act as snow accumulators.

Climate and Soil Requirements

Due to its C4 photosynthetic system, Proso millet is thermophilic like maize. Therefore shady locations of the field should be avoided. It is sensitive to cold temperatures lower than 10 to 13 degrees Celsius. Proso millet is highly drought resistant which makes it of interest to regions with low water availability and longer periods without rain. The soil should be lightly or medium-heavy. Due to its flat root systems, soil compaction must be prohibited. Furthermore Proso millet doesn't tolerate wetness due to dammed-up water.

Seedbed and Sowing

The seedbed should be fine crumbled like for sugar beet and rapeseed. In Europe proso millet is sowed between mid April and the end of Mai. 500g/are of seeds are required which comes up to 500 grains/m2. In organic farming this amount should be increased if a harrow weeder is used. For sowing the normal sowing machines could be used similar they are used for other crops like wheat. A distance between the rows among 16 to 25 centimeters is recommended if the farmer uses an inter row cultivator. The sowing depth should be 1.5 up to 2 cm in optimal soil or 3 to 4 cm in dry soil. Rolling of the ground after sowing is helpful for further cultivation. Cultivation in no till farming systems is also possible and often done in the US. The sowing then could be done two weeks later.

Field Management

Only a few diseases and pests are known but they aren't very important. Bigger problems are weeds. The critical phase is the juvenile development. The formation of the grains happens in the 3 up to 5 leaf stadium. After that all nutrients should be available for the proso millet. Therefore it is necessary to oppress the development of weeds. Inconventional farming herbicides could be used. In organic farming it's possible to use harrow weeders and inter row cultivators. For that special sowing parameters described in the chapter above are needed. For a good development of the plant fertilization with 50 to 75 kg nitrogen per hectare is recommended. Planting proso millet in a crop rotationafter maize should be avoided due to its same weed spectrum. Because proso millet is an undemanding crop, it also could be at the end of the crop rotation.

Harvesting and Postharvest Treatments

Harvest time is at the end of August until mid September. To determine the best harvest time isn't that easy because the ripeness of the grains isn't synchronized. The grains on the top of the panicle are ripe first while the grains in the lower parts need more time. That's why it's necessary to find a compromise and catch the date when the yield is highest. Harvesting could be done with a conventional combine harvester at moisture of the grains about 15-20 per cent. Usually proso millet is mowed at windrows first because the plants aren't dry like wheat. There they could wither which makes the threshing easier. Then the harvest is done with a pickup attached to the combine. The possible yield is between 2.5 and 4.5 tons per hectare under optimal conditions. Studies in Germany showed that even a higher yield could be reached.

Uses

Proso millet is one of the few types of millet not cultivated in Africa. In the United States, former Soviet Union and some South American countries, it is primarily grown for livestock feed. As a grain fodder, it is very deficient in lysine and needs complementation. Proso millet is also a poor fodder due to its low leaf: stem ratio and a possible irritant effect due to its hairy stem. Foxtail millet, having a higher leaf: stem ratio and less hairy stems, is preferred as fodder, particularly the variety called moha, which is a high quality fodder.

In order to promote millet cultivation, other potential uses have been considered recently. For example, starch derived from millets has been shown to be a good substrate for fermentation and malting with grains having similar starch contents as wheat grains. A recently published study suggested that starch derived from Proso millet can be converted to ethanol with an only moderately lower efficiency than starch derived from corn. The development of varieties with highly fermentable characteristics could improve ethanol yield to that of highly fermentable corn. Since Proso Millet is compatible with low input agriculture, cultivation on marginal soils for biofuel production could represent an important new market, for example for farmers in the High Plains of the US. The demand for more diverse and healthier cereal-based foods is increasing, particularly in affluent countries. This could create new markets for proso millet products in human nutrition. Protein content in proso millet grains is comparable with that of wheat, but the share of essential amino acids (leucine, isoleucine and methionine) is substantially higher in proso millet. In addition, health-promoting phenolic compounds contained in the grains are readily bioaccessible and high Calcium contents favor bone strengthening and dental health. Among the most commonly consumed products are ready-to-eat breakfast cereals made purely from millet flour as well as a variety of noodles and bakery products, which are, however, often produced from mixtures with wheat flour to improve their sensory quality.

Health Benefits of Proso Millet

- **Beneficial in anti-ageing:** Proso millet has high anti-oxidant content. Anti-oxidants are substances which remove free radicals from your body. Free radicals are produced in virtually every process in the body. These molecules damage the cells and fatigue them causing them to age. These are the reasons for wrinkles on skin, complexion loss etc. A healthy serving of Proso millets everyday can help you delay the ageing process.
- **Beneficial for nervous system:** Proso millet contains high amount of lecithin. Lecithin is a compound that indirectly affects the nervous system and keeps it functioning smoothly. Proso millet is also known to strengthen nervous system.
- **Beneficial in preventing pellagra and other Niacin dependent conditions:** Pellagra is a condition in which the skin becomes scaly and rough and starts falling off. This is caused due to the deficiency of niacin. Proso millet has a very high content of niacin. Thus, people whose diets are primarily meat based can avoid niacin deficiency by consuming daily portions of millet.
- **Useful for strengthening bones:** Proso millet contains high amounts of calcium which is essential for bone growth and maintenance. It actually contains one of the most absorbable sources of calcium found among food stuffs. This is especially useful for young children whose bones are in the growing phase and need plenty of calcium and for old people for whom large amounts of calcium are required for proper bone maintenance.
- **Beneficial in preventing cardiac diseases:** New research by the scientists in Seoul, South Korea suggests that regular consumption of millets can reduce the risk of heart diseases. Rats fed with millets for as short as a period of 4 weeks showed lower concentration of triglycerides as compared to non-millet rats. Millets also reduced the concentration of C-Reactive proteins which also helps in cutting heart disease risks.
- **Helps to lower cholesterol:** Proso millet has been shown to reduce the cholesterol levels in mammals. Apparently, Proso millet elevates the levels of High Density Lipoproteins in the body. The presence of HDLs increases the metabolism of cholesterol and helps bring it under control. Apart from this, Millet is also known to contain phyto chemicals like Phytic acid. Phytic acid is known to reduce cholesterol. This is an added benefit to the fact that it can also reduce heart disease risk.
- **Decreases risk of cancer:** Lignans are an essential phytonutrient present in millets that are very useful to humans. In the intestines of humans, the

bacteria act on them and convert them to mammalian lignans that are known to cut the risk of cancers like breast cancer.

- ☆ **Rich source of essential Phosphorus:** Millets are a very rich source of phosphorus for the body. A single cup of millet provides the 24 per cent of the body's daily requirement of phosphorus. Phosphorus is very essential for cell building and genetic replication in the body. Reduced risk of Diabetes: Recent research has shown that regular consumption of millet can reduce the risk of type 2 diabetes as it keeps a check on glucose and regulates insulin. This is believed to be because of the high content of magnesium in the millet.
- ☆ **Help prevent gall stones:** Millets are also known to prevent the formation of gallstones. This is because of the high content of insoluble fiber present in them which eradicates the possibility of excessive secretion of bile which is one major cause for gall stones.
- ☆ **Useful for people with coeliac disease:** Coeliac disease is a condition in which the individual is highly resistant and allergic to gluten. Gluten is found in almost every cereal grain apart from Common millet or Proso millet. Thus this is one of the foods which these people can consume safely.

Other Health Benefits

Apart from these, it has many other benefits too. It is beneficial for post-menopausal women suffering from signs of heart ailments, they combat high blood pressure, provide Iron, Vitamin B6, Zinc which are all essential for our day to day functioning. They are non-acid forming so they are very easy to digest. Millets are also high on fiber and satiate hunger quickly preventing people from overeating.

Proso Millet Uses

Food for People

Proso and other millets were an important part of the prehistoric diet 8,000-10,000 years ago. Proso millet porridge is a traditional food in Russian, German and Chinese cuisines. In Russia, it is eaten sweet (with milk and sugar added at the end of the cooking process) or savory with meat or vegetable stews. In China, it is eaten without milk or sugar and frequently with beans, sweet potato, or various types of squash. In Germany it is also eaten sweet. Apples are added to the boiling proso millet and honey is added afterward during the cooling process.

The nutritional value of proso millet is comparable to that of wheat. Proso millet is rich in B vitamins, especially vitamin-B6 and folic acid. The protein content is similar to that of wheat, but it contains no gluten and by itself is not suitable for yeast-leavened bread. When combined with wheat (or xanthan gum for those who have celiac disease), it can be used for bread. Alone, proso millet is suitable for flatbread. Since it's gluten-free, it can be used in foods for people with gluten intolerance or celiac disease.

The food industry in Europe and North America is interested in proso millet for its potential health benefits for humans as well as its mild flavor, light colour and gluten-free characteristics. Several food companies have started incorporating proso millet in their gluten-free recipes for people with celiac disease. Proso millet alone or in combination with other gluten-free grains is now used for making bread, flour, pastas, a couscous-like product and many other foods.

Proso Millet for Alcoholic Beverages

Millets are traditionally important grains used in brewing millet beer in some cultures, such as the Tao people of Orchid Island in Taiwan. The fermented millet is prepared in a large pot with hot water and people share the drink by sipping it through long straws. Millet is also the base ingredient for the distilled liquor rakshi in Nepal and the indigenous alcoholic drink tongba in Nepal and India. In Romania and Bulgaria, millet is used to prepare the fermented drink boza.

Gluten-free products are one of the fastest growing sectors in beverage production in Europe. The number one selling gluten-free beer in Germany is from malted proso millet due to its unique flavor and taste. Recently in the U.S. a few brewing companies in the Great Plains have started producing beer using proso millet as the grain source. Colorado Malting Co. (Alamosa, Colo.) is preparing malting millets from proso millet for commercial brewing companies in Colorado (*e.g.*, New Planet Beer, Eddyline Restaurant and Brewing Co. and Pagosa Brewing Co.) for production of gluten-free beer. In Nebraska the Modern Monks Brewery in Lincoln has brewed beer using proso millet.

Livestock and Poultry Feed

Proso millet has a nutritive value similar to that of other grains used for livestock feed and can be used for calves, growing and finishing cattle, dairy cows, sheep, swine and poultry. A hammer mill with a quarter-inch screen adequately cracks whole millet seeds for such uses.

Table 10.1: Composition of Digestible Nutrients (Per cent) in Proso Millet and other Grains Commonly Used for Livestock Feed

Grain	*Dry Matter*	*Digestible Protein*	*Total Digestible Nitrogen*	*Crude Protein*	*Fat*	*Fiber*
Corn	85.0	7.0	80.0	9.1	3.9	2.1
Barley	90.3	10.8	73.2	13.5	3.5	8.7
Oats	90.2	9.4	70.1	12.0	4.6	11.0
Milo	89.0	8.5	79.4	10.9	3.0	2.3
Rye	89.5	10.0	76.5	12.6	1.7	2.4
Wheat	89.4	11.3	79.6	13.5	1.8	2.8
Proso millet	90.4	8.4	76.9	11.9	3.4	8.1

Source: Feeds and Feeding, Abridged, Ninth Edition, F.B. Morrison, Morrison Company, Claremont, Ontario, Canada, 1961.

Amino acid content is generally similar to that of other grains and is an important factor in any feed. It is important to get a thorough feed analysis when proso millet is a substantial portion of any grain ration.

Feeding Calves

Combine proso millet with oats for starting calves on feed, as they will go on feed faster. Once started, proso millet alone can be used without any difference in efficiency.

Growing and Finishing Cattle

Proso millet has been shown to be equally as effective as corn, barley and wheat as an energy source in finishing rations. Adding some oats to proso millet may reduce a laxative effect in beef cattle. Like other grains, proso millet is deficient in calcium for ruminants and therefore supplemental calcium may be needed.

Dairy Cows

Adding as much as 40 percent proso millet to a grain mixture produced as much milk and a slightly higher body weight than pair-mate cows fed equal amounts of oats, corn and barley.

Sheep

As an energy source, proso millet is equal to corn, barley and wheat in lamb fattening rations. Ground proso millet gives considerably better results on fattening lambs than whole millet.

Swine

Proso millet-fed hogs performed the same as those fed corn or barley rations. The relative feeding value of proso millet is comparable (90-95 percent value) to corn when fed to swine.

Poultry

Proso millet can be used as a feed for multiple types of poultry: broiler chickens, laying hens and turkeys. Proso millet is comparable to yellow corn or commercial milo diets for mature hens (52 to 62 weeks of age) and laying hens (28 to 40 weeks of age) in either ground or whole (unground) form. Studies on turkey rations are similar to those of chicken. In some reports proso millet was found superior to corn and milo. Proso millet grain can be pelleted with other grains but as its percent increases, pellets tend to crumble.

Companion Animals

Proso millet also can be used in hamster and dog food.

Biofuel

Another potential use for proso millet is for ethanol production. Proso millet contains similar starch content as corn, the grain typically used in the U.S. for dry-grind, fuel ethanol industry. In a 2012 University of Nebraska–Lincoln study proso millet cultivars and six advanced breeding lines containing waxy starch were fermented with yeast and the resulting ethanol production was compared with normal corn and "highly fermentable" corn. Fermentation efficiencies in proso millet varieties ranged from 84 percent to 91 percent which compared to 97 percent in highly fermentable corn hybrid. Ethanol production with the experimental waxy line was nearly as high as the "highly fermentable" corn.

Dried distillers grains with soluble (DDGS) are an important co-product of ethanol production. More DDGS were recovered from proso millet lines than from the "highly fermentable" corn and the proso millet DDGS contained more protein than the corn DDGS. This suggests that the DDGS recovered from the proso millet-based fermentation can be used for animal feed similarly to that of corn. These results indicate that proso millet may be useful in fuel ethanol production. Breeding efforts to select proso millet varieties with "highly fermentable" (such as waxy type) characteristics could lead to ethanol production from proso millet that is as efficient as corn.

Table 10.2: Dry Distillers Grains with Solubles (DDGS) Yield and Composition

	DDGS (g/lb grain)	*Starch (Per cent)*	*Protein (Per cent)*	*Fat (Per cent)*	*Ash (Per cent)*
Corn					
Normal hybrid	214	14	17	8	4
Highly fermentable hybrid	163	4	23	11	5
Proso millet					
Non-waxy millet	201	4	30	9	8
Waxy millet	197	2	31	10	9

Proso Millet Products

Proso Millet Idle

Proso millet idli is a healthy breakfast recipe which is alternative to your daily and routine normal idli. Proso millet is white in color and had been in wide usage several thousand years back in India and China. It has rich in minerals and vitamins. Those who have skin allergy please avoid. You can also try proso millet dosa also with the same batter.

Ingredients

Solam (Proso millet) - 1 cup (100 g); Whole Urad Dal - 3/4 cup; Rice-1 cup; Salt - 4 tsp.

Preparation

Wash and soak urad dal, Rice and proso millet over night separately. Grind them, mix well, add salt and allow it for fermentation (8 hours aprx). Make idli as usual.

Other Products

Several food products can be prepared from proso millet as other millets.

Foods and Alcoholic Beverages Prepared from Proso Millet

(a) *Awaokoshi*, candied millet puffs, a specialty of Osaka, Japan (Source: Mugu-shisai, GFDL); (b) Millet with mushrooms (Source: Kathy Steger); (c) White millet soup (Source: Jean Hediger); and (d) Tongba, a millet-based alcoholic brew in Nepal and India.

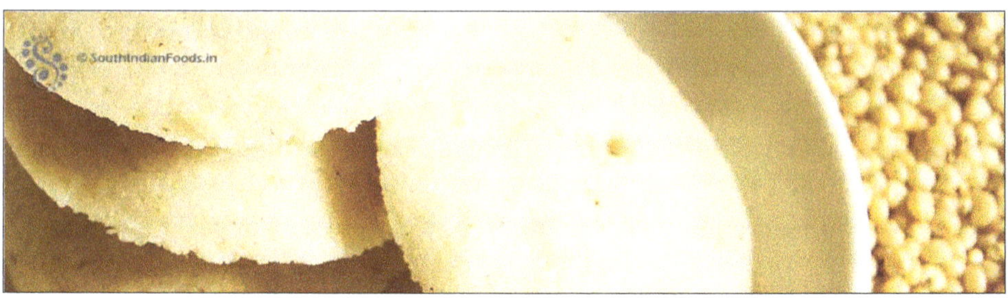

Proso *idle*

Proso Millet *Pakoda*

References

Daniel Zohary and Maria Hopf, *Domestication of plants in the Old World*, third edition (Oxford: University Press, 2000), p. 83.

Li, L.M. (2007). *Fighting Famine in North China: State, Market and Environmental Decline, 1690s-1990s*. Stanford University Press. ISBN 9780804753043.

Merkblatt für den Anbau von Rispenhirse im biologischen Landbau, www.biofarm.ch, http://www.biofarm.ch/assets/files/Landwirtschaft/Merkblatt Biohirse_Version per cent 2012_2010.pdf(23.11.14)

National Research Council (1996-02-14). "Ebony". *Lost Crops of Africa: Volume I: Grains*. Lost Crops of Africa **1**. National Academies Press. p. 260. ISBN 978-0-309-04990-0. Retrieved 2008-07-18.

Pearl Millet and Other Millets, Wayne W. Hanna, David D. Baltensperger, Annadana Seetharam (2004).

Producing and marketing proso millet in the Great Plains, U. Nabraska-Lincoln Extension.

Rose, D.J., Santra, D.K., 2013. Proso millet (*Panicum miliaceum* L.) fermentation for fuel ethanol production. *Industrial Crops and Products*, 43: 602-605.

Saleh, A.S.M., Zhang, Q., Chen, J., Shen, Q., 2012. Millet Grains: Nutritional Quality, Processing and Potential Health Benefits. *Comprehensive Reviews in Food Science and Food Safety*, 12: 281-295.

Taylor, J.R.N., Schober, T.J., Bean, S.R., 2006. Novel food and non-food uses for sorghum and millets. *Journal of Cereal Science*, 44: 252-271.

The Plant List: A Working List of All Plant Species. Retrieved 8 January 2015.

USDA GRIN Taxonomy. Retrieved 8 January 2015.

Zohary and Hopf, *Domestication*, p. 86

Chapter 11

Rice: Coarse Millet (*Oryza sativa*)

Parboiling of Paddy

Rice parboiling process consists of five major steps-soaking, steaming, drying, conditioning and cooling. During soaking the water should penetrate the centre of the grain. The water content of paddy after soaking increased to about 50-55 per cent on dry basis. During the steaming process the rice is gelatinized. Finally the paddy is dried to desirable moisture content so that it toughens enough and does not break easily during milling. A few Traditional Methods of the traditional premilling treatments to improve the milling, nutritional, cooking and keeping qualities are:

A Tapa

This treatment originated in Bengal. The paddy is soaked in water at room temperature for 24 hours and then dried in the sun (hence the name *atapa* or sun-dried). The dried paddy is then milled by traditional methods. The relative breakage of rice is more in this process.

Balam

This treatment also originated in Bengal. It is slightly better than *atapa*. Here the paddy is sprinkled with water, which inflates the grain. When the paddy is dried in the sun and milled, the hull is easier to remove.

Josh

This treatment was developed in Larkhana, Pakistan. Large earthen pots filled with paddy and water is placed on 15′cm layer of hull which is used as fuel. The pots remain on the fire overnight. The next day the water is drained and the paddy is placed on shallow iron pans and heated over fire for one hour. The paddy is then dried in sun.

Sela

This treatment was originated in Saharan pur, India. Paddy is soaked in water at room temperature for 24-48 hours and then gently roasted in hot sand (80-90°C). The roasted paddy is then further dried in sun and milled.

Siddha

This treatment originated in Bengal. Paddy is soaked in water at room temperature for 20 hours and then boiled for few minutes. It is finally dried in sun, In this method, rice from over-soaked paddy becomes coarse in appearance and rice from over-dried paddy shows poor milling quality.

Parboiling

Parboiling is the latest premilling treatment which improves the quality of rice. The traditional parboiling process in India is carried out in different ways. The principal methods are single-boiling and double-boiling process.

The double boiling method (Dobhapa): This involves double steaming in its sequence of operations. The vertical steaming kettles, generally two in number, are made of mild steel plates. The size of each kettle is generally 700-900 mm in diameter and 1.2-2 meters in height with about 600 mm deep conical bottom having a 300 mm diameter flanged outlet, fitted with a sliding valve. The flat.top cover is fitted with a 300 mm diameter central opening for feeding the raw paddy. The kettles are provided with steam pipe extending vertically through the centre about half way down through the top cover. When the direct steam from the boiler at 6 to 7 kg/cm^2 is used, a 20 to 25 mm diameter steam pipe is supplied. The soaking is done in large masonry water tanks constructed on the ground floor. These are generally two in number working each in line with a steaming kettle. Each tank usually of 23 meters in depth hold's 20 to 35 tons of paddy. The water level in the tank is maintained such as to cover the paddy completely during the soaking period.

The process involves filling the dry paddy in the steaming kettle and opening the steam valve. During steaming the top opening of the kettle is 'covered with a gunny bag. Depending on the size of the batch the steam starts blowing out through the bag on the top in three to eight minutes. Steam is turned off and the paddy is discharged from the bottom and then quickly dumped into the water in the soaking tank. The temperature in the tank rises, as more and more paddy is dumped. Finally at the end the temperature of soak water is around 70 to 80° C. But because of large exposed surface, the temperature of soak water drops to 50-60° C within two hours. The paddy is allowed to soak for 24 to 72 hours after which the water is drained off. The colour of the soak water turns brawn. In most cases signs of fermentation have been observed within 18 hours. These result in a bad odour and brown colour in the milled rice. The soaked paddy is then lifted back into the steaming kettles for the second steaming or actual parboiling operation. The procedure followed is the same as is done in case of the first steaming. The parboiled paddy is discharged from the kettles and allowed to dry in the drying yard or in a mechanical drier.

The single boiling method: The paddy is soaked in cold water in the cement tanks for a few days and then steamed in the usual manner. The soaking time is generally more in the case of the single boiling process. In both these methods, during prolonged soaking of the paddy, the rice prepared out of it produces a bad odour, as a result of fermentation during soaking. These difficulties are eliminated by the improved methods of parboiling developed in India at the Central Food Technological Research Institute, Mysore, Jadavpur University and Calcutta and in certain other parts of the world. In the modern methods of parboiling, the long steeping and steaming cycle using law temperatures is replaced by those with short cycles using high temperatures and pressures and the process is carried out either in a batch plant or a semi-continuous or continuous plant. The treatment will depend on the variety and quality of paddy and the characteristics of the final product desired. The choice of any parboiling technique will also depend on the initial investment, running costs, local conditions and amount of automation desired.

Parboiling Method of CFTRI, Mysore

The system developed by CFTRI (Central Food Technological Research Institute) was primarily aimed at improving the yield and quality of rice with a lower capital investment. The soaking and steaming is Jane in the same mild steel cylindrical tank. Steam enters through the perforated pipe at the centre and there are more perforated pipes arranged radically at the bottom of the tank. The base of the tank is cone-shaped and is closed at the bottom by a water tight hatch. At the side of the hatch there is a valve for draining of the steeping water. During parboiling the tank is filled with water heated by steam injection to 85° C. The paddy is then poured inside the tank. The temperature of water drops to 70-75° C. After two to three and a half hours steeping, the water is drained off. Pressurized steam is then passed through the perforated pipes until the husks just begin to crack open. After steaming the hot paddy is unloaded through the bottom hatch and then spread over the drying floor.

Rice Milling

Rice milling machinery used in different countries range from crude hand pounding equipment and small scale hullers to highly sophisticated and capital intensive units. However, the rice milling machinery can be broadly classified into two groups: traditional and modern rice milling machinery (Figure 11.1).

Traditional Rice Milling Machinery

Traditional rice mills include hand pounding equipments, single huller and battery of hullers, sheller-cum-huller and sheller mills.

Hand Pounding

A variety of implements are used for the purpose of hand pounding, the more common being; (a) mortar and pestle, (b) Dhenki and (c) hand stone (*chakki*).

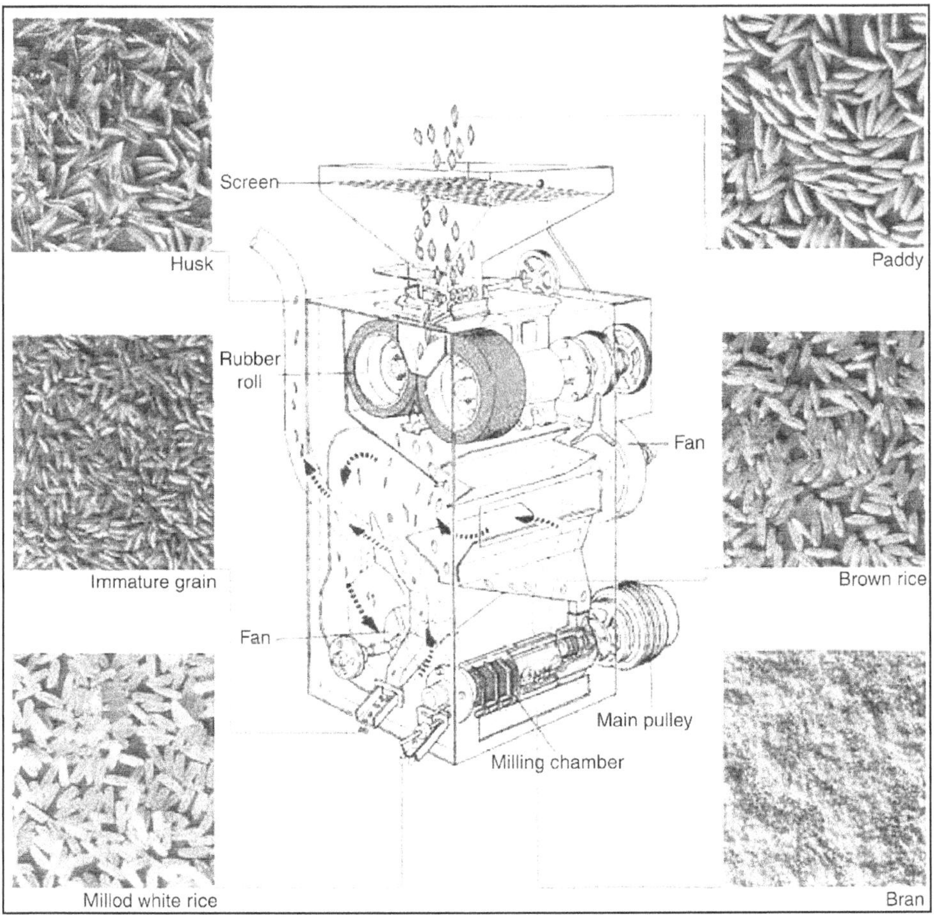

Figure 11.1: Rice Pearler Combined with Husker.

Single Huller

An iron bar called knife or blade is protruded to resist the movement of grain in the milling chamber and to control the degree of whitening. The blade can be moved forward and backward manually to adjust the clearance. Paddy is fed at one end of the upper half and discharged from the other end of the half of the housing. Paddy handling capacities of the huller from 0.25 to 0.75 tons/hr are common. Generally in a single huller paddy is husked and whitened in a single operation by applying high pressure resulting in a high degree of whitening and high percentage of broken in the processed rice. The total and head yields can be slightly increased by milling paddy in a number of passes. Single hullers of small.capacity are located in the village for custom milling: But rice mills consisting of two to six hullers are located in tons market areas and cities. The total rice out turns is about 56 and 64 per cent for raw and parboiled paddy respectively.

Huller Mill

The huller mill consists of battery of hullers, sieves for cleaning paddy, reciprocating sieves for removing brokens, etc. and bucket elevators. Sometime, aspirator is also employed for removing husk. The machines are arranged in proper sequence. Sun dried raw or parboiled paddy is first cleaned to remove foreign materials. The clean paddy is fed to the firs, huller, used mainly for husking. Then the mixture of husked rice, paddy and husk IS fed in equal amount to the second and third hullers who are operated with narrower clearance between the blade and rotating ribbed roll. In the second and third hullers complete husking and whitening takes place. The mixture of whitened rice, husk and some bran obtained from these two hullers are separated in the sieve sifter.

Sheller Mill

As far as basic milling operations are concerned, the sheller mill is almost identical to a modern mill except that the husking operation is done by the under runner disc husker (sheller) in place of modern rubber roll husker. This type of mill consists of: (a) a cleaner, (b) one or more disc shellers, (c) aspirator (to remove husk), (d) one or more paddy separators and (e) one or more cone type rice whitener. Commonly, the capacities and total power consumption vary from 1 to 2 tons/hr and 35 to 50 BHP respectively. Among all traditional rice mills, total and head yields are highest in sheller mill. However, in huller and sheller-cum-huller and sheller mills certain amount of husked rice, smaller and lighter paddy escape along with the husk due to defective design of the aspiration system and small brokens are also lost with the bran. Moreover, immature paddy cannot be separated in these mills and is mostly lost in the pile of husk.

Modern Rice Milling Machinery

The modern rice milling machinery can be divided into two major groups (Figure 11.1): (1) rice milling machinery developed in Japan and (2) rice milling machinery developed in Europe. Both European and Japanese modern rice milling machines have been described in this chapter. The major operations performed by modern rice mills are as follows: (1) storage, (2) cleaning, (3) husking, (4) separation, (5) whitening and (6) grading.

Cleaning

The paddy procured from the farmer is cleaned with the help of paddy cleaners. The removal of impurities from the grains is essential to protect the subsequent milling machinery from unusual wear and tear and to improve the quality of the final product. If the procured paddy contains excessive amounts of foreign materials or if paddy is parboiled prior to milling then the procured paddy is sometimes pre-cleaned in an open double sieve type pre-cleaner or enscalper installed outside the mill room.

Husking

The purpose of a modern husking machine is to remove husk from paddy without damage to the bran layer and rice kernel. Husking machines are known by different names such as huskers, dehuskers, shellers and sometime hullers also.

Impact Type Paddy Husker

The working principle of the impact or centrifugal type of husker is based on the utilization of impact and frictional force for husking of paddy. In the impact type husker, paddy is thrown against a rubber wall by a rotating disc. The impact on the rubber wall due to the centrifugal force of the rotating disc causes cracking of hulls with a minimum damage to the kernel. Relations between husking percentage and rpm of the disc, husking percentage and angle of inclination of the husking plate with japonica varieties of paddy are shown s that the husking percentage increased sharply from 2,000 rpm and reached its maximum at 4,300 rpm. But the head yield decreased as rpm increased. Therefore, optimum rpm of the disc for different varieties of paddy is very important. It was also observed that the most important factor affecting husking ratio was not the diameter of the rotating disc but the circumferential speed.

Rubber Roll Husker

When paddy is passed between two resilient surfaces rotating at different speeds in different directions, its husk will be split and stripped off. If paddy is allowed to pass through the gap (smaller than the thickness of paddy) between two rubber rolls rotating at different speeds then it makes contact with the two rolls for different periods of time. The contact of the faster revolving roll is longer than the slower revolving roll. As a result the paddy is sheared and compressed and its husk is stripped off and removed.

Both rolls have the same diameter. Diameter, of the rubber rolls varies from 150 of 250 mm depending on the capacity of the husker. The wear of the rubber is considerable and with the reduction of the roll diameter, capacity is also reduced. The main reason for the capacity reduction is the decrease in the relative speed of the, two rolls. In general only 85 to 90 per cent of the paddy fed is husked due to variation in moisture content, size and degree of maturity of grain, eccentricity of the revolving rubber, rolls and uneven wearing of rubber roll surfaces. The clearance between the two rolls can be adjusted manually. A sophisticated pneumatic device for automatic adjustment of the clearance has been introduced in place of manual adjustment. Nowadays rubber roll huskers are equipped with a blower to blow air on the rubber rolls' surface and bring down their temperature rise due to friction between husk and roller during husking.

When the husker is in operation for a certain period, the faster revolving roll wears more than the slower revolving roll. As a result the diameter of the former becomes smaller than the latter and the difference in peripheral velocity between the two rolls becomes less. Though the speed ratio is kept constant, yet it causes lowering of husking efficiency. To encounter this difficulty application of high pressure is not the proper means as it would break the grains. However, the rubber rolls are to the interchanged at a regular interval of time.

Separation

The husked rice separated from the mixture of paddy and husked rice by the paddy separator. The paddy separator consists of a number of identical inclined trays with dimples over the surface. These dimpled trays are in reciprocating motion. When a mixture of husked rice and paddy is delivered at the upper comer of the tray, the husked rice being heavier occupies the bottom layer and comes in contact with the dimpled tray, while the paddy being lighter floats on the husked rice. The size of the dimples is slightly bigger than the brown rice but smaller than the paddy. The downward movement of brown rice is thus partially restricted by the dimples while the paddy is free to move downward. With the reciprocating movement of the tray, the husked rice, in contact with the tray picks up the movement and moves to the side, while the paddy flow downward fast by gravity. At the same time, paddy and husked rice gradually move to the right side of the tray separating the husked rice at the upper part and the paddy at lower part and leaving the mixture at the central part. The tray overflow is, therefore, received by the three compartments divided by flaps. The two flaps divide the whole receiving chamber into three compartments. The position of the flaps is adjustable so that pure brown rice can be received in the upper compartment, paddy in the lower compartment and the mixture of paddy and brown rice in the middle compartment of the chamber.

Whitening

The term whitening refers to the operation of removal of germ, pericarp, tegmen and aleurone layers from husked rice kernels. It is also called polishing or pearling or scouring. There are three major kinds of whitening machines used in the modern rice processing industry. They are: (1) vertical abrasive whitening cone, (2) the horizonontal abrasive whitening roll and (3) the horizontal metallic friction type roll.

Vertical Whitening or Pearling Cone

Basically, the machine consists of an inverted cast iron frusteconical rotor covered with abrasive material mounted on a vertical spindle, revolving inside a crib. The crib is lined with steel wire cloth or perforated metal sheets and it is equipped with rubber brakes which are placed vertically and spaced equally and protruded into the gap between cone and crib. The pressure inside the machine can be adjusted by pushing in or pulling out the rubber brakes. The peripheral speed of the cone should be 13 m/sec. The husked rice enters the gap and is dragged along by the rough surface of the rotating cone. The rubber brakes tend to stop it and cause it to pile up against their side. While pressed up against the brakes, the grains undergo a strong swirling and revolving movement because of their oval shape and smooth surface. Each grain is scoured by the abrasive surface of the cone. It also rubs against the surrounding grains and the rough lining of the crib as well. The grains meet almost same conditions all the way round the cone until they sink lower and lower by gravity and are finally discharged at the bottom of the cone.

The bran is finally ground due to scouring and rubbing and escapes through the lining of the crib and regularly removed. As the vertical abrasive roll is an inverted truncated cone, the peripheral speed at the upper part is higher than the lower part. But the density and pressure at the lower part are higher than at the upper part. As a result, the grinding action is predominating at the upper part while frictional force is stronger at the lower part. In the strict sense this machine is a combination of abrasive and friction types of machines, generally, a series of two to four whitening cones are used to whiten rice successively in two to four passes.

Grading

After rice milling and processing the grading is done on the basis of grain size and shape as well as colour using various types of machines available in the market.

Health Benefits of Rice

The health benefits of rice include its ability to provide fast and instant energy, regulate and improve bowel movements, stabilize blood sugar levels and slow down the aging process, while also providing an essential source of vitamin B1 to the human body. Other benefits include its ability to boost skin health, increase the metabolism, aid in digestion, reduce high blood pressure, help weight loss efforts, improve the immune system and provide protection against dysentery, cancer and heart disease. Rice is a fundamental food in many cultural cuisines around the world and it is an important cereal crop that feeds more than half of the world's population.

The various benefits of rice can be found in more than forty thousand varieties of this cereal that is available throughout the world. The two main categories are whole grain rice and white rice. Whole grain rice is not processed very much, so it is high in nutritional value, whereas white rice is processed so that the bran or outer covering is removed, leaving it with less nutritional value. People choose different styles of rice for particular flavours, depending on their culinary needs, the availability and the potential for healthy benefits as well.

Rice can also be defined by the length of each grain. Indian or Chinese cuisines specialize in long grained rice, whereas western countries prefer short or medium length grains.

- ☆ **Great Source of Energy:** Since rice is abundant in carbohydrates, it acts as fuel for the body and aids in the normal functioning of the brain. Carbohydrates are essential to be metabolized by the body and turned into functional, usable energy. The vitamins, minerals and various organic components increase the functioning and metabolic activity of all your organ systems, which further increases energy levels.
- ☆ **Cholesterol Free:** Eating rice is extremely beneficial for your health, simply because it does not contain harmful fats, cholesterol or sodium. It forms an integral part of balanced diet. Any food that can provide nutrients without having any negative impacts on health is a bonus! Low levels of fat, cholesterol and sodium will also help reduce obesity and the health conditions associated with being overweight. Rice is one of the most

widely used and eaten foods in the world because it can keep people healthy and alive, even in very small quantities.

- **Blood Pressure Management:** Rice is low in sodium, so it is considered one of the best foods for those suffering from high blood pressure and hypertension. Sodium can cause veins and arteries to constrict, increasing the stress and strain on the cardiovascular system as the blood pressure increases. This is also associated with heart conditions like atherosclerosis, heart attacks and strokes, so avoiding excess sodium is always a good idea.
- **Cancer Prevention:** Whole grain rice like brown rice is rich in insoluble fiber that can protect against many types of cancer. Many scientists and researchers believe that such insoluble fibers are vital for protecting the body against the development and metastasis of cancerous cells. Fiber specifically is beneficial in defending against colorectal and intestinal cancer. However, besides fiber, rice also has natural antioxidants like vitamin C, vitamin-A, phenolic and flavonoid compounds, which also act as or stimulate antioxidants to scour the body for free radicals. Free radicals are by products of cellular metabolism that can do serious damage to your organ systems and cause the mutation of healthy cells into cancerous ones. Boosting your antioxidant levels is a great idea and eating more rice is a wonderful way to do that.
- **Skin care:** Medical experts say that powdered rice can be applied topically to cure certain skin ailments. On the Indian subcontinent, rice water is readily prescribed by ayurvedic practitioners as an effective ointment to cool off inflamed skin surfaces. The phenolic compounds that are found in rice, particularly in brown or wild rice, have anti-inflammatory properties, so they are also good for soothing irritation and redness. Whether consumed or topically applied, substance derived from rice tends to relieve a number of skin conditions. The antioxidant capacity also helps delay the appearance of wrinkles and other premature signs of aging that can affect the skin.
- **Alzheimer's disease:** Brown rice is said to contain high levels of nutrients that stimulate the growth and activity of neurotransmitters, subsequently helping to prevent Alzheimer's disease to a considerable extent. Various species of wild rice have been shown to stimulate neuroprotective enzymes in the brain, which inhibit the effects of free radicals and other dangerous toxins that can cause dementia and Alzheimer's disease.
- **Diuretic and Digestive Qualities:** The husk part of rice is considered to be an effective medicine to treat dysentery and some people say that a three month old rice plant's husks are said to have diuretic properties. Chinese people believe that rice considerably increases appetite; cures stomach ailments and reduce all digestive problems. As a diuretic, rice husk can help you lose excess water weight, eliminate toxins from the body like uric acid and even lose weight, since approximately 4 per cent of urine is actually made up of body fat. The high fiber content also increases bowel

Rice Processed Products

Cooked Rice

Rice Cookies

Rice Crackers

Rice Extruded

Rice Sticks

Rice Pullava

Rice Gel Cake

Rice Cake

Rice Oil

Rice Gel Kurdai

movement regularity and protects against various types of cancer, as well as reducing the chances of cardiovascular diseases.

☆ **Rich in Vitamins:** Rice is an excellent source of vitamins and minerals like niacin, vitamin D, calcium, fibber, iron, thiamine and riboflavin. These vitamins provide the foundation for body metabolism, immune system health and general functioning of the organ systems, since vitamins are commonly consumed in the most essential activities in the body.

☆ **Cardiovascular Health:** Rice bran oil is known to have antioxidant properties that promote cardiovascular strength by reducing cholesterol levels in the body. We have already spoken about the cardiovascular

Biryani Varieties

Egg Biryani

Chicken Biryani

Shrimp Biryani

Cooked Meat Biryani

Vegetable Biryani

Hyderabadi Biryani

Kaadai Biryani

Mushroom Soya Biryani

Pulao Varieties

Kashmiri Pulao

Zafrani Pulao

Jeera Pulao

Peas Pulao

Pulao

Mixed Vegetable Pulao

South Indian Variety Rice

Curry Leaf Rice

Curd Rice

Tomato Rice

Mint Coriander Rice

Mango Rice

Lemon Rice

Fllu Sadam

Khuska Biryani

Coconut Rice

Vegetable Rice

Carrot Rice

Asparagus Pulao

Capsicum Rice

Corn Rice

Gooseberry Rice

Spinach Rice

Methi Pulao

Okra Rice

Cauliflower Pulao

Potato Pulav

Vangi Bath

benefits of fibber and low levels of fat and sodium. Wild rice and brown rice varieties are far better than white rice in this category, since the husk of the grain is where much of the nutrients are; the husk is removed in white rice preparation.

☆ **Resistant starch:** Rice abounds in resistant starch, which reaches the bowels in an undigested form. This type of starch stimulates the growth of useful

Rice with Lentils (Kitchadi and Pongal Varieties)

Puli Pongal

Ven Pongal

Sambhar (Kadamba) Sadam

Bisi Bele Bath

Sweet Rice Dishes

Zarda Rice

Rice Pudding

Sweet Pongal

Phirni (Rice Kheer)

Nei Payasam

Akkaravadisal

Other Rice Dishes

Chicken Fried Rice

Turkey Fried Rice

Payaru Kanji

Tri Colour Rice Dhokla

bacteria that help with normal bowel movements. Also, this insoluble rice is very useful in reducing the effects of conditions like Irritable Bowel Syndrome (IBS) and diarrhoea.

According to the International Rice Research Institute in the Philippines, the nutritional value of rice needs to be improved even more so that it benefits mankind. Rice, being the most dominant cereal crop in most of the world can improve the lives of millions of people who consume it.

Efforts are currently being made to increase the micro-nutrient value of rice by combining traditional methods of growing crops with modern biotechnology. The institute further states that the development of rice with high iron and zinc compounds could be possible through bio-fortification. That can also induce high quality yields, which could be eagerly accepted by farmers as well as rice consumers for healthier, longer lives.

References

Anderson, R. A. (1962). A note on wet-milling of high amylose corn containing 75 per cent-amylose starch, *Cereal Chem.*, 39, 406-408.

Anderson, R. A. (1963). Wet-milling Properties of grains: Bench- Scale Study, *Cereal Sci. Today*, 8, No.6, 190-92, 195, 221.

Anon, 1964, Joint United States-Canadian tables of feed composition, *Natl. A cad. Sci.*, Nath. Res. Council Publication 1232.

Anonymous, (1982). Prevention of Food Adulteration Act, Government of India Publication, New Delhi.

Athalye (1992). Plastics in Packaging. Tata McGraw-Hill Publishing Company Limited, New Delhi.

Atkins, T. A. and Geddes, W. F. (1939). The relationship between protein content and strength of flours in gluten enriched flours, *Cereal Chem.*, 16, 223-31.

Autrey, H. S. (1953). Effect of variables upon milling yields, *Rice J. Annual.*

Autrey, H. S., Grigorieff, W. W., Attschul, A. M. and Freeman, E. E. (1955). Effect of milling conditions on breakage of rice grains, *J. Agr. Food Chem.*, 3(7), 593.

Bender, A. E. (1973). Composition of infant foods, In Nutritional and Dietetic Foods, International Textbook Co. Ltd., London, U. K., p. 64.

Board, P.W. (1988). Quality Control in Fruit and Vegetable Processing. FAO Food and Nutrition Paper 39. Food and Agriculture Organisation of the United Nations, Rome.

Borasio, L. and Gariboldi, P. (1957). Illustrated glossary of rice processing machines, Food. and Agricultural Organization of the United Nations, Rome.

Brautlecht, CiA, (1953). *Starch, Its sources, Production and Uses*, Reinhold Publishing Corp., New York.

Brekke, 0 L. and Kwolek, W. F. (1969). Com dry milling: Cold tempering and degermination of com of various initial moisture contents, *Cereal Chern..* 46, 545-49.

Brekke, O. L. (1970). Dry milling artificially dried com: Roller milling of degerminator stock at various moistures, *Cereal Science Today*, 15, 37-42.

Chandra, P. K. (1972). Methodology and Procedures for.evaluating performance of commercial rice mills, Unpublished M. Tech. Thesis, L I. T., Kharagpur.

Chavan, J. K. and Kadam S. S. (1989). Nutritional improvement of cereals by fermentation. *CRC in Food Sci. and Nut*. 28: 349-399.

CSIR, (1980). The Wealth of India: Raw Materials, Councill of Scientific and Industrial Research, New Delhi.

Dauthy, M.E. (1995). Fruit and Vegetable Processing. FAO Agricultural Services Bulletin 119. Food and Agriculture Organisation of the United Nations, Rome.

Desrosier, N.W. and Desrosier, J.N. (1987). The Technology of Food Preservation. CBS Publishers and Distributors, New Delhi.

Desrosier, N.W. and Tressler, D.K. (1977). Fundamentals of Food Freezing. The AVI Publishing Company, Inc., Westport, Connecticut Distributors, New Delhi.

Elias, D. G. and Scott, R. A. (1957). British flour milling technology, *Cereal Sci. TorJay*, 7, 180-84.

Fennema, O.R. Ed.(1976). Principles of Food Science, Part-l. Food Chemistry, Marcel Dekker Inc, New York.

Frazier, W.e. and Westhoff, D.C. (1983). Food Microbiology. Tata McGraw-Hill Publishing Company Limited, New Delhi.

Garibaldi, F. (1974). Rice Milling Equipment Operation and aintenance, Agricultural Services Bulletin No. 22, FAO, Rome.

Goel, A. K., Kumar, R. and Mann, S. S. (2007). Postharvest management and value addition. Daya Publishing House, Delhi.

Gopalan, C, Rama Sastri, B.V. and Bala Subramanian, S.C. (1982). Nutritive Value of Indian Foods. National Institute of Nutrition, R.CM.R., Hyderabad.

Hardenburg, R. E., A.E. Watada and C. Y. Wang, (1986). The Commercial Storage of Fruits, Vegetables and Florist and Nursery Stocks, Agricultural Handbook 66, US Dept. Agric. Washington D.C.

Harris, RS. and Karmas, E. (1975). Nutritional Evaluation of Food Processing. 2nd Edition. AVI Publishing Co. Westport, Connecticut.

Haung, M. T, O. Toshihiko, C. T Ho and R.T. Rosen, (1994). Food Phytochemicals for Cancer Prevention, I. Fruits and Vegetables ACS Symposium Series 546, American Chemical Soc. Washington, D. C.

Hogan, J. T. and Deobald, H. J. (1995), Measurement of the degree of milling of rice, *Rice* 1, 68 (10), 10.

Ilyas, S. M. (2003). Importance of postharvest processing and value- addition for additional income and employment generation. Presented at National Symposium on Resource Management for Eco-friendly Crop Production, held at CSAUA and T, Kanpur, February, 26-28.

Ilyas, S. M., Wanjari, O. D. and Goyal, R. K. (2002). Prospects of income and employment generation through agro-processing and value-addition. Presented at Conference on Emerging Opportunities of Agro and Food Processing Industries in Jharkhand held at Indian Institute of Coal Management, Ranchi during October, 8-9.

Ilyas, S.M. and Coyal, R.K. (2002). Quality assurance in food processing-impact of WTO, In: Proc. National Seminar on Development of Food Processing Industries in U.P., held at Lucknow during January 16-17.

Jacob, T. (1987). Poisons in our Food. Publications Division, Ministry of Information and Broadcasting, Government of India, New Delhi.

Jones, CR. (1940). The production of mechanically damaged starch in milling as a governing factor in the diastatic activity of flour, *Cereal Chem.*, 17, 133-69.

Karel, M., O.R. Fennema and D.S. Lund, Eds. (1975). Principles of Food Science Part II. Physical Principles of Food Preservation. Marcel Dekker, Inc. New York.

Kazanas, N. and Fields, M. L. (1981). Nutritional improvement of sorghum by fermentation. *J. Food Sci.* 46: 819-821.

Kester, E. B. and Matz, S. A. (1970). Rice Processing, In: Cereal Technology, ed. by S. A. Matz, Avi Pub. Co., Westport, Conn.

Koga, Y. (1969). Drying, husking and milling in Japan IV and V., *Farming,* Japan.

Lockwood, J. F. (1952). *Flour Milling,* Third ed., Northern Publishing Co. Ltd., Liverpool.

Lockwood, J. F. (1960). *Flour Milling,* The Northern Publishing Co., Liverpool.

Macrae, R., R. K. Robinson and M. J. Sadler, Eds, (1993). Encyclopaedia of Food Science, Food Technology and Nutrition, Academic Press, London.

Matz, S. A. (1959). The Chemistry and Technology of Cereals as Food and Feed, Avi Publishing Co., Westport, Conn.

Miller, O. H. and Burns, E. E. (1970). Starch characteristics of selected grain sorghum as related to human food. *J. Food Sci.* 35: 666-668.

Misra, J.B. (1983). A simple arrangement for the accurate determination of the specific gravity of potato tubers. J. *Indian Potato Assoc.* 10: 121-128.

Modi B. S. (1972). Factors affecting the performance of paddy separator, Unpublished *M. Tech. Thesis,* IIT, Kharagpur.

Mudambi, S.R and Rao, S. (1986). Food Science. Wiley Eastern Limited, New Delhi.

Nagy, S. and P.E. Shaw, Eds. (1980). Tropical and Subtropical Fruits AVI, Westport, CT.

Novellie, L. (1977). Beverages from sorghum and millets in "Proc. of Symp. On Sorghum and Millets for Human Food". International Association for Cereal Chemistry. Symposium. Vienna. 1976. Tropical Products Institute, London. Pp. 73-77.

Odunfa, S. A., Nordsrom, J. and Adeniran, S. A. (1994). Development of starter cultures for nutrient enrichment of ogia West African fermented cereal gruel. Report submitted to HBVC research grants program. USAID, Washington, U.S.A.

Paulus, K. O. (1984). Modelling in Industrial Cooking. In: *Thermal Processing and Quality of Foods.* Zeuthen *et al.* (eds.), Elsevier Applied Science Publishers, London.

Potter, N. N. (1984). *Food Science.* 3rd Edition. The AVI Publishing Company, Inc., Westport, Connecticut.

Potter, N.N. (1973). *Food Science.* The AVI Publishing Company, Inc., Westport Connecticut.

Potter, N.N. and Hotchkiss, J.H. (1996). *Food Science.* 5th Edition. CBS Publishers and Distributors, New Delhi.

Primo, E., Barber, S., Tortosa, E., Camacho, J., Ulldemolins, J., Jimenez, A and Vega, R, (1970). Chemical composition of the byproducts obtained in the different steps of the rice milling process (In Spanish), Rev. Agroquim Teenol, Alimentos 10': 244.

Quick, Graeme R. (1980). An In-depth Look at Farm Fuel Alternatives. Power Farming Magazine, Feb. 10-17 pp.

Rai, Mangla and Mauria S. (1999). Coarse cereal: Not for the poor alone. *The Hindu Survey of Indian Agriculture.* pp. 55-58.

Reichert, R. D. and Young, C. G. (1976). Dehulling cereal grains and grain legumes for developing countries. I. Quantities comparison between attrition and abrasive type mills. *Cereal Chem.* 53: 829-839.

Salunkhe, D.K., S.S. Kadam and S. J. Jadhav, (1991). Potato: Production, Processing and Products, CRC Press, Soca Raton.

Satake Engineering Co. Ltd. (1973). Rice Milling Machinery, Technical Note No. 601, Extension and.Training Institute, Satake Engineering Co. Ltd., Tokyo, Japan.

Schoch, T. J. (1941), Physical aspects of starch behavior, *Cereal Chem.,* 18, 121-28.

Schrimshaw, N. S. (1974). Nutritional requirements for high protein foods, Proc. Int. Symp. Protein Foods and Concentrates, Central Food Technol. Res. Institute, Mysore, India, pp. 13-18.

Scott, J. H. (1951), *Flour Milling Processes,* Second ed., Chapman and Hall, London.

Seetharam A. (1997). Coarse cereals: Waiting for incentives. The Hindu Survey of Indian Agriculture. pp. 55-59.

Shakuntala Manay, N. and Shadaksharaswamy, M. (1987). Foods: Facts and Principles. Wiley Eastern Limited, New Delhi.

Shibano, M. (1973). Construction and Function of Abrasive Roll Type Rice Whitening Machine, Technical note, JRMA, Japan.

Shibano, M., 1973, Construction and Function of Friction Type Rice Whitening Machine Technical note, JRMA, Japan.

Shreve, N. R. (1956). The Chemical Precess Industries, McGraw-Hili Book Co. Inc., Tokyo.

Singh, G. (1997). Data book on Mechanization and Agro-Processing Since Independence, CIAE, Bhopal-462038.

Smith, L. (1944), *Flour Milling Technology,* Third ed., Northern Publishing Co., Liverpool.

Srivastasva, M (1985). Essentials of Food and Nutrition Vol. II (Applied Aspects). The Bangalore Printing and Publishing Company Limited, Bangalore.

Srivastava, R.K and Singh, S. (1987). Utilization of Mushrooms for Preparation of Different Types of Preserved Products. Government Fruit Preservation and Canning Institute,.U.P., Lucknow.

Stermer, R A, (1968). Environmental conditions and stress cracks in milled rice, *Cereal Chem.,* 45, 365.

Stivers, T. *E.,* Jr. (1955), American corn milling systems for degermed products, *Assoc. Operators Millers Bull.,* 2168-79.

Tauro, P., Kapoor, KK and Yadav, KS. (1986). An Introduction to Microbiology. Wiley Eastern Limited, New Delhi.

Thompson, R. A. and r-oster, G. H. (1963). Stress cracks and breakage in artificially dried corn, U. S. Dept. Agr., AMS, 631.

Chapter 12

Sorghum: Coarse Millet (*Sorghum bicolour* L.)

Introduction

Sorghum (*Sorghum bicolour* L. Moench) is one of the major cereal crops among the all millets grown mostly in dryland. It is grown as rainfed crop in both season *Kharif* (mansoon) and *Rabi* (winter). It is mainly grown in the Deccan plateau, Central and Western India apart from a few patches in northern India. Almost entire grain produce is used for human consumption in India. It is nutritionally superior to other fine cereals such as rice and wheat and hence it is known as nutritious cereal. Nutritionally sorghum grain contains 4.4 to 21.1 per cent protein, 2.1 to 7.6 per cent fat, 1.0 to 3.4 per cent crude fiber, 57.0 to 80.6 per cent total carbohydrates, 55.6 to 75.2 per cent starch and 1.3 to 3.5 per cent total minerals (ash). Sorghum also provides 350 Kcal energy, calcium, phosphorus, potassium, carotene and thiamin as well as antioxidants through phenolics and various types of tannins. It is mainly consumed as *Bhkari* (*roti*) *i.e.* un-leavened pancake in various states of India. Apart from the traditional products like *bhakari, bhatwadi, papad, popped grains, kurda, high fiber cookies, biscuits, flakes, thalipeeth, upama, rawa idali, dosa, uttappa, chiwada, chakali, papadi, ambil, shankarpale, cookies and cakes* are also prepared from sorghum and consumed as snack food items. Sorghum grains are not only good source of nutrients but it also contains special constituents such as phyto-chemicals, dietary fiber as well as resistant starch, which are more essential to the human nutrition. However, the grain sorghum consumption has remained restricted to the weaker sections of the society due to poor nutritional quality of grain and inferior quality of the products such as Bhakari as well as very low price of this food item as compared to the other cereal grains such as wheat, rice and other millets.

Sorghum is of African origin. A large variety of wild and cultivated sorghum are grown in the tropics and subtropics of the world. In India, sorghum constitutes an important article of food, after rice and wheat. The sorghum grain is small and rounded, varying in colour from off-white to white to varying shades of red, yellow or brown. The grain size varies, the weight ranging from 7.0 to 61g/1000 grains, with most sorghums weighing 20 – 30 g/1000 grains. The chemical composition of grain sorghum is similar to that of maize. Generally, sorghum has more protein than maize, a lower fat content and about the same amount and proportions of carbohydrate components. The proximate analysis of Indian sorghum grain indicates moisture, 11.9; protein 10.4; fat 1.9; fiber 1.6; carbohydrates 72.6 and minerals 1.6; minerals present in the grain are calcium, magnesium, potassium and iron.

In comparison with maize, sorghum grain contains approximately the same quantities of riboflavin and pyridoxine but more pantothenic acid, nicotinic acid and biotin. Nicotinic acid occurs in the grain in available form.

Starch is the major carbohydrate of the grain. The other carbohydrates present are simple sugars, cellulose and hemicelluloses. The amylose content of starch varies from 2 1 to 28 per cent. Starch from waxy varieties contains little amylose. Both waxy and regular starches contain free sugars upto 1.2 per cent. Sucrose being the major constituent (0.85 per cent) followed by glucose (0.09 per cent), fructose (0.09 per cent), maltose and stachyose. Sorghum grain contains no detectable amount of glucoside, but on germination within 3 days old seedling gives 3.5 per cent Dhurin which leads to the poisoning of animals consuming such sorghum seedlings.

The percentage of different protein fractions to the total protein of sorghum grown in India is albumin 5; globulin 6.3; prolamin 46.4 and glutelin 30.4. Prolamin and glutelin are principally present in the endosperm. Amino acid analysis of various protein fractions shows that there is better distribution of all essential amino acids in globulins than in prolamins. Sorghum protein is superior to wheat protein in biological value and digestibility. A vegetarian diet based on some varieties of sorghum is somewhat better than a rice-based diet. Sorghum lipids mostly consist of triglycerides, which are rich in the unsaturated fatty acids, oleic and linoleic, their percentage being 33 and 47 respectively.

The nutritional quality of grain sorghum can be improved by malting, fermentation and by mixing of flour from other cereal grains or pulses. Malting or germination of sorghum grain have been reported to improve the nutritional quality in terms of increased soluble proteins, free amino acids, reducing sugars, digestibility of proteins and starch. Such type of grain processing treatments increases dietary fiber content in the final processed products. The higher proportion of dietary fiber of grain sorghum confer many advantages such as nutritional, digestive and physiological benefits such as hypocholesterolemic and hypoglycemic effects, incidence of colon cancer, constipation and gastro-intestinal complications etc. A large number of diabetic patients are regularly consuming sorghum bhakari with fenugreek or bitter gourd *subji* through every day meals. However, the bhakari prepared from malted sorghum flour has been found to contain higher proportion of dietary fiber with superior nutritive and digestive values with additional physiological benefits. Natural sorghum roti consumption has been restricted to only weaker section of the society due to its coarse nature and palatability. This type of problems can be overcome by using malting and fermentation technology for sorghum.

Sorghum is undoubtedly the staple cereal of Botswana and is grown widely in the country. Statistical surveys indicated that 96 per cent population of southern and Southeastern districts of Botswana mainly consuming sorghum as a major food. Several varieties of sorghum and millet, both indigenous and exotic are grown in the country. All these varieties have proved to be of good food value and may be use for different industrial applications. The development of these varieties and their subsequent use has been steered by a number of institutions in the country and abroad, notably the ministry of Agriculture, SADC/ICRISAT. The Botswana Technology Center (and its international collaborators-CIRAD, NRI and IICT) and others. An identified constraint that continues to higher the national objective of food security is postproduction losses. These losses are believed to be high in magnitude although there is no loss assessment studies carried out in the country. These losses are caused by a number of agents, mainly insects and postproduction practices.

In many parts of the world sorghum has traditionally been used in several food products and various food items, like porridge, unleavened bread, cookies, cakes, couscous and malted beverages are made from this versatile grain. The traditional food preparation of sorghum is quite varied according to the region. Boiled sorghum is one of the simplest uses and small, corneous grains are normally desired for this type of food product. The whole grain may be ground in to flour or decorticated before grinding to produce a fine particle product or flour, which is then used in various traditional foods. Sorghum is a good source of lactobacilli, which is used in souring of foods used in traditional fermented foods and drinks. Nigerians produce "*Ogi*" a traditional fermented thin porridge. Fermented breads called "*masa*" and unleavened bread called "*waina*" are also produced in Nigeria. "*Tuwo*" is a stiff porridge produced in Nigeria from adding boiling water to sorghum flour and continuously stirring. Couscous can be made from sorghum. The grain is also used in making starch, dextrose, syrup, gur, vodka, malted flour for weaning foods, beers and other alcoholic beverages.

Sorghum is playing very important role in animal feed production in various countries like the USA, Mexico, South America and Australia. The nutritional value of good quality sorghum has an average feeding value of 96 to 98 per cent of corn. Sorghum can be processed to further improve its food as well as feed value and techniques such as grinding, crushing, steaming, steam flaking, popping and extruding have all been used to enhance the grain for food and feeding purpose. The products are then fed to beef or dairy cattle; egg laying hens, poultry and pigs are also used in pet foods.

Industrial uses for sorghum include wallboard and biodegradable packaging materials. In addition to the wallboard grain sorghum can be processed interchangeably with corn for the production of ethanol-many ethanol producing factories within the United States use grain sorghum as their principal grain due to their favourable geographic location. A very valuable co-product created from the processing of grain sorghum for ethanol is dry distillers grain. Grain sorghum having high protein animal feed supplement generates additional revenues for the ethanol processing plant.

Benefits of Sorghum Consumption

1. Sorghum food provides non-glutinous flours which is useful for avoiding damage to the lining of the intestine and make easy for nutrients to be absorbed.
2. It avoid celiac disease problems.
3. It also avoid dermatitis herpetiforms disease.
4. Sorghum is non-acid forming food.
5. Sorghum foods are a soothing and easy to digest.
6. Sorghum foods are least allergenic and mostly digestible.
7. It is good source of dietary fibers.
8. It also good source of vitamins and minerals.
9. It also provides antioxidants for controlling cancers.
10. Malted sorghum flour can be used for anti-constipation drinks.
11. It provides sufficient amount of dietary fiber in human diet.
12. Selected sorghum genotypes can be provide enough amount of lysine to meet human requirement.
13. Sorghum diet is very good to the people suffering from the Jaundice.
14. Sorghum has very low glycimic index so it helps to human health.
15. Sorghum helps in loose and control of body weight.
16. It increases body's sensitivity to insulin.
17. It also helps in diabetes control.
18. It reduces risk of heart diseases.
19. It also reduces blood cholesterol levels.
20. It also reduces hunger and keeps fuller for longer time.
21. It also helps to manage polycystic ovary syndrome in women.

Anti Nutritional Factors and Mycotoxins in Sorghum

Sorghum contains polyphenols, which are generally associated with grain pigmentation. Polyphenols commonly referred to as tannins, interfere with the bioavailability of nutrients. It has been reported that sorghum grains contain substantial quantities of condensed tannins and trace amounts of hyrdrolyzale tannins. Brown sorghums with high tannins are known to be resistant to bird depredation and have reduced preharvest germination and grain moulding.

Sorghum grain is known for its hardness compared to other food grains. The hardener of the grain is due to higher content of protein prolamin. Prolamin content varies from 36.0 to 51.0 per cent. The antinutritional factors present in sorghum grain are polyphenols and phytic acid. Polyphenols are the secondary metabolites produced and the presence of polyphenols/tannins in diet result in poor digestibility of proteins as they bind the proteins present in grain and make them unavailable for the intestinal absorption. Mostly polyphenols present in colored sorghums. In white sorghum, flavan 3-ols or flavan 4-ols (monomers of polyphenols) are present in very low quantity. Apart from flavan 4-ols, phytic acid is also present in sorghum in the form of 6-inositol phosphate. Phytic acid usually forms insoluble compounds with minerals like calcium, iron, magnesium and zinc thus making them unavailable.

Presently the new research investigations on polyphenols and phytic acid consider these compounds as health factors and consumption of these factors increase immunity in animal and human systems against several diseases. Due to grain mould the kharif grain gets severely affected by *Fusarium* and *Aspergillus* fungi, which produce harmful toxins. These mycotoxins namely aflatoxins and fumonisins cause deleterious effects to human and animal health. Hence the grain that is used for food preparation and consumption should be free of these toxins. However these toxins have a safety limit beyond which they cause ill health. The safety limit for aflatoxins is 20 ppb and for fumonisins it is 200 ppb as per the CODEX committee. Aflatoxin contamination was relatively less compared to Fumonisin contamination.

Nutritional Enrichment of Sorghum with other Cereal/Pulse Grain

Since the need for nutritive food at low cost is in demand and majority of the population cannot afford to buy variety of food grains for their bread and *roti* preparation, there is a great necessity to make a readymade *atta* with all the possible ingredients in a desired combination. The concept of multigrain flour is more ideal not only for nutritional and cost benefits but to improve the texture and shelf life of the flour. Efforts have been made to make composite flour with sorghum, wheat, maize, ragi, bajra and soybean keeping sorghum up to 50-60 per cent. The kneading quality of the composite flour was very good. Also the roti made of the composite flours was very soft and good in taste. The dough has good viscosity and the mean diameter of the roti made from composite flours was 24.16 cm. as compared to sorghum grain flour, which was 22.39 cm. The composite flour is also ideal for making different foods listed in this book. The composite flour is very useful for children with malnutrition and under nutrition. The composite flour is also ideal to include in the mid day meal programme as the school children need complete balanced diet.

The substitution of sorghum flour to poor vegetarian diets based on sorghum and finger millet increased the protein content of the diet and markedly improved the rate of weight gain in experimental trials. Consumer acceptability tests indicated that groundnut flour could be incorporated up to a level of 20 percent with sorghum in familiar food preparations. Soybean flour when incorporated in sorghum flour, it was found that the blend of 70 to 30 percent was found to be the best. This blend had 19 percent protein and 2.26 PER, which is very close to that of casein (2.5). The blend was used in popular jowar recipes such as *roti* and *laddu,* which were found to be highly acceptable by the people. This blend also satisfied the I.S.I. specifications for infant foods. Therefore, preparations of roti and laddu from sorghum-soy bean blend seem to be quite promising in bridging the protein-calorie gap.

Comparison with other Cereals

Grain sorghum is mostly used for human food as well as in small quantity for animal feeds. The utilization of sorghum grains in the world is shown in Table 12.1. From this table it is conformed that most of the sorghum grains are used as food in the Asia and Africa. Developed countries are using very small quantity of sorghum grains as a food and remaining mostly using for animal feeds.

Table 12.1: Sorghum Utilization (Average million tons)

Region	*Food*	*Feed*	*Other Uses*	*Total*
Africa	8.0	0.4	2.3	10.7
Asia	15.1	6.3	2.1	23.5
Central America	0.3	8.4	0.2	8.9
South America	–	4.6	0.3	4.9
North America	–	12.6	0.1	12.7
Europe	–	1.4	–	1.4
USSR	–	2.3	0.3	2.6
Oceania	–	0.4	–	0.4
World	23.4	36.4	5.3	65.1
Developing countries	23.2	15.6	4.8	43.6
Developed countries	0.2	20.8	0.5	21.5

Sorghum grains are rich in fiber and minerals apart from having a sufficient quantity of carbohydrates, proteins and fat. The comparison of sorghum grain with other cereals is given in Tables 12.2–12.7.

Table 12.2: Average Composition of Cereal Grains

Grains	*Moisture (Per cent)*	*Crude Protein (Per cent)*	*True Protein (Per cent)*	*Oil/Ether Extract (Per cent)*	*Crude Fiber (Per cent)*	*CHO (NFE) (Per cent)*	*Ash (Per cent)*
Barley	15.0	9.0	8.5	1.5	4.5	67.4	2.6
Sorghum	11.0	9.6	8.6	3.8	1.9	71.3	2.4
Maize	13.0	9.9	9.4	4.4	2.2	69.2	1.3
Millet	13.0	10.5	9.9	3.9	8.1	60.7	3.8
Oats	13.0	10.4	9.5	4.8	10.3	58.4	3.1
Rice	11.4	8.3	7.2	1.8	8.8	64.7	5.0
Rice polished	13.0	6.7	6.4	0.4	1.5	77.6	0.8
Rye	13.0	11.6	10.7	1.7	1.9	69.8	2.0
Wheat	13.0	12.2	11.0	1.9	1.9	69.3	1.7

Table 12.3: Amino Acid Composition of Sorghum, Wheat and Rice Grains (g/16 g Nitrogen)

Amino Acid	*Sorghum*	*Wheat*	*Rice*
Lysine	2.60	2.72	3.68
Histidine	2.06	2.08	2.08
Arginine	4.20	4.64	7.68
Aspartic acid	7.21	3.40	4.85
Threonine	3.00	2.88	3.68
Serine	3.66	4.30	4.56
Glutamic acid	20.30	32.50	11.69
Proline	6.25	11.60	5.27
Glycine	3.45	3.20	6.41
Alanine	8.64	2.00	3.23
Cystine	0.97	2.24	1.44
Valine	4.39	4.48	6.08
Methionine	1.39	1.44	2.40
Isoleucine	3.83	3.52	4.80
Leucine	12.27	6.56	8.00
Tyrosine	3.25	2.88	4.64
Phenylalanine	4.55	4.48	4.48
Trptophan	1.12	1.12	1.28
Total	93.14	96.04	86.25
Energy (Kcal)	349	341	345

Table 12.4: Mineral Composition of Sorghum, Wheat and Rice Flour (mg/100 g)

Mineral	*Sorghum*	*Wheat*	*Rice*
Sodium	21.00	19.3	2.2
Potassium	537.00	315.0	7.4
Calcium	25.00	48.0	10.0
Phosphorus	526.00	183.0	160.0
Magnesium	212.00	132.0	90.0
Iron	8.48	4.90	0.70
Zinc	3.91	2.20	1.40
Copper	0.86	0.51	0.14
Manganese	3.50	2.29	0.59

Table 12.5: Vitamins Content in Sorghum, Wheat and Rice Grains

Vitamin	Sorghum	Wheat	Rice
β-Carotene (μg/100g)	47.00	64.00	–
Thiamine (mg/100g)	0.37	0.45	0.06
Riboflavin (mg/100g)	0.13	0.17	0.06
Niacin (mg/100g)	3.10	4.30	1.90
Total B6 (mg/100g)	0.21	0.57	–
Folic acid (μg/100g)	20.00	35.80	8.00

Table 12.6: Oxalic Acid, Phytic Phosphorus and Dietary Fiber in Sorghum, Wheat and Rice Grains

Constituent	*Sorghum*	*Wheat*	*Rice*
Oxalic acid (mg/100g)	10.00	8.00	3.00
Phytic Phosphorus (mg/100g)	172.00	238.00	83.00
Phytic P as percent of total P	77.00	80.00	52.00
Total dietary fiber (per cent)	12.69	11.40	–

Table 12.7: Synergistic Effects of Legume (Green gram) on PER of Sorghum

Source of Protein	*PER*
Casein	2.87
100 per cent Sorghum	0.93
90 per cent Sorghum + 10 per cent green gram	1.11
80 per cent Sorghum + 20 per cent green gram	1.36
70 per cent Sorghum + 30 per cent green gram	1.75
50 per cent Sorghum + 50 per cent green gram	1.56

Grain Structure and Quality

Many scientists have described the kernel structure of sorghum. The kernel characteristics of sorghum described by using light, fluorescence and electron microscopy. The sorghum kernel can be divided into the following main and sub-components:

A. Seed Coat

1. Pericarp
    a. Epicarp
        1. Epidermis
        2. Hypodermis

b. Mesocarp
c. Cross-cell layer
d. Tube-cell layer

2. Testa (not present in all grain types)

B. Embryo or Germ

1. Embryonic axis
2. Scutellum

C. Endosperm

1. Based on hardness (structure)
   a. Peripheral
   b. Corneous, vitreous, horny or flinty
   c. Intermediate
   d. Floury or soft
2. Based on chemical composition
   a. Waxy
   b. Non-waxy
   c. Sugary

Sorghum grain processing involves converting shelled grain into consumable or suitable for further processing. The primary processing events involves cleaning, dehulling (decorticating), pounding and milling into fine powder. Then comes secondary processing, which involves turning raw material into food that includes blending, roasting, fermentation, frying or cooking etc.

Sorghum grain require the removal of the outer layer before any further processing because the milling and sensory qualities of sorghum for *roti/bhakari* making are influenced by the kernel characteristics. More recently, Chavan and Salunkhe (1984) have described the kernel characteristics and their implications in the nutritional and processing properties of sorghum. The sorghum kernel can be divided into seed coat, embryo or germ and endosperm (Figure 12.1). The seed coat contributes 3 to 10 per cent of dry weight of the kernel. It is made up of four sub-layers: epicarp, mesocarp, cross-cell layers and tube cell layers. The epicarp is subdivided into epidermis and hypodermis. Many times, epidermis contains pigments depending upon the genetic background of the seed. If the kernel is enclosed in pigmented glumes, these pigments migrate to the endosperm, especially, during humid weather or if crop receives rains at late maturity.

The presence of pigments (red, brown, lemon yellow or black) in seed coat or in endosperm poses a serious problem of their separation by pearling or milling which affects the quality of final product. The layer beneath the pericarp is called testa. The testa layer is complete in some sorghum grains, partial, sporadically around the pericarp in some while totally absent in others. These layers contain

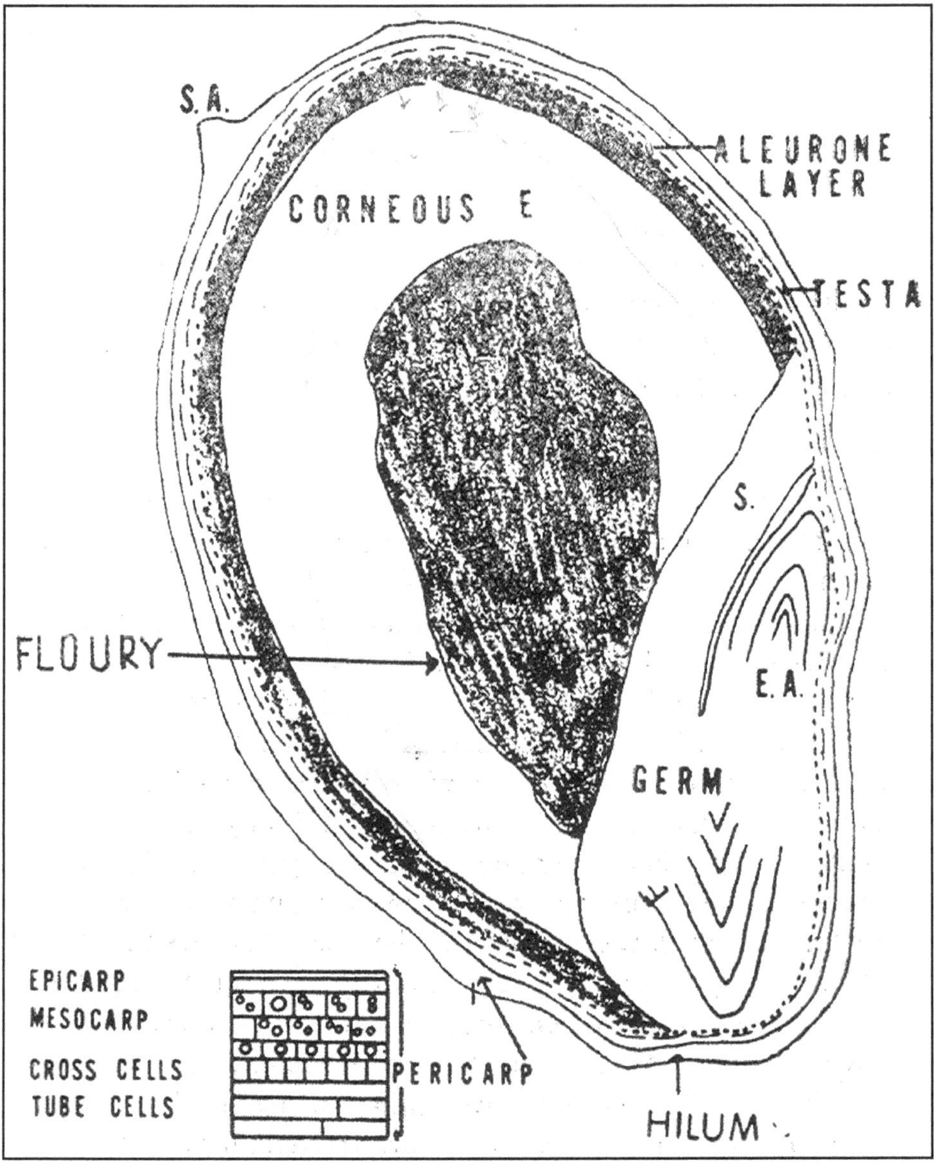

Plate 12.1: Sorghum Grain Structure.

the polyphenolic pigments called tannins. Due to its fibrous irritating nature and presence of pigments, the seed coat and testa are considered as undesirable parts of the grain from processing and utilization point of view. In the traditional milling process, part of the seed coat and testa are ground to powder, which is not fully separated by shifting the flour through a common sieve leaving considerable amount of bran with tannins in the resultant flour. Therefore, these components need to be effectively removed during milling to improve the quality and acceptability of sorghum products. The embryo or germ contributes to about 8 to 12 per cent of the dry weight of the kernel. It contains oil globules, protein bodies and few starch granules. The germ of some sorghum cultivars is more deeply embedded inside the endosperm and is difficult to remove. The germ contains more proteins that rich in lysine. Hence, a deep placement of embryo is desirable to avoid its loss during pearling or decortications.

The endosperm is the largest component contributing about 81 to 85 per cent to the dry weight of the kernel. It consists of aleurone layer, peripheral (hardest), corneous (hard), intermediate (semi hard) and floury (soft) portions. The aleurone is a single layer of block like rectangular cells and contains a large amount of minerals, water-soluble vitamins, autolytic enzymes, oil and protein bodies. The remaining portions of the endosperm contain mostly of nutrients such as proteins, starch, oil and minerals of the sorghum grain. The highest protein concentration is in the peripheral region and decreases towards floury area. The major proteins in the outer portions are prolamine, a lysine deficit protein while the proteins in floury region are relatively rich in lysine. The relative proportions of the corneous to floury endosperm within the sorghum kernel (endosperm texture) are known to influence processing properties of sorghum. Usually corneous endosperm is associated with more hardness, less breakage during pearling, easy removal of bran, resistance to pest infestation during storage and higher yields during milling. The physical properties of the grain influence *bhakari* making quality of sorghum. In general, pale yellow grain with intermediate corneous endosperm and with thin pericarp without testa produced *bhakari/roti* with acceptable quality. Waxy and floury grains produce poor quality *bhakari/roti*. The detailed physical properties and chemical composition of various types of grains with typical endosperm character are given in Tables 12.8 and 12.9. The protein concentration is highest in peripherial endosperm and decrease towards floury. The major protein in the outer portions is prolamin, a lysine deficit protein. This results in the lower concentration of lysine in the peripherial and corneous while the floury endosperm is higher in the lysine content.

Various in the structural features of the sorghum grain are large. Such differences in relative proportions of the seed coat, germ and endosperm structure and types affect the nutritional and processing quality of the grain. Identification of the cultivars early in the breeding programme for specific and uses can be possible by selecting the lines with certain physical characters such as pericarp thickness, presence or absence of testa, endosperm texture, placement and proportion of germ, hardness and shape of the grain.

The outer layers of certain sorghum varieties seed contain tannins, which are slightly toxic or have a bitter taste. Traditionally the processing of sorghum and millet has been carried out by grinding the whole grain between two plates of stones or by pounding grain using a pestle and mortar. The latter method is the commonest. The grain is then winnowed using levels to remove the bran. Pounding and winnowing are repeated several times for getting good quality flour. The objective of hand pounding is thus twofold, the first is to remove the bran and the second is to produce the good quality flour, but this is time consuming and backbreaking system.

Table 12.8: Kernel Characteristics, Physical Properties and Proximate Analysis of Grain Sorghum with different Endosperm Structures

Parameter	Corneous		Intermediate		Floury	
	Ref: 1	2	1	2	1	2
1000 kernel weight (g)	20.0	19.0–28.5	24.0	24.5–48.7	22.0	11.6–28.8
Hardness, per cent	33.5	17.9–63.8	15.7	1.2–44.1	7.2	0.1–30.8
Density (g/cm²)	1.4	1.31–1.4	1.37	1.32–1.38	1.2	1.22–1.36
Protein, per cent	12.9	13.8–17.8	11.8	12.4–15.6	16.6	12.6–14.6
Lipid, per cent	2.8	2.8–4.1	2.6	3.2–4.2	3.4	3.3–4.2
Ash, per cent	1.6	–	1.3	–	1.9	–
Starch, per cent	73.4	61.5–70.5	75.2	68.1–71.4	66.7	62.3–67.5
Amylose (per cent in total starch)	29.0	–	29.0	–	28.5	–

Source: 1. Sullins and Rooney, 1974; 2. Karim and Rooney, 1972.

Table 12.9: Kernel Characteristics, Physical Properties and Chemical Composition of Sorghum Grains with different Endosperm Types

Parameter	Waxy	Normal	Sugary	Yellow	High Lysine
1000 kernel weight (g)	27.6–34.6	24.5–48.7	22.4–33.1	26.6–59.2	19.3
Hardness, per cent	4.3–43.5	1.2–44.1	5.0–29.1	16.4–29.6	–
Density (g/cm²)	1.32–1.33	1.32–1.38	1.29–1.30	1.32–1.36	–
Protein, per cent	13.3–15.7	12.4–15.6	14.7–17.3	11.4–14.0	11.7
Lipid, per cent	3.4–3.7	3.2–4.2	5.2–5.4	2.7–3.7	–
Starch, per cent	68.8–72.6	68.1–71.4	60.9–64.1	69.7–75.9	–
Amylose per cent(in total starch)	0.79–4.8	27.72	26.80	29.82	–
Lysine (g/100g protein)	–	2.34	–	–	3.51

Source: Karim and Rooney, 1972; Figures from Miller and Burns, 1970; Sullins *et al.*, 1975.

The improved methods of processing sorghum and millets improve two stages, which use a dehuller and a hammer mill. The dehuller is used to remove the bran from the sorghum before it is milled. A dehuller has an abrasive disc or stones set on a horizontal shaft rotating at a high speed inside a casing. The grinding action of the spinning stones and the friction of the other grains rub off the bran. After primary processing, cereal products, flour or whole grain are further processed in the home and by small cottage industries into final products. Commonly final products from sorghum and millets include foods with porridge, fermented drinks and weaning foods. There is still a great potential for sorghum to be processed into snack foods, breakfast cereals, whole grain foods and baked products.

Milling

Dry milling separates the grain in to three components, germ, endosperm and seed coat. Milling techniques practiced mostly depend on the end use of the products (Plate 12.2). Pounding in a mortar and pestle to remove outer bran, with dry or slightly wet grain, does traditional milling (Plate 12.3). After pounding, the outer bran is removed by winnowing and the endosperm is pulverized in the same pestle and mortar or ground as small *chakkai*. Pounding is a very laborious and time-consuming process. Hence the usage of hand pounding is not advantageous to practice in any food enterprise. The dry milling process starts with the cleaning of grains. The cleaned grain is conditioned by addition of water, to soften the endosperm and milled by the conventional roller mills to separate the endosperm, germ and bran from each other. The endosperm is recovered in the form of grits, with the minimum production of flour. Yields of various fractions from the dry milling process are grit, 76.7; bran 1.2; germ 11 and fiber 10 per cent. Bran and germ are further processed as in the case of maize by dry milling for the preparation of oils and feeds.

Milling of Sorghum

Milling process of sorghum is called pearling and decortications. In this case cleaned grains are wetted by spraying water for 2 – 3 min and immediately milled in a rice huller to remove a major part of the coarse fiber, pigment, with minimum degree of cracking of the grain. A maximum of 12 per cent polishing can be carried out. This type of milling can give products rich in protein (up to 27 per cent) and which are also high in fat and give high yield of ash, but are low in fiber. These products are used in the preparation of food products of high protein content.

Wet Milling of Sorghum

Wet milling of sorghum is carried out by methods similar to that of wet milling of maize. Clean sorghum seeds are steeped for 48 hrs in warm water (50 °C). Steeping of water softens the kernel and assists separation of the hull, germ and fiber from each other. After steeping, the steep water is drained off and the sorghum is coarsely ground in degerminating mills to free the germ from the grain. Then the ground material flows down separating through in which hulls and grits settle, while the germ overflows. The degerminated material ground in to fine flour.

However, the milling of sorghum is more difficult than that of maize because of the small size and spherical shape of the sorghum kernel and the dense high-protein peripheral endosperm layer. Manufacture of starch is the main purpose of wet milling. However, some of the pigments of the pericarp and sub coat of the grain leach out and stain the starch. Thus grains with dark coloured outer layers are not satisfactory for wet grinding. Sorghum starch is indistinguishable from cornstarch and can be used interchangeably with cornstarch in most industrial applications. Sorghum oil obtained from germ fraction, after refining is used for salads and general cooking. Kernel residues containing bran and gluten are processed as cattle feed.

The starch of waxy sorghums consists almost entirely of amylopectins. These types were developed for starch in the wet milling process. There is some evidence that waxy sorghums may be more digestible. Sticky millet grown in Burma and Hawaii (var. Splendidum) had a more glutinous endosperm and was used to make cakes and sweet meats.

Sorghum can be milled for starch and oil etc. using very similar (wet milling) technology to that of maize. Products were starch (regular or waxy), germ oil and livestock feed. The starch was used in the same ways as maize starch, *e.g.*, paper and board, in the food industry etc. Some was converted to sugar, syrup and dextrin. There were two livestock feeds, gluten feed (21 per cent protein) and gluten meal (41 per cent protein) product.

Improved Milling Methods

Improved milling methods include roller, milling, pearling, impacting, grinding and air classification. The roller milled products are reported to have a higher production costs partly due to lower extraction rate. This may not therefore be suitable for adoption of millets commercially. The pearling technique has been applied successfully to sorghum and other millets.

For pearl millet, extraction rates varied between 80 and 90 per cent for similar dehulling times. For elusions only 3 min dehulling is required with extraction rate of 90 per cent. Mechanical dehulling improved physical appearance and functional properties. Dehulling effectively removed the colour of all varieties of sorghum. The detail procedure of mechanical dehulling process is shown in following flow chart. Mechanical dehulling reduces the drudgery. It converts millets into convenient ready-to-cook grain and improves the quality of flour.

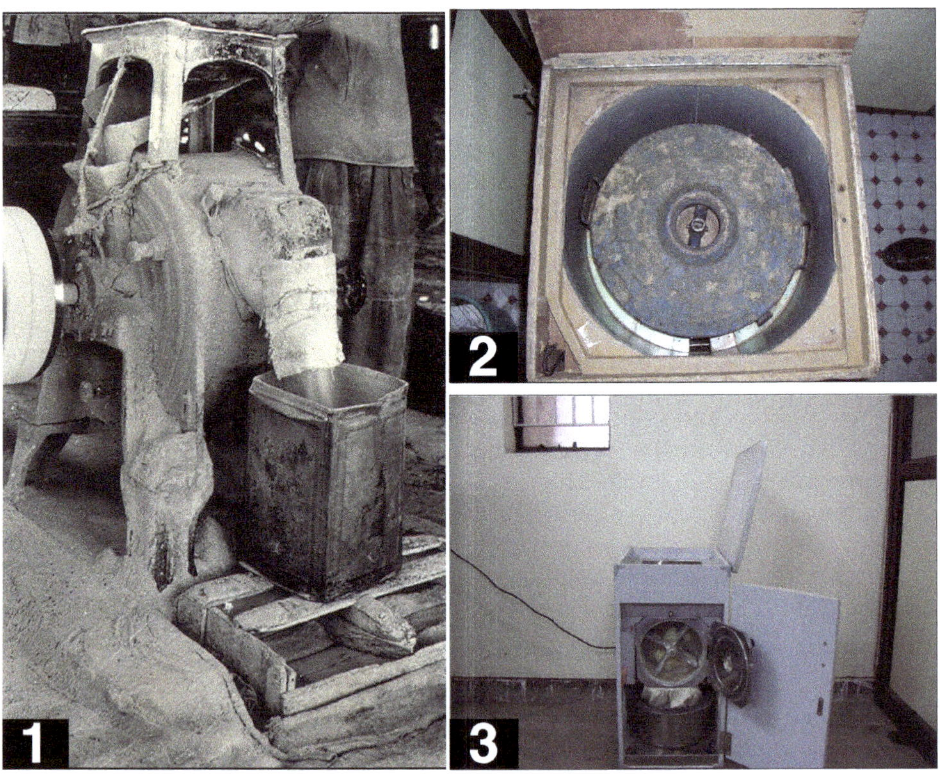

Figure 12.2: Flour Mills.
1. Vertical mill; 2. Horizontal mill and 3. Pin mill.

Figure 12.3: The Stone Mill and the Pestle-Mortar System Used for Milling Cereals in India.

Roller Milling

Basically it involves a series of rolls. First, the grooved or corrugated set of rollers break opens the grains. Subsequently, the exposed endosperm is crushed between a series of smooth reduction rollers and freed from the toughened bran. Screening separates the fine grains and the coarse fraction is further fed to the next set of rollers after every pass. The bran is generally, removed by screening and aspiration. The utility of dry milling largely depends upon the maximum yield of endosperm and the use of germ and bran as valuable byproducts. The break flour obtained from the first set of rollers is mostly floury endosperm. It is low in protein, oil, fiber and ash; amounts to 10-15 percent of the grain; behaves like starch and appears specky. Five fractions of grain sorghum obtained by specific gravity separators (Table 12.10) and found that the separation of germ more difficult from waxy sorghum than that of non-waxy.

Attrition Milling

The attrition mill fitted with saw-tooth blade plates, which provide the dehulling/pearling action (Figure 12.4). The grains were further abrased by a drum rotating in a cylindrical screen. Finally, the hulls were separated from the kernels by an air separator. Most attrition type dehullers comprise two stone or metal discs, either or both of them rotate around a vertical or horizontal axis. The attrition is provided by introducing metal pins or blades into the surface of either or both of the rotors or the rotor and stator.

Table 12.10: Yield and Proximate Composition of Dry-Milled Non-Waxy and Waxy Sorghum

Fraction	*Yield*	*Protein*	*Oil*	*Fiber*	*Ash*
		Non-Waxy sorghum			
Whole grain	100	9.6	3.4	2.2	1.5
Grits	67	9.6	0.6	0.8	0.5
Bran	12	8.9	5.5	8.6	2.4
Germ	11	15.1	20.0	2.6	8.2
Fines	10	7.1	2.4	1.3	1.0
		Waxy sorghum			
Whole grain	100	9.8	3.2	2.2	1.6
Grits	65	9.8	1.0	0.8	0.6
Bran	15	7.1	4.8	11.3	2.1
Germ	14	15.3	18.7	3.0	8.5
Fines	6	6.5	2.3	1.5	1.0

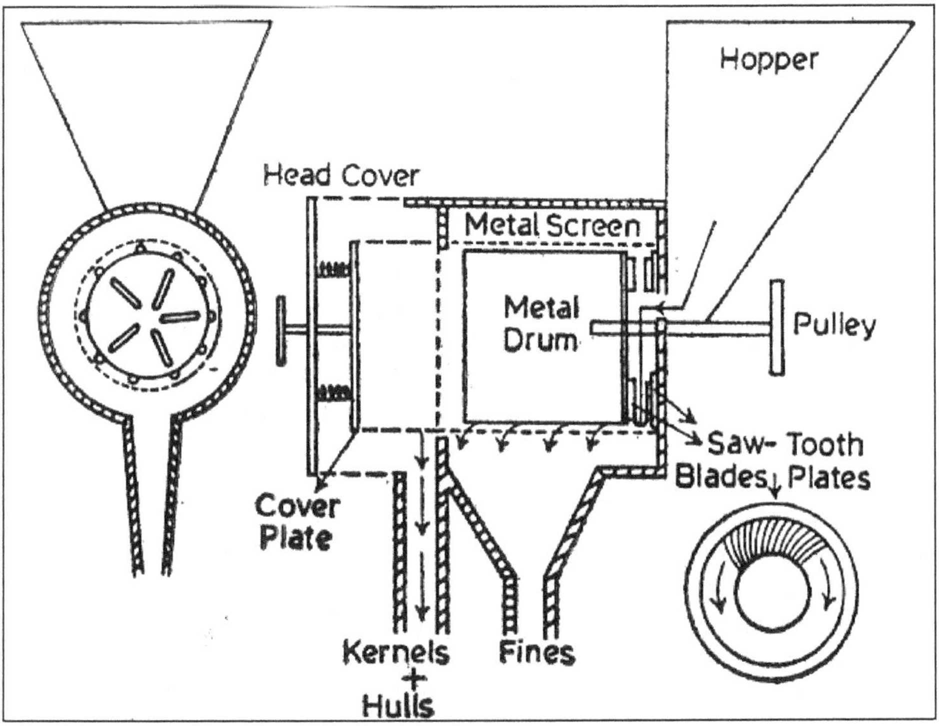

Figure 12.4: Schematic Diagram of the Attrition Mill.
***Source*: Reichert and Young, 1976.**

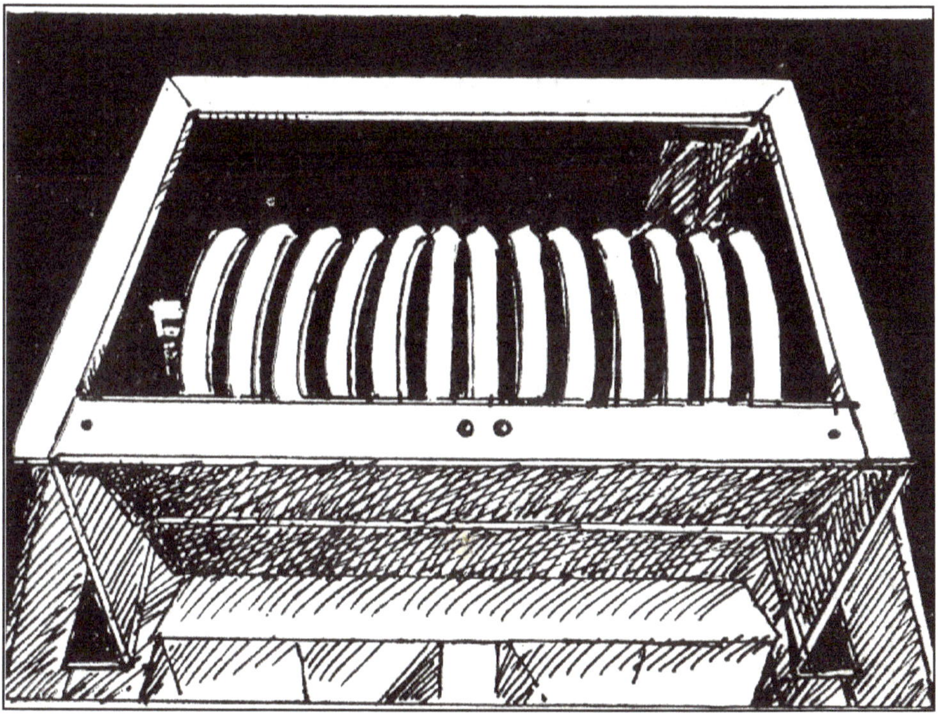

Figure 12.5: Dry Milling of Grain Sorghum.
***Source*: Reichert and Young, 1976.**

Figure 12.6: Dehulling/Pearling Machine.

Abrasive Milling

The abrasive mill consists of 13 carborundum stones (12-inch diameter), which can be driven at a speed up to 200 rpm (Figure 12.5). In continuous operation, grains are fed through a hopper at one end and released after stone action through an overflow outlet at the other end. The amount of kernel removed as fine is determined by the retention time in the mill, which in turn, depends upon the grain-feeding rate. After discharge from the mill the grain was passed through the air separator on the attrition mill to remove the fines. In addition to efficiency, other factors such as size, maintenance requirements and convenience favour the use of abrasive over attrition type mill for a village scale milling operation.

Clean sorghum

↓

Dry the grain thoroughly (8 -10 hrs)

↓

Grade

↓

Weigh 5 kg of grain and fill the dehuller

↓

Run the dehuller for 6-8 min

↓

Remove the dehulled grains

↓

Winnow

↓

Husk | **Dehulled/pearled sorghum grains**

Figure 12.7: Mechanical Dehulling of Sorghum Grains.

Sorghum Semolina Preparation

Improvement in primary process itself has helped in getting fine sorghum flour. Use of appropriate milling procedures and sieves help in getting sorghum semolina. Fine quality sorghum flour can be substituted in place of Bengal gram flour and wheat flour or wheat semolina. The process developed for sorghum flour and semolina is shown in Figure 12.7. In a similar manner, pearl millet, ragi and maize flour and semolina can be obtained.

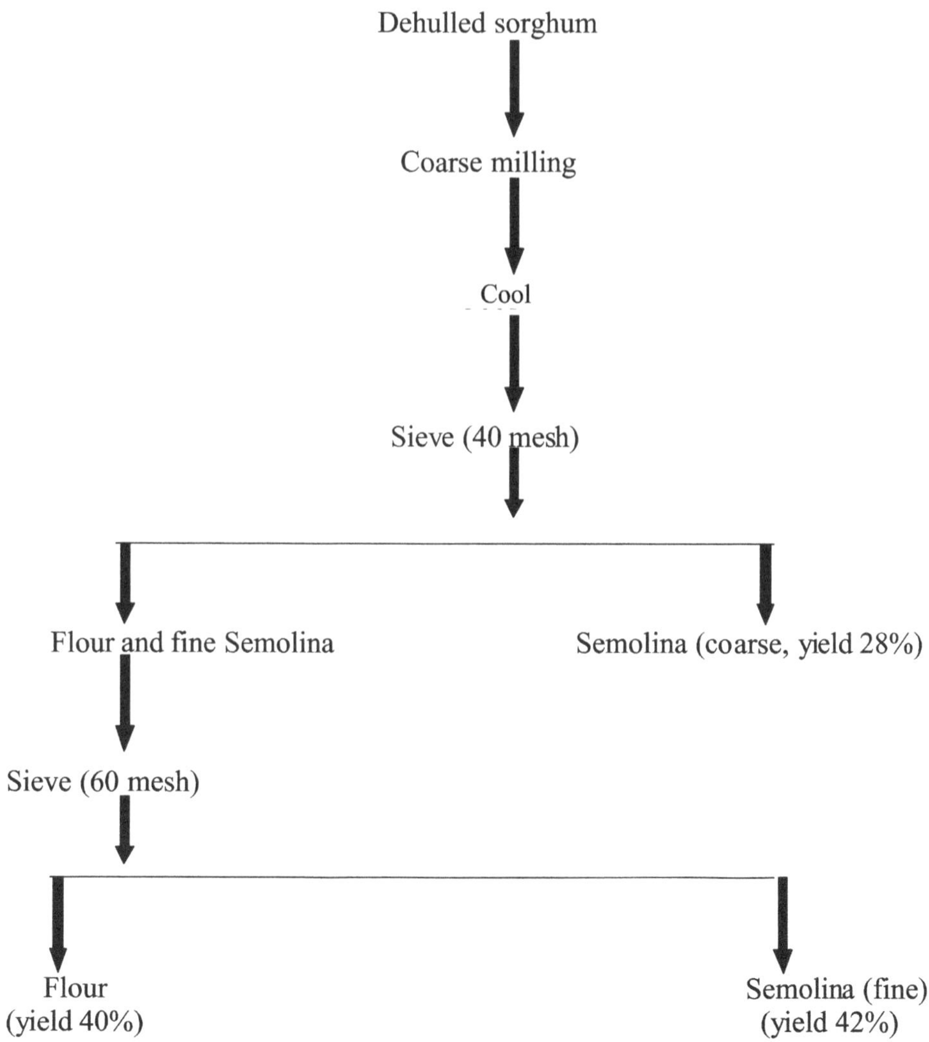

Figure 12.7: Processing of Dehulled Sorghum in to Semolina.

Preparation of Sorghum *Flakes* and their Products

Preparations of sorghum flakes are very easy and it has very high potential in human nutrition. The sorghum flakes have very high nutritional status because it is not producing acidity after consumption. Therefore, sorghum flakes can replace the rice flakes in various food products. Therefore, it is very essential to increase the processing industry in this field. Some of the work done at Mahatma Phule Krishi Vidyapeth, Rahuri is presented here. Ten sorghum varieties *viz.*, Phule Revati, Phule Vasudha, Phule Chitra, Phule Yashoda, Phule Maulee, Phule Anuradha, CSV-22, CSV-18, Selection-3, M 35-1 (Maldandi) and five hybrids *viz.*, CSH-15-R, SPH-1620, SPH-1647, SPH-1664 and SPH-1665 were obtained from the All India Co-ordinated Sorghum Improvement Project, Mahatma Phule Krishi Vidyapeeth, Rahuri for sorghum flakes preparation.

Figure 12.8: Sorghum Roaster for *Flakes*.

Figure 12.9: Sorghum Flakes making Machine.

Preparation of *Flakes* from Sorghum Grains

Flakes making machines (sorghum grain roaster and flakes maker) were used for the preparation of sorghum flakes from the Directorate of Sorghum Research, Rajendranagar, Hyderabad. The detail procedure is given in the following flowchart (Figure 12.10) recovery of flakes in Table 12.11.

Preparation of Sorghum *Flakes* and their Products

Sorghum grains
↓
Cleaning
↓
Sun drying (5-6 hr)
↓
Washing (2-3 times)
↓
Conditioning (12 hr in closed stainless steel drum)
↓
Roasting (340 °C for 1 min)
↓
Flaking in flaking machine
↓ ⟶ Husk separated
Sieving
↓ ⟶ Broken seperated
Flakes
↓
Packaging storage

Figure 12.10: Flow Chart for Preparation of Sorghum Flakes.

Nutritional Constituents of Sorghum Grains and *Flakes*

Nutritional/Chemical analysis of sorghum grains and *flakes* for protein, total sugar, crude fiber and starch were done using standard methods (Tables 12.11–12.13). These sorghum flakes were used for the preparation of various food products and their organoleptic evaluation and nutritional quality also studied for consumers' acceptance.

Preparation of Sorghum Flakes Chiwada

Recipe for the preparation of sorghum flakes chiwada: flakes 50g, oil 10g, turmeric powder 0.5g, chanadhal 5g, peanuts 10g, mustard and cumin 0.2g, chilli powder 2g, curry leaves 2-3 leaves and salt 1g. Chiwada was prepared as per method shown in Figure 12.11.

Table 12.11: Recovery of *Flakes* from different Genotypes of Sorghum

Genotype	*Flakes (Per cent)*	*Broken (Per cent)*	*Husk (Per cent)*
Variety			
Phule Revati (RSV-1006)	53.55	18.32	28.13
Phule Vasudha (RSV-423)	54.60	17.46	27.94
Phule Chitra (SPV-1546)	62.42	10.79	26.79
Phule Yashoda (SPV-1359)	58.78	14.35	26.87
Phule Maulee (RSLG-262)	63.43	13.56	23.01
Phule Anuradha (RSV-458)	64.14	8.45	27.41
CSV-22	51.73	18.41	29.86
CSV-18	50.62	17.52	31.86
Selection-3	51.89	18.35	29.76
Maldandi (M 35-1)	62.32	9.79	27.89
Hybrid			
CSH-15R	61.04	17.49	21.47
SPH-1620	53.37	22.40	24.23
SPH-1647	60.03	14.32	25.65
SPH-1664	61.76	15.25	22.99
SPH-1665	58.55	19.04	22.41
Range	50.62-64.14	8.45-22.40	21.47-31.86
Mean	57.88	15.71	26.41
SE ±	0.436	0.315	0.298
CD at 5 per cent (n=3)	1.261	0.911	0.863
CV per cent	2.264	6.027	3.395

Sorghum flakes

↓

Fry in sufficient quantity of oil

↓

Fry cumin, mustard, Chanadahl, groundnut, curry leaves, chilli powder, turmeric powder in other pan

↓

Add fried flakes in it and mix well

↓

Serve it while hot.

Figure 12.11: Flow Chart for Preparation of Sorghum Chiwada.

Table 12.12: Nutritional Composition of Sorghum Grains

Genotype	*Crude Protein, Per cent*	*Starch, Per cent*	*Total Sugars, Per cent*	*Crude Fiber, Per cent*
Variety				
Phule Revati (RSV-1006)	9.47	65.57	1.93	2.78
Phule Vasudha (RSV-423)	10.16	69.01	1.63	2.84
Phule Chitra (SPV-1546)	9.74	61.79	1.82	3.21
Phule Yashoda (SPV-1359)	9.49	67.66	2.32	2.82
Phule Maulee (RSLG-262)	9.83	61.27	1.93	3.41
Phule Anuradha (RSV-458)	9.13	62.62	1.93	3.16
CSV-22	10.42	60.38	1.95	3.20
CSV-18	9.45	66.92	1.85	2.76
Selection-3	10.39	61.07	2.12	3.18
Maldandi(M 35-1)	10.45	68.93	1.83	2.92
Hybrid				
CSH-15R	8.75	63.56	1.74	2.83
SPH-1620	8.30	63.71	1.75	2.72
SPH-1647	8.16	63.32	1.45	2.72
SPH-1664	7.81	64.65	1.83	2.56
SPH-1665	8.16	63.55	1.41	2.59
Range	7.81-10.45	61.07-69.01	1.41-2.32	2.56-3.41
Mean	9.31	64.26	1.78	2.91
SE ±	0.014	0.137	0.010	0.021
CD at 5 per cent (n=3)	0.043	0.398	0.031	0.063
CV per cent	0.480	0.643	1.823	2.245

Table 12.13: Nutrient Content in Sorghum *Flakes*

Genotype	*Crude Protein, Per cent*	*Starch, Per cent*	*Total Sugars, Per cent*	*Crude Fiber, Per cent*
Variety				
Phule Revati (RSV-1006)	8.83	65.11	1.06	2.14
Phule Vasudha (RSV-423)	9.25	67.68	1.54	2.34
Phule Chitra (SPV-1546)	8.45	61.26	1.62	2.93
Phule Yashoda (SPV-1359)	8.70	66.34	1.87	2.54
Phule Maulee (RSLG-262)	8.11	60.81	1.73	3.12
Phule Anuradha (RSV-458)	8.19	61.55	1.63	3.04
CSV-22	9.12	59.95	1.74	3.10
CSV-18	8.82	65.52	1.64	2.62
Selection-3	9.62	60.56	1.88	2.95
Maldandi(M 35-1)	9.71	67.87	1.74	2.86

Contd...

Table 12.13–*Contd...*

Genotype	*Crude Protein, Per cent*	*Starch, Per cent*	*Total Sugars, Per cent*	*Crude Fiber, Per cent*
Hybrid				
CSH-15R	8.53	63.35	1.59	2.76
SPH-1620	8.13	63.42	1.59	2.67
SPH-1647	7.92	63.17	1.22	2.63
SPH-1664	6.89	64.62	1.53	2.49
SPH-1665	7.34	63.06	1.20	2.52
Range	6.89-9.71	59.95-67.87	1.06-1.88	2.14-3.12
Mean	8.50	63.62	1.53	2.71
SE ±	0.008	0.009	0.007	0.008
CD at 5 per cent (n=3)	0.023	0.027	0.022	0.024
CV per cent	0.285	0.045	1.498	0.950

Preparation of Sorghum Poha

Recipe for sorghum poha: Flakes 50g, peanuts 10g, oil 10g, onion 10g, coriander leaves 2g, salt 1g, green chilli 2g, cumin 1g and mustard 1g. Preparation procedure is as follows (Figure 12.12).

Sorghum flakes

↓

Sprinkle small quantity of water on sorghum flakes to moisture it

↓

Soak for 1-2 min.

↓

Fry cumin, mustard, groundnut, curry leaves, green chilli (chopped), turmeric powder in oil

↓

Add moistured flakes in it and mix well

↓

Serve it while hot.

Figure 12.12: Flow Chart for Preparation of Sorghum *Poha*.

Organoleptic Evaluation of *Chiwada* and *Poha*

Organoleptic evaluation of *Chiwada* and *Poha* for colour and appearance, flavour, texture, taste and overall acceptability was carried out using standard method of Amerine *et al.* (1965). For this 10 semi-trained judges were used and 1 to 9 point hedonic scale was used for rating the quality of the sorghum product (Table 12.14).

Table 12.14: Overall Organoleptic Evaluation of *Chiwada* and *Poha* Prepared from *Flakes* of different Genotypes of Sorghum

Genotype	*Overall Acceptability for Chiwada*	*Overall Acceptability for Poha*
Variety		
Phule Revati (RSV-1006)	7.8	7.6
Phule Vasudha (RSV-423)	8.0	8.0
Phule Chitra (SPV-1546)	7.8	7.8
Phule Yashoda (SPV-1359)	8.0	8.0
Phule Maulee (RSLG-262)	7.4	7.0
Phule Anuradha (RSV-458)	7.6	7.2
CSV-22	7.8	7.8
CSV-18	7.6	7.8
Selection-3	7.8	8.2
Maldandi (M 35-1)	6.4	6.8
Hybrid		
CSH-15R	7.2	7.6
SPH-1620	7.4	7.8
SPH-1647	7.0	6.6
SPH-1664	7.6	7.8
SPH-1665	7.8	7.6
Range	6.4-8.0	6.6-8.2
Mean	7.52	7.54
SE ±	0.129	0.146
CD at 5 per cent (n=5)	0.366	0.415
CV per cent	8.618	9.737

Figure 12.13: Sorghum *Flakes*.

Figure 12.14: Sorghum *Chiwada* from Sorghum *Flakes.*

Figure 12.15: Sorghum *Poha* from Sorghum *Flakes.*

Preparation of Sorghum Semolina and their Products

Grain semolina is a very famous product which is utilized for preparation of various food products. Mostly wheat semolina is mostly utilized for preparation of various food products. But now a day's people are getting allergy of wheat gluten and some people are unable to digest wheat gluten. Sorghum is a non-gluten food item. Therefore, preparation of sorghum semolina is very important and its utilization in various food products. This sorghum semolina can be utilized in various food products where wheat semolina is used. Sorghum semolina has various health benefits over wheat semolina. Therefore, it is very essential to increase the processing industry in this field. Some of the work done at Mahatma Phule Krishi Vidyapeth, Rahuri, Maharashtra is presented here.

The grains of ten sorghum varieties *viz.*, Phule Revati, Phule Vasudha, Phule Chitra, Phule Yashoda, Phule Maulee, Phule Anuradha, CSV-22, CSV-18, Selection-3, Maldandi and five hybrids *viz.*, CSH-15-R, SPH-1620, SPH-1647, SPH-1664 and SPH-1665 were procured from the All India Co-ordinated Sorghum Improvement Project, Mahatma Phule Krishi Vidyapeeth, Rahuri for the preparation of sorghum semolina.

Semolina (*rava*) from sorghum grains was prepared as per the following procedure (Figure 12.16) and using pulverizer (Figure 12.17).

Sorghum grain
↓
Cleaning
↓
Reduction by break rolls
↓
Purification
↓ ⟶ **Flour**
Granular finished product (semolina/*rava*)
↓
Packaging
↓
Storage

Figure 12.16: Flow Chart for Preparation of Sorghum Semolina (*Rava*).

Nutritional constituents of sorghum grains and semolina: The nutritional constituents from the sorghum grains and semolina were evaluated for protein, total sugar, crude fiber, starch and minerals using standard methods (Table 12.15 and 12.16).

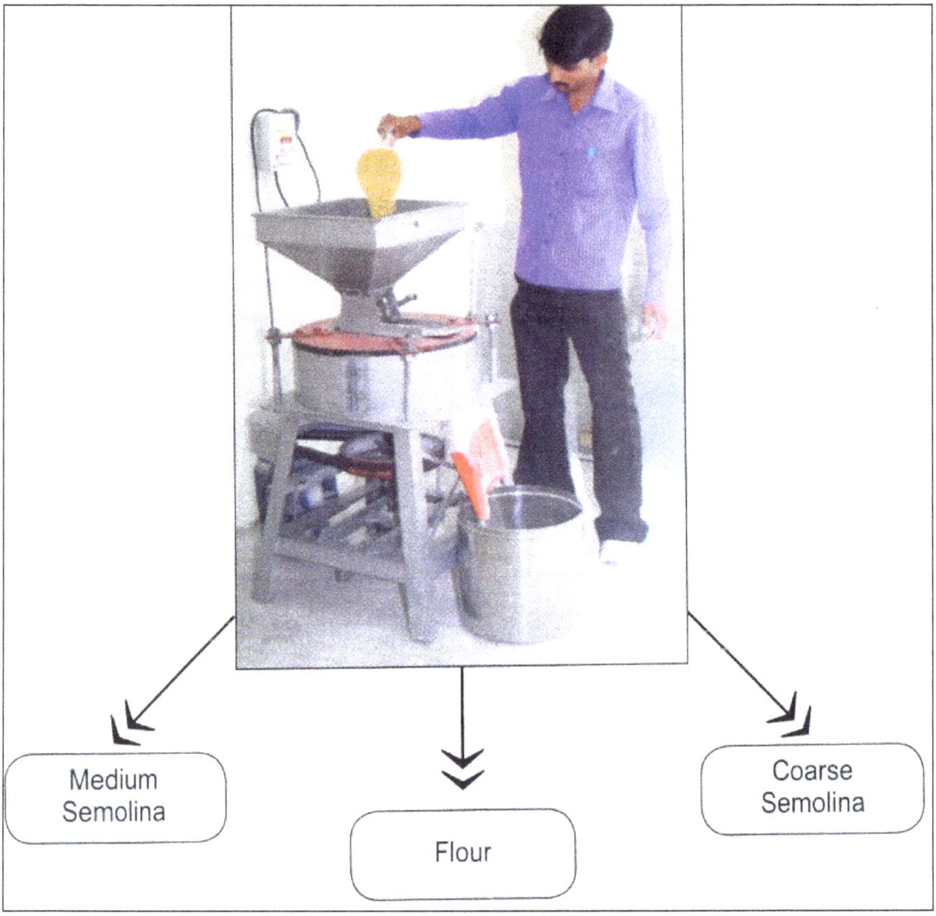

Figure 12.17: Milling of Sorghum Grain using Pulverizer for Semolina/*Rava*.

Preparation of Sweet (Shira) from Sorghum Semolina

Recipe for preparation of sorghum sweet (*shira*): Sorghum semolina 50g, sugar 50g, cashewnut 10g, almond 10g, cardamom 02g, vegetable ghee 25g and water 100ml (as per requirement) (Figure 12.18).

Preparation of *Upma* from Sorghum Semolina

Sorghum semolina 50g, gram flour ½ spoon, blackgram flour ½ spoon, peanuts 20g, tomato 10g, onion 15g, green chilli 2g, mustard 0.2g, curry leaves 2-3 leaves, oil 10g, salt 1g and water 100ml (as per requirement) (Figure 12.19).

Preparation of *Idli* from Sorghum Semolina

Sorghum semolina 50g, blackgram (dhal) 25g, oil ½ spoon, salt 2g and water 100ml (as per requirement) (Figure 12.20).

Table 12.15: Nutritional Composition of Sorghum Grains

Genotype	*Crude Protein, Per cent*	*Starch, Per cent*	*Total Sugars, Per cent*	*Crude Fiber, Per cent*
Variety				
Phule Revati (RSV-1006)	9.47	65.57	1.93	2.78
Phule Vasudha (RSV-423)	10.16	69.01	1.63	2.84
Phule Chitra (SPV-1546)	9.74	61.79	1.82	3.21
Phule Yashoda (SPV-1359)	9.49	67.66	2.32	2.82
Phule Maulee (RSLG-262)	9.83	61.27	1.93	3.41
Phule Anuradha (RSV-458)	9.13	62.62	1.93	3.16
CSV-22	10.42	60.38	1.95	3.20
CSV-18	9.45	66.92	1.85	2.76
Selection-3	10.39	61.07	2.12	3.18
Maldandi (M 35-1)	10.45	68.93	1.83	2.92
Hybrid				
CSH-15R	8.75	63.56	1.74	2.83
SPH-1620	8.30	63.71	1.75	2.72
SPH-1647	8.16	63.32	1.45	2.72
SPH-1664	7.81	64.65	1.83	2.56
SPH-1665	8.16	63.55	1.41	2.59
Range	7.81-10.45	61.07-69.01	1.41-2.32	2.56-3.41
Mean	9.31	64.26	1.78	2.91
SE ±	0.014	0.137	0.010	0.021
CD at 5 per cent (n=3)	0.043	0.398	0.031	0.063
CV per cent	0.480	0.643	1.823	2.245

Table 12.16: Sorghum Semolina Recovery and their Nutrient Contents

Genotype	*Semolina Recovery, Per cent*	*Flour, Per cent*	*Nutrient Content in Semolina*			
Variety			*Crude Protein, Per cent*	*Starch, Per cent*	*Total Sugars, Per cent*	*Crude Fiber, Per cent*
Phule Revati (RSV-1006)	49.24	50.76	7.61	64.16	1.05	2.59
Phule Vasudha (RSV-423)	49.16	50.84	8.35	64.30	1.16	2.76
Phule Chitra (SPV-1546)	48.90	51.10	6.73	61.25	1.37	3.15
Phule Yashoda (SPV-1359)	49.40	50.60	6.37	64.76	2.12	2.75
Phule Maulee (RSLG-262)	48.91	51.09	6.64	60.75	1.43	3.26
Phule Anuradha (RSV-458)	48.73	51.27	7.24	61.83	1.68	3.13
CSV-22	46.51	53.49	8.06	59.93	1.46	3.12

Contd...

Table 12.16–*Contd...*

Genotype *Variety*	*Semolina Recovery, Per cent*	*Flour, Per cent*	*Nutrient Content in Semolina*			
			Crude Protein, Per cent	*Starch, Per cent*	*Total Sugars, Per cent*	*Crude Fiber, Per cent*
CSV-18	48.44	51.56	7.14	65.05	1.33	2.63
Selection-3	50.61	49.39	7.43	60.51	1.91	3.12
Maldandi (M 35-1)	49.57	50.43	7.41	66.43	1.75	2.87
Hybrid						
CSH-15R	54.29	45.71	6.61	63.09	1.61	2.78
SPH-1620	51.32	48.68	6.64	63.04	1.60	2.65
SPH-1647	50.96	49.04	7.26	63.21	1.24	2.63
SPH-1664	52.69	47.31	5.43	64.20	1.42	2.44
SPH-1665	52.38	47.62	6.08	63.20	1.19	2.48
Range	46.51-54.29	45.71-53.49	5.43-8.35	59.93-66.43	1.05-2.12	2.44-3.26
Mean	50.08	49.92	7.00	63.04	1.45	2.82
SE ±	0.479	0.475	0.011	0.009	0.007	0.008
CD at 5 per cent (n=3)	1.385	1.373	0.033	0.028	0.022	0.024
CV per cent	2.877	2.858	0.500	0.046	1.631	0.893

Sorghum semolina (rava)

↓

Rosting till it becomes brown

↓

Place deep fry pan on gas

↓

Add ghee and heat to warm

↓

Add sufficient water in it and boil it

↓

Add semolina, sugar, salt etc.

↓

After cooking spread almond, cashewnut

↓

Serve it while hot.

Figure 12.18: Flow Chart for Preparation of Sorghum Sweet (*Shira*).

Sorghum semolina

↓

Rosting till it becomes brown

↓

Cut onion, chilli, coriander leaves, curry leaves

↓

Place deep fry pan on gas

↓

Add oil and heat to warm

↓

Add mustard, onion, chilli, curry leaves, Gram dhal powder, black gram dhal powder

↓

Fry thoroughly

↓

Add sufficient water in it and boil it

↓

Add semolina, salt (cook 15-20 min.)

↓

After cooking spread coriander leaves on it

↓

Serve it while hot

Figure 12.19: Flow Chart for Preparation of Sorghum *Upama.*

Organoleptic Evaluation of Semolina Products

Organoleptic evaluation of sweet (*shira*), *Upma* and *Idali* for colour and appearance, flavour, texture, taste and overall acceptability was carried out using standard method of Amerine *et al.* (1965). For this 10 semi-trained judges were used and 1 to 9 point hedonic scale was used for rating the quality of the sorghum product for consumers acceptance (Table 12.17).

Supplementing Sorghum Flour with Wheat Maida

Effect of incorporation of sorghum flour in wheat maida on physical and sensory properties of bread, buns, cookies and biscuits:

Bread, sweet buns, cookies and orange biscuits were prepared from the composite flour of wheat and sorghum. The products were evaluated for their

Soak black gram dhal for 8-10 hr

↓

Then drain out the water

↓

Grind into fine paste

↓

soak samolina for 1-2 hr

↓

Add grinded black gram dahl, salt in soaked semolina

↓

Mix thoroughly (Use mixer)

↓

Keep one night for fermentation

↓

Apply teaspoon oil to the Idali mould and pour Idali batter in it

↓

Steam it in Idali cooker for 10-15 min

↓

Take out and remove Idali from the moulds

↓

Serve hot with chuteny

Figure 12.20: Flow Chart for Preparation of Sorghum *Idali.*

Figure 12.18: Sorghum Semolina.

Table 12.17: Overall Organoleptic Evaluation of Sweet, *Upama* and *Idali* Prepared from Semolina of different Genotypes of Sorghum

Genotype	*Overall Acceptability for Sweet (Shira)*	*Overall Acceptability for Upama*	*Overall Acceptability for Idali*
Variety			
Phule Revati (RSV-1006)	6.6	7.6	7.6
Phule Vasudha (RSV-423)	8.2	8.2	8.2
Phule Chitra (SPV-1546)	7.6	7.4	7.4
Phule Yashoda (SPV-1359)	8.2	8.0	8.2
Phule Maulee (RSLG-262)	7.2	7.4	7.4
Phule Anuradha (RSV-458)	6.6	7.2	7.0
CSV-22	6.8	8.0	7.8
CSV-18	7.4	7.4	7.2
Selection-3	7.4	8.0	7.6
Maldandi (M 35-1)	7.6	6.8	7.0
Hybrid			
CSH-15R	7.2	7.0	7.4
SPH-1620	7.6	7.8	7.4
SPH-1647	7.4	7.0	7.0
SPH-1664	7.2	7.4	7.6
SPH-1665	7.4	7.8	7.4
Range	6.6-8.2	6.8-8.2	7.0-8.2
Mean	7.36	7.52	7.48
SE ±	0.101	0.128	0.117
CD at 5 per cent (n=5)	0.288	0.363	0.333
CV per cent	6.928	8.549	7.871

Figure 12.19: Sorghum Sweets (*Shira*).

Figure 12.20: Sorghum *Upma.*

Figure 12.21: Sorghum *Idali.*

physical and sensory properties. The results revealed that sorghum flour with 80 per cent extraction could be mixed with maida to the extent of 20 per cent (w/w) to prepare acceptable bread and buns and to the extent of 30 per cent (w/w) to prepare acceptable cookies and biscuits without using any other additives (Figures 12.22 and 12.23 and Tables 12.18–12.20).

Table 12.18: Sensory Properties of Bread, Sweet Buns, Cookies and Orange Biscuits Prepared from Wheat : Sorghum Blends

Product/Blend	*Flavour*	*Crust Colour*	*Crumb or Texture*	*Taste (Sweetness)*	*Overall Acceptability*	*Spread Factor (w/t)*
Bread wheat:sorghum						
100:0	8.00	8.50	9.50	9.50	8.88	–
90:10	7.50	8.00	6.50	7.50	7.38	–
80:20	7.50	7.00	6.50	7.00	7.00	–
70:30	5.50	6.50	4.50	5.00	5.38	–
S. E. ±	0.24	0.41	0.36	0.28	0.32	–
CD at 5 per cent	0.78	1.33	1.18	0.90	1.05	–
Sweet buns						
100:0	8.00	8.00	8.00	8.00	8.00	–
90:10	8.00	8.00	8.00	8.00	8.00	–
80:20	7.50	8.00	6.50	8.00	7.50	–
70:30	6.00	6.50	6.00	6.50	6.26	–
S. E. ±	0.13	0.13	0.13	0.13	0.13	–
CD at 5 per cent	0.41	0.41	0.41	0.41	0.41	–
Cookies						
100:0	7.70	7.25	7.35	7.30	7.40	6.45
90:10	6.75	6.75	7.25	7.00	6.94	6.62
80:20	6.25	6.25	6.50	6.75	6.44	6.70
70:30	5.85	6.50	6.50	6.75	6.40	6.71
60:40	3.50	6.00	5.00	6.00	5.13	6.48
50:50	3.00	6.00	5.50	5.75	5.06	6.47
40:60	3.00	6.00	5.25	5.50	4.94	6.44
S. E. ±	0.44	0.30	0.35	0.35	0.36	0.013
CD at 5 per cent	1.32	0.90	1.04	1.06	1.08	0.040
Orange biscuits						
100:0	7.50	7.75	7.25	7.50	7.50	8.40
90:10	6.75	7.50	7.25	7.50	7.25	8.61
80:20	6.25	6.75	6.50	7.50	6.75	8.70
70:30	6.20	6.75	6.25	7.50	6.68	8.80
S. E. ±	0.35	0.29	0.26	0.24	0.29	0.007
CD at 5 per cent	1.15	0.94	0.86	0.77	0.93	0.023

Table 12.19: Physical Properties of the Bread and Sweet Buns Prepared from Wheat : Sorghum Blends

Product/Blend	*Volume (ml)*	*Vol./Wt. (Ratio)*	*Weight of Bread (g) during Storage Period*			
			O hr	*24 hrs*	*48 hrs*	*72 hrs*
Bread Wheat : Sorghum						
100:0	960	2.65	362.00	346.50 (4.3)	346.00 (4.4)	344.80 (4.8)
90:10	945	2.60	363.00	347.50 (4.3)	347.00 (4.4)	345.50 (4.8)
80:20	905	2.46	367.80	353.00 (4.0)	349.00 (5.1)	346.00 (5.9)
70:30	900	2.43	370.00	356.00 (3.8)	351.00 (5.1)	349.50 (6.5)
Sweet buns Wheat : Sorghum						
100:0	140.50	3.07	45.70	45.00 (1.5)	44.80 (2.0)	43.90 (3.9)
90:10	140.00	2.99	46.90	46.00 (1.9)	45.70 (2.6)	45.00 (4.1)
80:20	132.00	2.78	47.50	47.00 (1.1)	46.60 (1.9)	45.20 (4.8)
70:30	130.00	2.68	48.50	48.00 (1.0)	47.00 (3.1)	46.80 (3.5)

Figures in brackets indicate percent loss in weight during storage period.

Table 12.20: Protein and Ash Contents of the Bakery Products Prepared from Wheat : Sorghum Blends (Dry weight basis)

Products	*Protein (Per cent)*	*Ash (Per cent)*
Bread		
Sorghum 100 per cent	5.60	0.89
Wheat : Sorghum		
100:0	11.80	2.83
90:10	11.18	2.63
80:20	11.42	2.55
70:30	9.51	2.40
Sweet buns Wheat : Sorghum		
100:0	9.13	1.80
90:10	9.10	1.75
80:20	8.00	1.72
70:30	7.90	1.71
Cookies Wheat : Sorghum		
100:0	6.19	2.09
90:10	5.33	1.74
80:20	4.95	1.64
70:30	4.95	1.54
60:40	4.57	1.40
50:50	4.19	1.33
40:60	4.10	1.30
Orange biscuits Wheat : Sorghum		
100:0	7.23	1.87
90:10	6.85	1.50
80:20	6.60	1.39
70:30	6.20	1.31

Dehydrated Food Products from Sorghum

The snack food products can be prepared from various cereals, pulses and some oilseeds. But mostly the *papads* are prepared from black gram and finger millets are famous in the society. The consumers are mostly accepting these snack products. Now-a- days a very good quality *papads* can be prepared from sorghum grains. Therefore commercially *papads* from sorghum can be produced. Therefore, it is very essential to increase the processing industry in this field. Some of the work done at Mahatma Phule Krishi Vidyapeth, Rahuri is presented here. Usually *papads* are prepared with cereal and pulse combination. But *papads* can be prepared from dehulled sorghum flour (Figure 12.24).

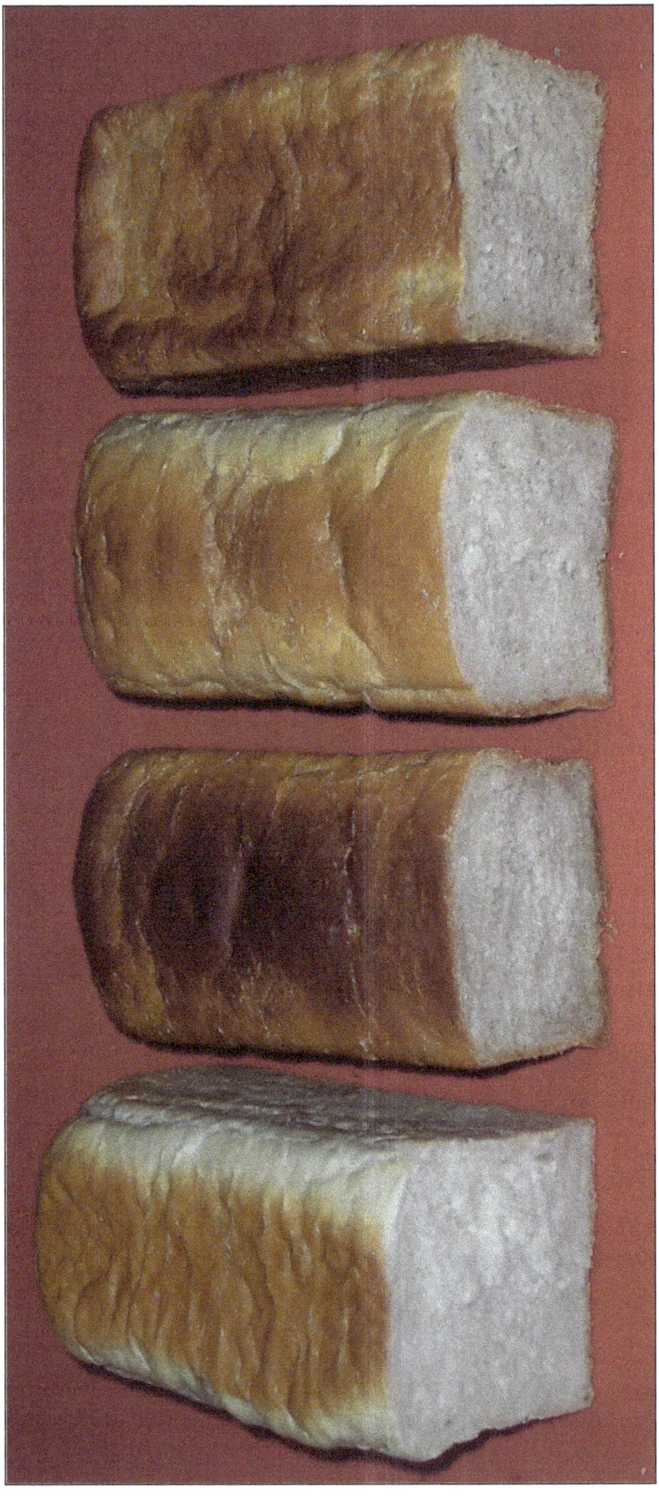

Figure 12.22: Sorghum Bread.

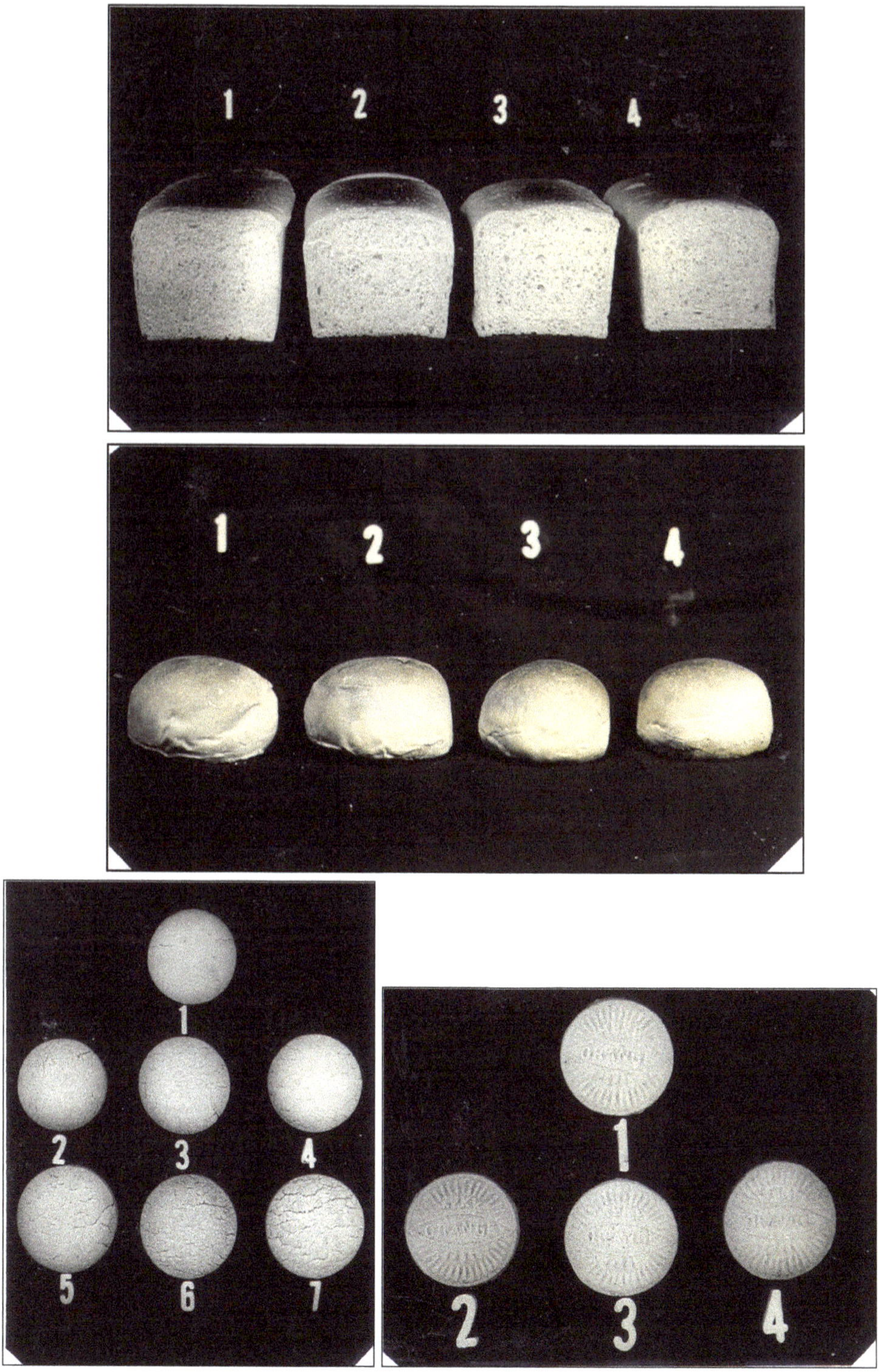

Figure 12.23: Bread, Buns, Cookies and Biscuits Prepared using Sorghum Flour.

Dehulled sorghum flour (25 g)

↓

Add chilli powder (7.5 g), salt (12.5g), cumin seeds (5g) and mix well

↓

Add lemon juice (1 tsp)

↓

Add the mixture to boiling water (500 ml) and cook for 10-15 min

↓

Remove from fire and knead well

↓

Make into small round balls

↓

Roll into papads

↓

Dry under sun or in dryer

↓

Papads

↓

Store in polythene bags

Figure 12.24: Processes for Preparation of Sorghum Papads.

Figure 12.25: Sorghum Papads.

Mostly the local genotypes of sorghum *viz., Pandhari Chikni* and *Dadar* local are used for preparation of *papads* in the cottage industry but at very small scale. The new genotype RPASV-3 identified for the preparation of sorghum *papads* showed slightly higher level of protein, fiber and pehnolics content. The total sugar, fat and ash content are less than other cultivars studied. *Papads* prepared from the genotype RPASV-3 had good sensory qualities like texture/crispiness, taste and overall acceptability and also good frying quality like puffiness, crispiness and expansion as compared to other genotype. *Papads* of genotype RPASV-3 having low oil absorption capacity (17.9 per cent) than other genotype and especially from market *papads* (black gram and finger millet *papads*). Due to less moisture percentage (2-3 per cent) in sorghum *papad* there was no any microbial contamination. The product remains safe throughout the storage period. From the present study it is concluded that the genotype RPASV-3 is best suited for the *papad* snack product preparation. The nutritional composition and other results are presented in Tables 12.21–12.29.

Table 12.21: Various Formulas for Papad Ingredients

Ingredients	*Formula No. 1*	*Formula No. No. 2*	*Formula No. No. 3*	*Formula No. No. 4*	*Formula No. 5*
Salt	Suhana papad	8	16	24	32
Papadkhar	masala pack	2	5	8	10
Black pepper	used (Market	1.20	6	4	1.20
Sesame seed	pack)	20	50	50	20
Asafoetida		0.30	1.4	4	0.50
Cumin seed		–	–	50	–
Water required (ml)		1000	1000	1000	1000

Table 12.22: Nutritional Composition of Sorghum Grains

Genotype	*Crude Protein (Per cent)*	*Lipids (Per cent)*	*Starch (Per cent)*	*Total Sugars (Per cent)*	*Fiber (Per cent)*	*Ash (Per cent)*	*Phenolics (Per cent)*
Varieties							
RPASV-3	11.27	1.63	58.96	1.12	3.61	2.36	1.63
Pandhari chikni	10.28	1.74	61.56	1.49	3.15	2.84	1.54
Dadar local	10.33	1.90	61.44	2.27	3.33	2.49	1.34
M-35-1	11.37	1.68	61.31	2.47	3.17	2.33	1.18
Phule Revati	11.36	1.21	58.56	1.46	3.36	2.47	1.24
Hybrids							
SPH-1620	10.43	1.65	60.44	1.76	3.56	3.01	0.92
CSH-15R	10.77	1.52	60.13	1.78	3.71	3.07	0.94
Range	10.28-11.37	1.21-1.90	58.56-61.56	1.12-2.47	3.15-3.71	2.33-3.07	0.92-1.63
Mean	10.83	1.61	60.34	1.76	3.40	2.65	1.26
SE ±	0.46	0.20	1.12	0.43	0.21	0.10	0.25
CD at 5 per cent	N.S.	0.59	N.S.	1.31	N.S.	0.30	N.S.

N.S.: Non-Significant.

Table 12.23: Amino Acid Content in Sorghum Grain (g/16 g N) Genotype

Genotype	*Cys*	*Glu*	*Gly*	*His*	*Ile*	*Leu*	*Lys*	*Met*	*Phe*
Variety									
RPASV-3	0.87	20.54	2.61	1.83	3.79	11.67	2.61	1.43	4.90
Pandhari chikni	1.14	20.82	3.24	1.98	3.36	13.46	2.89	1.55	4.59
Dadar local	0.70	21.49	2.31	2.09	4.09	13.75	2.61	1.36	5.02
M-35-1	0.72	21.32	2.29	2.12	3.94	13.46	2.69	1.34	5.00
Phule Revati	0.86	21.08	2.36	2.09	3.72	12.57	2.61	1.37	5.12
Hybrid									
SPH-1620	0.88	21.25	2.76	2.03	3.10	12.21	2.55	1.36	5.69
CSH-15R	0.87	21.37	2.48	2.13	3.76	13.28	2.66	1.35	5.07
Range	0.70-1.14	20.54-21.49	2.29-3.24	1.83-2.13	3.10-4.09	11.67-13.75	2.55-2.89	1.34-1.55	4.59-5.69
Mean	0.86	21.12	2.578	2.038	3.68	12.91	2.66	1.39	5.05
S.E. +	0.13	0.31	0.312	0.09	0.313	0.716	0.102	0.06	0.30
CD at 5 per cent	0.39	N.S.	N.S.	0.29	0.94	N.S.	0.30	0.20	0.91
Genotype	*Pro*	*Ser*	*Thr*	*Trp*	*Tyr*	*Ala*	*Arg*	*Asp*	*Val*
Variety									
RPASV-3	5.30	3.57	3.08	1.15	2.87	7.75	3.82	8.59	4.13
Pandhari chikni	4.96	4.12	3.30	0.76	3.44	5.48	5.69	9.04	3.33
Dadar local	6.12	3.14	2.96	1.05	2.77	7.84	5.21	6.69	5.15
M-35-1	6.12	3.05	2.92	1.12	2.68	8.03	5.23	6.65	5.35
Phule Revati	5.99	3.08	2.99	1.19	2.84	7.95	5.08	7.05	3.98
Hybrid									
SPH-1620	6.24	3.47	2.98	1.15	2.75	8.05	5.10	7.08	3.83
CSH-15R	6.18	3.17	2.95	1.08	2.87	8.02	5.40	6.56	4.21
Range	4.96-6.24	3.05-4.12	2.92-3.30	0.76-1.19	2.68-3.44	5.48-8.05	3.82-5.69	6.56-9.04	3.33-5.35
Mean	5.84	3.37	3.02	1.07	2.88	7.58	5.07	7.38	4.28
S.E. +	0.46	0.35	0.12	0.13	0.23	0.86	0.54	0.93	0.66
CD at 5 per cent	N.S.	N.S.	0.36	0.40	0.70	N.S.	1.64	N.S.	2.00

N.S.: Non-significant.

Procedure for the Preparation of Sorghum *Pops*

The corn *pops* are very famous all over the world. It is used as a snack food product. Similar sorghum *pops* also can be prepared and make it available to the consumers. Now a day's some farmers and small cottage industries preparing sorghum *pops* commercial and selling in the market. Therefore, it is very essential to increase the processing industry in sorghum processing field. Some of the work done at Mahatma Phule Krishi Vidyapeeth, Rahuri Maharashtra is presented here.

Table 12.24: Mineral Composition of Sorghum Grains (mg/100 g)

Genotype	*Ca*	*Fe*	*Mn*	*Mg*	*P*	*K*	*Cu*	*Na*	*Zn*
Variety									
RPASV-3	21.77	4.98	1.77	238	513	564	0.90	23.63	4.12
Pandhari chikni	30.19	4.14	2.41	221	505	512	1.09	29.55	4.73
Dadar local	36.44	4.36	2.82	225	494	506	1.02	17.99	3.87
M-35-1	34.72	4.88	2.83	221	491	515	1.03	17.67	3.86
Phule Revati	34.75	4.15	2.76	223	498	556	0.90	25.54	3.75
Hybrid									
SPH-1620	34.50	4.74	3.04	224	517	535	0.93	24.38	3.68
CSH-15R	34.42	4.11	2.87	222	499	531	1.04	22.65	3.79
Range	21.77-36.44	4.11-4.98	1.77-3.04	221-238	491-517	506-564	0.90-1.09	17.67-29.55	3.68-4.73
Mean	32.39	4.43	2.64	224	502	531	0.98	23.05	3.97
S.E. +	4.68	0.34	0.39	5.54	8.97	20.60	0.07	3.87	0.33
CD at 5 per cent	14.05	N.S.	1.19	N.S.	N.S.	N.S.	N.S.	11.6	1.00

N.S.: Non significant.

Table 12.25: Organoleptic Evaluation of Various *Papads* Prepared by using Five Formulas

Formula No.	*Colour and Appearance*	*Texture*	*Flavour*	*Taste*	*Overall Acceptability*	*Ranking*
Formula No. 1	7.7	7.4	8.0	8.2	7.8	3
Formula No. 2	8.2	8.4	8.3	8.5	8.3	2
Formula No. 3	7.2	7.1	7.6	7.8	7.4	4
Formula No. 4	6.6	6.7	7.0	7.2	6.8	5
Formula No. 5	8.4	8.7	8.5	8.8	8.6	1

* 1 to 9 hedonic scale (Amerine *et al.*, 1965).

Pops can be prepared from sorghum grains, but it required special type of sorghum grains (special variety of sorghum), which gives very attractiveness and fully open *pops*. First clean sorghum grains. Wash it thoroughly. Sprinkle the grains with 2 per cent hot water so as to attain 14 – 16 per cent moisture and cover it for 4 – 6 hrs. Place iron skillet on gas and place salt in the skillet. After sufficient heating place the soaked sorghum grains and allow it to puff for about 5 min. (maximum pop grains). If the popping machine is available the soaked grains can be popped in the machine by adjusting their temperature and moving road distance in the machine. Remove the popped grains by sieving through perforated scoop by separating from salt. As per the requirement mix spices with the popped grains. Serve it when it is hot, it tastes crunchy and crispy. It can be packed immediately into HDPE pouch by vacuum packaging and store for about 6 months.

Table 12.26: Physical Characteristics of Sorghum, Blackgram and Finger Millet Papads

Name of Genotype	*No. of Papad/kg*	*Colour of Papad*	*Thickness of Papad (mm)*	*Diameter of Papad (cm)*		*Expansion (Per cent)*	*Weight of Papad (g)*		*Increase Weight (Per cent)*	*Bulk Density (in box) (g/cm³)*
				Before Frying	*After Frying*		*Before Frying*	*After Frying*		
RPASV-3	65	Red brown	0.059	16.5	34.54	22.2	16.2	19.1	17.90	0.045
Pandhari Chikani	62	Faint white	0.060	16.4	29.88	21.3	18.9	23.8	25.93	0.042
Dadar Local	65	Faint yellow	0.056	14.1	20.57	17.0	16.3	21.3	30.67	0.043
M-35-1	65	Faint yellow	0.057	15.8	27.21	20.1	19.4	23.4	20.62	0.043
Phule Revati	64	Yellowish	0.061	14.9	29.53	19.3	20.1	25.2	25.37	0.044
SPH-1620	57	Yellowish	0.058	16.4	10.97	18.2	20.0	25.2	26.00	0.043
CSH-15R	56	Yellowish	0.061	16.1	20.49	19.4	20.4	25.1	23.03	0.044
Finger millet (market sample)	–	Faint yellow	0.058	11.2	23.2	107.14	3.6	14.6	30.55	0.044
Black gram (market sample)	–	Yellowish	0.057	13.6	15.1	11.03	6.7	10.1	50.75	0.043

Table 12.27 Organoleptic evaluation of sorghum papad*

Genotype	*Colour and Appearance*	*Texture/ Crispiness*	*Flavour*	*Taste*	*Overall Acceptability*
Variety					
RPASV-3	7.2	7.8	7.0	7.8	7.6
Pandhari Chikni	7.4	7.2	7.6	7.6	7.4
Dadar local	7.0	6.4	7.0	6.8	6.6
M-35-1	6.6	6.6	6.8	6.2	6.4
Phule Revati	6.4	6.8	7.0	6.6	6.6
Hybrid					
SPH-1620	7.4	7.2	7.0	7.0	7.1
CSH-15R	7.0	7.2	6.8	7.2	7.0
Range	6.4-7.4	6.4-7.8	6.8-7.6	6.2-7.8	6.4-7.6
Mean	7.0	7.02	7.02	7.07	6.85
S.E. +	0.35	0.43	0.25	0.52	0.44
CD at 5 per cent	N.S.	N.S.	0.74	1.55	N.S.

N.S.: Non significant*; 1 to 9 hedonic scale (Amerine *et al.*, 1965).

Table 12.28: Chemical Composition of Sorghum, Finger Millet and Blackgram Papads

Name of Genotype	*Protein (Per cent)*	*Oil (Per cent)*
RPASV-3	11.14	23.59
Pandhari chikni	10.20	30.48
Dadar Local	10.11	24.82
M-35-1	11.35	32.50
Phule Revati	11.25	32.18
SPH-1620	10.30	35.42
CSH-15R	10.33	25.42
Range	10.11-11.35	23.59-35.42
Mean	11.88	29.20
Finger millet *papad* (market sample)	9.10	57.72
Black gram *papad* (market sample)	22.00	34.35
S.E. +	3.64	8.55
CD at 5 per cent	N.S.	N.S.

N.S.: Non significant.

Table 12.29: Comparison of Sorghum Papad with Black Gram and Finger Millet Papad*

Name of Genotype	*Colour and Appearance*	*Texture*	*Flavour*	*Taste*	*Overall Acceptability*	*Ranking*
Sorghum papad (RPASV-3)	8.5	8.2	8.0	8.2	8.23	2
Finger millet papad	8.8	8.4	8.3	8.5	8.50	1
Black gram papad	8.0	7.8	7.6	7.8	7.80	3

* Semi-trained 10 judges used and 1 to 9 hedonic scales (Amerine *et al.*, 1965).

Figure 12.26: Sorghum *Pops.*

RPOSV-3 (Phule Panchami) having very good popping characteristics as well as nutritional quality as compared to the other local genotypes (Tables 12.30–12.37). It can be stored one year without any adverse effect in plastic bags with proper sealing. The economical returns from *pop* sorghum are very good for commercialization of the *pop* sorghum.

Table 12.30: Popping Quality of Popped *Rabi* Sorghum Variety RPOSV 3 (Phule Panchami)*

Popping Quality Parameters	*Genotypes*				
	RPOSV 3	*Local 1*	*Local 2*	*M 35-1*	*Maize*
Popping (per cent) on gain number basis	98	82	63	59	98
Expansion volume (ml/g)	6.5	4.8	4.6	2.6	19.6
Colour of pop grains	White	White	White	White	Creamy white
Size of popped grains	Extra large white fully opened and excellent	Large white	Small whitish	Small whitish, yellow	Extra large, fully opened, creamy white, excellent

* Results are means of 10 determinations.

Table 12.31: Popping Percentage (on wt. basis) of Various *Rabi* Sorghum Genotypes*

Genotype	*Popped Grains Per cent*	*Un-popped Grains Per cent*	*Processing Losses Per cent*	*Grain Price (Rs./Kg)*
RPOSV 3	87.4	8.8	3.8	20
Local 1	73.2	18.6	8.2	20
Local 2	56.6	36.4	7.0	20
M 35-1	38.0	53.8	8.2	18
Maize	82.6	4.0	13.4	52
Range	38.0-87.4	4.0-53.8	3.8-13.4	–
Mean	67.56	24.32	8.12	–
S.E. ±	18.14	18.45	3.09	–
C.D. at 5 per cent (n=3)	54.42	55.34	9.28	–

*Results are means of three determinations.

Table 12.32: Chemical Constituents of Sorghum *Pop* Grains (RPOSV-3; Phule Panchami)*

Parameter	*RPOSV 3*	*Local 1*	*Local 2*	*M 35-1*	*Maize*	*S.E.±*	*CD at 5 per cent*
Starch (per cent)	75.1	70.1	74.6	64.4	69.7	3.88	11.66
Protein (per cent)	8.65	8.15	7.03	10.47	9.03	1.12	3.38
Soluble protein (per cent)	1.00	1.13	0.93	1.43	0.59	0.27	0.82
Total sugars (per cent)	1.25	1.11	0.87	1.92	0.93	0.37	1.13
Free amino acids (mg/100g)	48.70	51.36	38.13	83.40	23.67	19.74	59.23
Hectoliter wt. (Kg/hl)	81.11	79.36	77.76	79.21	142.08	25.11	75.33

*Results are means of three determinations.

Table 12.33: Nutritional Constituents of *Pops* Prepared from *Pop* Sorghum Variety RPOSV-3*

Parameter	*Starch (Per cent)*	*Protein (Per cent)*	*Soluble Protein (Per cent)*	*Total Sugars (Per cent)*	*Free Amino Acids (mg/100g)*
RPOSV 3	75.8	8.52	0.85	1.05	23.17
Local 1	70.9	8.01	1.01	0.95	21.16
Local 2	75.1	6.93	0.86	0.71	13.05
M 35-1	65.0	10.23	1.33	1.70	33.11
Maize	70.3	9.00	0.55	0.78	13.12
Range	65.0-75.8	8.01-10.23	0.55-1.33	0.78-1.70	13.05-33.11
Mean	71.42	8.54	0.92	1.04	20.72
S. E. ±	3.88	1.09	0.25	0.35	7.43
C. D. at 5 per cent	11.66	3.27	0.76	1.06	22.30

*Results are means of three determinations.

Table 12.34: Organoleptic Evaluation of Sorghum Pops RPOSV-3 (Phule Panchami)*

Genotype	*Colour and Appearance*	*Flavour*	*Crispness*	*Taste*	*Overall Acceptability*	*DMR Rank*
RPOSV 3	8.6	7.4	8.0	8.4	8.1	2
Local 1	7.4	7.2	6.8	7.6	7.3	3
Local 2	6.4	6.8	6.6	7.4	6.8	4
M 35-1	6.2	7.0	5.6	6.8	6.4	5
Maize	7.8	8.0	8.2	8.6	8.2	1
Range	6.2-8.6	6.8-8.0	6.6-8.2	6.8-8.6	6.4-8.2	–
Mean	7.28	7.28	7.04	7.76	7.36	–
S.E. ±	0.89	0.41	0.95	0.66	0.70	–
C.D. at 5 per cent	2.67	1.24	2.87	1.99	2.12	–

* Ten semi-trained judges were used for the sorghum *pops* organoleptic/sensory quality evaluation and 1 to 9 point hedonic scale used for scoring. DMR = Duncan multiple range rank.

Table 12.35: Nutritional Composition of Sorghum Grains (Phule Panchami), Pops, Pop Grits and Pops' Flour*

Parameter	*Grain*	*Pops*	*Pops' Frits*	*Pops' Flour*	*SE ±*	*CD at 5%*
Crude protein, per cent	8.65	9.04	8.37	10.82	0.95	2.86
Soluble proteins, per cent	1.00	1.59	1.20	1.15	0.21	0.65
Lipids, per cent	1.52	1.17	1.21	0.85	0.23	0.71
Crude fiber, per cent	3.94	2.05	2.67	3.12	0.68	2.07
Total sugars, per cent	1.25	1.03	1.15	1.05	0.08	0.26
Starch, per cent	75.01	75.26	67.35	66.33	4.18	12.56
Ash, per cent	2.86	2.72	2.64	2.53	0.12	0.36
Mineral elements						
Calcium, mg/100g	14.51	25.81	27.29	26.95	5.29	15.90
Potassium, mg/100g	517.57	254.35	215.53	197.98	129.33	388.01
Phosphorus, mg/100g	524.78	510.81	519.15	516.73	5.01	15.04
Sodium, mg/100g	29.93	13.11	11.90	8.61	8.27	24.82
Iron, mg/100g	9.49	6.48	5.31	7.76	1.55	4.66
Magnesium, mg/100g	219.58	217.02	207.13	211.87	4.79	14.38
Zinc, mg/100g	4.28	4.19	4.00	4.24	0.10	0.32
Copper, mg/100g	0.84	0.76	0.75	0.79	0.03	0.11
Manganese, mg/100g	2.05	2.07	2.07	1.81	0.11	0.33
Vitamins						
β-Carotene, μg/100g	52.75	3.78	–	–	–	–
Thiamine, mg/100g	0.45	–	–	–	–	–
Riboflavin, mg/100g	0.32	0.34	0.36	0.35	0.01	0.04

Contd...

Table 12.35–*Contd...*

Parameter	*Grain*	*Pops*	*Pops' Frits*	*Pops' Flour*	*SE ±*	*CD at 5%*
Niacin, mg/100g	1.63	–	–	–	–	–
Folic acid, µg/100g	31.26	9.54	9.29	9.77	9.40	28.23
Antinutritional factors						
Free amino acids, mg/100g	93.09	89.04	85.70	87.50	2.72	8.18
Phenolics, g/100g	1.72	1.65	1.17	1.65	0.21	0.66
Oxalic acid, mg/100g	14.48	–	–	–	–	–
Phytic phosphorus, mg/100g	65.45	39.79	42.16	42.68	10.40	31.23

*Results are means of three determinations. - Not detected

Table 12.36: Amino Acid Composition of Sorghum Grains (Phule Panchami), *Pops*, *Pop* Grits and *Pops'* Flour (g/16 g Nitrogen)*

Parameter	*Grain*	*Pops*	*Pops' Frits*	*Pops' Flour*	*SE ±*	*CD at 5%*
Alanine	7.15	6.44	6.10	5.39	0.63	1.90
Arginine	4.89	3.30	4.72	3.73	0.35	1.08
Aspartic acid	8.72	6.07	5.33	7.58	1.31	3.95
Cysteine	1.22	0.92	0.91	0.77	0.16	0.49
Glutamic acid	20.11	23.28	24.45	24.89	1.86	5.60
Glycine	3.89	3.56	3.39	2.44	0.53	1.62
Histidine	1.89	2.49	2.77	2.72	0.34	1.05
Isoleucine	3.23	3.95	4.04	4.73	0.53	1.59
Leucine	11.86	15.46	16.72	17.86	2.25	6.76
Lysine	2.75	1.71	1.63	1.66	0.46	1.41
Methionine	1.49	1.58	1.63	1.72	0.08	0.25
Phenylalanine	4.32	4.67	4.58	5.18	0.31	0.94
Proline	5.28	5.30	5.34	3.96	0.58	1.75
Serine	4.24	4.09	3.89	3.66	0.21	0.65
Threonine	3.19	3.33	3.42	3.58	0.14	0.42
Tryptophan	1.05	0.77	0.65	0.51	0.19	0.60
Tyrosine	3.48	4.57	4.91	4.57	0.53	1.62
Valine	2.46	2.40	2.33	2.39	0.04	0.14

*Results are means of three determinations.

Sorghum Pop Laddu

First prepare sorghum pops as mentioned in the preparation of sorghum pops. The good quality sorghum pops then used for the preparation of sorghum pop laddu.

Table 12.37: Economical Returns from *Pop* Sorghum

Variety	*Grain Yield Kg/ha*	*Net Grain Yield Used for Popping (kg/ha)*	*Market Rate Rs/Kg*	*Total Price of Grain (Rs)*	*Processing Cost (Rs)*	*Cost of Grain with Processing (Rs)*	*Market Price of Processed Product (Rs)*	*Net Expected Return (Rs)*
RPOSV 3	1376	1202	18/-	24768/-	18824/-	43592/-	15,6260/-	1,12,668/-
M 35-1	1712	–	20/-	34240/-	–	34240/-	34240/-	Nil

Note:

1. Processing cost 76 per cent on raw material cost: (Ref: Bakery products book, 2008; J. K. Chavan and V. A. Dhotre).
2. Sorghum pop recovery from 1 kg grain produces 26 packets of 500 ml/25g each.
3. Present market price of sorghum pop packet is Rs. 5/each (*i.e.* Rs. 130/kg grain).
4. Un-popped grains 8.8 per cent, popping losses 3.8 per cent (*i.e.* total loss is 12. 6 per cent).

Ingredient	*Quantity*
Sorghum pops	1000 g
Jaggary/Sugar	600 g
Water	250 g

Procedure for the Preparation of Sorghum *Pop* Laddu

Dissolve sugar/jaggery in warm water and then heat on medium flame until it get sticky consistency and then add sorghum pops and mix it very well. All pops must be sticky with jaggery/sugar syrup. While it is warm taking small quantity and press with two hands in such a way that it forms a round shape laddu and keep for cooling and hardening. Then pack into plastic bags or plastic container and store in dry clean place. It can remain in good condition for about 15 days.

Figure 12.27: Sorghum *Pop* Laddu.

Sorghum *Pop Grits* Laddu

First prepare sorghum pops as mentioned under preparation of sorghum pops. Then prepare sorghum pops grits using mixer on small scale, for large scale grinding mill can be used.

Ingredient	*Quantity*
Sorghum pop grits	1000 g
Jaggary/Sugar	600 g
Water	250 g

Procedure for the Preparation of Sorghum Pop Grits Laddu

Dissolve sugar/jaggery in warm water and then heat on medium flame until it get sticky consistency and then add sorghum pops grits and mix it very well. All pops grits must be sticky with jaggery/sugar syrup. While it is warm taking small quantity and press with two hands in such a way that it forms a round shape laddu and keep for cooling and hardening. Then pack into plastic bags or plastic container and store in dry clean place. It can remain in good condition for about 15 days.

Sorghum Pop Grits Chikki

First prepare sorghum pops as mentioned under preparation of sorghum pops. Then prepare sorghum pops grits using mixer on small scale, for large scale grinding mill can be used.

Ingredient	*Quantity*
Sorghum pop grits	1000 g
Jaggary/Sugar	600 g
Fat	30 g
Water	250 g

Figure 12.28: Sorghum Pop Grits Chikki.

Sorghum Food Products

Sorghum Roti

Papads

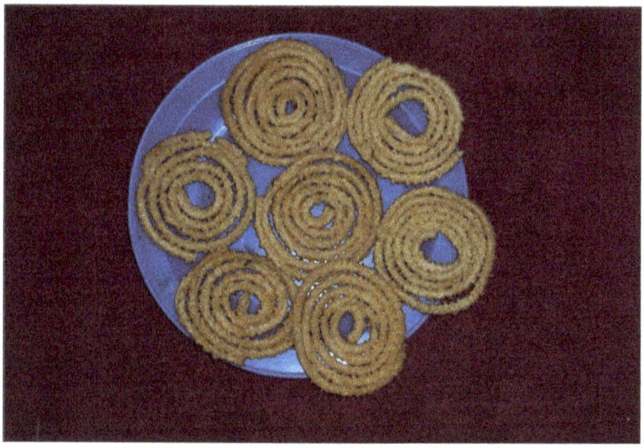

Sorghum Chakali

Noodles

Sweet Khir

Cookies

Upma

Pakoda

Sweet Shira

Idli

Dosa

Uttapa

Laddoo

Masala Chakali

Shev

Papadi

Plain Papadi

Shankar Pali

Cake

Sorghum Starch

Sorghum Ethanol Industry and Storage

Sorghum Ethnic Cafe

Procedure for the Preparation of Sorghum Pop Grits Chikki

Take water in a frying pan, warm it and then add jaggery/sugar, heat till it becomes like a soft paste. Add sorghum pop grits, mix well and then transfer the mixture to a greased flat plate and spread evenly. Allow to cool and then cut into attractive pieces. Store in airtight container. It can remain more 30 days in good condition at room temperature.

References

Amerine, M. A, Pangborn, R. M. and Rossler, E. B. (1965). Principles of sensory evaluation of foods. Academic Press, New York, pp. 350-376.

Anderson, R. A. (1962). A note on wet-milling of high amylose corn containing 75 per cent-amylose starch, *Cereal Chem.*, 39, 406-408.

Anderson, R. A. (1963). Wet-milling Properties of grains: Bench- Scale Study, *Cereal Sci. Today*, 8, No.6, 190-92, 195, 221.

Anon, 1964, Joint United States-Canadian tables of feed composition, *Natl. A cad. Sci.*, Nath. Res. Council Publication 1232.

Anonymous, (1982). Prevention of Food Adulteration Act, Government of India Publication, New Delhi.

Au, P. M. and Fields, M. L. (1981). Nutritive quality of fermented sorghum. *J. Food Sci.* 46: 652-654.

Autrey, H. S. (1953). Effect of variables upon milling yields, *Rice J. Annual.*

Badi, S. M. and Hoseney, R. C. (1976). Use of sorghum and pearl millet flours in cookies. *Cereal Chem.* 53: 733-738.

Bankar, J. R., Chavan, J. K. and Kadam, S. S. (1986). Effect of grain conditioning on physico-chemical properties of sorghum bhakari. *J. Maharashtra Agric. Univ.,* 11: 221-223.

Bender, A. E. (1973). Composition of infant foods, In Nutritional and Dietetic Foods, International Textbook Co. Ltd., London, U. K., p. 64.

Beta, T., Rooney, L. W. and Waniska, R. D. (1995). Malting characteristics of sorghum cultivar. Cereal Chemistry 72 : 533-538.

Bhat, C.M., Sharma, R.N. and Sehgal, S. (1982). A Manual on Food Preservation at Home. Directorate of Publication, Haryana Agricultural University, Hissar.

Bishop, W. and Dustin, J. (1951). Topnotch wet com mill house, *Food Ind.,* 23, No.3, 121-24, 211-12.

Bose, T K. and M. K. Mitra, Eds. (1986). Fruits of India Tropical and Subtropical. Naya Prokash, Culcutta.

Brautlecht, CiA, (1953). *Starch, Its sources, Production and Uses,* Reinhold Publishing Corp., New York.

Brekke, O. L. (1970). Dry milling artificially dried com: Roller milling of degerminator stock at various moistures, *Cereal Science Today,* 15, 37-42.

Brekke, O. L., Weinecke, L. A., Boyd, J. N. and Griffin, E. L., Jr. (1963). Com dry milling: Effects of first-temper moisture, Screen perforations and rotor speed on Beall egerminator throughput and products, *Cereal Chem., 40,* 423-29.

Chavan, J. K. and Kadam S. S. (1989). Nutritional improvement of cereals by fermentation. *CRC in Food Sci. and Nut.* 28: 349-399.

Chavan, J. K. and Salunkhe, D. K. (1984). Structure of sorghum grain in nutritional and processing quality of Sorghum. *Qual. Plant. Pl. Foods Human Nutr* 29: pp.21-31.

Chavan, J. K., Chavan, U. D. and Nagarkar, V. D. (1989). Effect of malting and fermentation on nutritional quality of sorghum. *J. Maharashtra Agric. Univ.,* 14: 246-247.

Chavan, U. D. and Chavan, J. K. (1991). Utilization of malted sorghum, mung bean and black gram in preparation of bhakari. *J. Maharashtra Agric. Univ.,* 16: 141-142.

Chavan, U. D. and Patil, J. V. (2010). Grain Processing of Sorghum. IBDC publishers, Lucknow. pp.10-15.

Chavan, U. D., Patil, J. V. and Shinde, M. S. (2009). Nutritional and Roti Quality of Sorghum Genotypes. *Indonesian Journal of Agricultural Science* 10(2): 80-87.

Chavan, U.D. (1987). "Effect of Fermentation on the Nutritional Quality of Grain Sorghum (*Sorghum bicolour* L. Moench). "*M. Sc. (Agric.) Thesis,* Mahatma Phule Agriculture University Rahuri, Maharashtra, INDIA. pp.i-95.

Chavan, U.D. (1987). "Effect of Fermentation on the Nutritional Quality of Grain Sorghum (*Sorghum bicolour* L. Moench). "M. Sc. (Agric.) Thesis, Mahatma Phule Agriculture University Rahuri, Maharashtra, INDIA. pp.i-95.

Chavan, U.D. and Chavan, J.K. (1992). Utilization of groundnut flour in preparation of sorghum bhakari. *J. Maharashtra Agric. Univ.*, 17 (20) : 346-347.

Chavan, U.D., Chavan, J.K. and Kadam, S.S. (1988). Effect of fermentation on soluble proteins and *in-vitro* protein digestibility of sorghum, green gram and sorghum + green gram blends. *J. Food Sci.* 53 : 1574-1575.

Chevassua-Agnes, S., Favier, J. C. and Joseph, A. (1976). Traditional technology and nutritive value of sorghum beer from camaoon (in French). *Cah. Nutr. Diet.* 11: 89-104.

Dayakar Rao, B., Seetharama, N., Elangovan, M., Vilas, A. Tonapi and Ratnavathi, C.V. (2007). Changing Scenarioof Millets Cultivation in India. Souvenir, Farm Fest-2007, 23-25th February at Uppalam, Puducheri.

Dayakar, Rao B. and Singh, J. P. (2010). Status paper on millets in India. Background paper prepared for *National Seminar on Millets*, organized by Directorate of Millets Development, Jaipur and Directorate of Sorghum Research, Hyderabad.

Deshpande, S. P., Borikar, S. T., Ismail, S. and Ambekar, S. S. (2003). Genetic studies for Improvement of Quality Characters in *Rabi* sorghum using land races. ISMN. 44.2003.

Desikachar, H. S. R. (1975). Processing of maize, sorghum and millet for food uses. *Journal of Scientific and Industrial Research* 34(4): 231-237.

Desikachar, H. S. R. (1977). Processing of sorghum and millets for versatile food uses in India. In Sorghum and millets for human food. Pp. 41-45. *Symp. Proc. IACC*, Vienna, 1976. Tropical Products Institute, London.

Eggum, B. O., Monowar, L., Knudsen, B. K. E., Munck, L. and Axtell, D. (1983). Nutritional quality of sorghum and sorghum foods from Sudan. *Journal of Cereal Science* 1: 127-137.

Elias, D. G. and Scott, R. A. (1957). British flour milling technology, *Cereal Sci. TorJay*, 7, 180-84.

Ezaki, H. (1973). Paddy husker, Group training course-Fiscal, Institute of Agricultural Machinery, Japan.

FAO, (1995). Food and Agricultural Organization, Sorghum and Millet in Human Nutrition. FAO : Food and Nutrition Series. No. 27, ISBN – 92-51-03381-1.

Fennema, O.R. Ed.(1976). Principles of Food Science, Part-l. Food Chemistry, Marcel Dekker Inc, New York.

Frazier, W.e. and Westhoff, D.C. (1983). Food Microbiology. Tata McGraw-Hill Publishing Company Limited, New Delhi.

G.O.I. (2011). *Agriculture statistics at a glance*. Directorate of Economics and Statistics, New Delhi.

Goel, A. K., Kumar, R. and Mann, S. S. (2007). Postharvest management and value addition. Daya Publishing House, Delhi.

Gopalan, C, Rama Sastri, B.V. and Bala Subramanian, S.C. (1982). Nutritive Value of Indian Foods. National Institute of Nutrition, R.CM.R., Hyderabad.

Hahn, R. R. (1969). Dry milling of grain sorghum. Cereal Sci. Today. 14: 234-237.

Hall, A. J. (2000). Sorghum utilization and the livelihoods of the poor in India : A review of findings and recommendations, International Crops Res. Institute for the Semi-Arid Tropics, ICRISAT, Patancheru, p. 41.

Harris, RS. and Karmas, E. (1975). Nutritional Evaluation of Food Processing. 2nd Edition. AVI Publishing Co. Westport, Connecticut.

Houston, D. F. (1973). Western Marketing and Nutrition Research Division, Agricultural Research Service, U. S. Department of Agriculture, Berkeley, California.

Ibrahim, E., Abdel, R. and Magdi, A. O. (2010). Changes in Chemical Composition x Available Corbohydrates and Amino Acids Content during Soaking and Germination of Saudi Sorghum Cultivars. *J. Saudi. Soc. for Food and Nutrition* 5 : 2.

Ilyas, S. M., Wanjari, O. D. and Goyal, R. K. (2002). Prospects of income and employment generation through agro-processing and value-addition. Presented at Conference on Emerging Opportunities of Agro and Food Processing Industries in Jharkhand held at Indian Institute of Coal Management, Ranchi during October, 8-9.

Ilyas, S.M. and Coyal, R.K. (2002). Quality assurance in food processing-impact of WTO, In: Proc. National Seminar on Development of Food Processing Industries in U.P., held at Lucknow during January 16-17.

Jacob, T. (1987). Poisons in our Food. Publications Division, Ministry of Information and Broadcasting, Government of India, New Delhi.

Jones, CR. (1940). The production of mechanically damaged starch in milling as a governing factor in the diastatic activity of flour, *Cereal Chem.*, 17, 133-69.

Karel, M., O.R. Fennema and D.S. Lund, Eds. (1975). Principles of Food Science Part II. Physical Principles of Food Preservation. Marcel Dekker, Inc. New York.

Karim, A. and Rooney, L. W. 1972. Characterization of pentosans in sorghum grain. *J. Food Sci.* 37: 369-371.

Kazanas, N. and Fields, M. L. (1981). Nutritional improvement of sorghum by fermentation. *J. Food Sci.* 46: 819-821.

Kent-jones, D. W. and Amos, A. J. (1957). *Modern Cereal Chemistry*, Fifth Ed., Northern Publishing Co. Ltd., Liverpool.

Kerr, R. W. (1950). *Chemistry and Industry of starch*, 2nd Edition, Academic Press, New York.

Kleih, U., Bala Ravi, S. and Dayakar Rao, B. (2000). Industrial utilization of sorghum in India. Institute for the Semi-Arid Tropics, ICRISAT, Patancheru, pp. 38.

Klopfenstin, C. F. and Hoseney, R. C. (1995). Nutritional properties of sorghum and millets. In : Sorghum and Millets : Chemistry and Technology. American Assoc. of Cereal Chemists, USA, pp. 125-168.

Koga, Y. (1969). Drying, husking and milling in Japan IV and V., *Farming,* Japan.

Kordylas, J.M. (1991). Processing and Preservation of Tropical and Subtropical Foods. Macmillan Education Ltd., Houndmills, Basingstoke, Hampshire.

Lockwood, J. F. (1952). *Flour Milling,* Third ed., Northern Publishing Co. Ltd., Liverpool.

Lockwood, J. F. (1960). *Flour Milling,* The Northern Publishing Co., Liverpool.

Macrae, R., R. K. Robinson and M. J. Sadler, Eds, (1993). Encyclopaedia of Food Science, Food Technology and Nutrition, Academic Press, London.

Matz, S. A. (1959). The Chemistry and Technology of Cereals as Food and Feed, Avi Publishing Co., Westport, Conn.

Mayande, V. M. (1987). Pigeonpea (*Caianus cajan* L.) milling Technology: Biochemical and Engineering Studies, unpublished M. Tech. Thesis, G. B. Pant University of Agric. and Technology, Pantnagar.

Miller, O. H. and Burns, E. E. (1970). Starch characteristics of selected grain sorghum as related to human food. *J. Food Sci.* 35: 666-668.

Mosse, J., Huet, J. C. and Baudet, J. (1988). The Amino Acid Composition of Whole Sorghum Grain in Relation to its Nitrogen Content. American Association of Cereal Chemists, Inc. Vol. 65, No. 4, 1988 271. Robertson S.K., Perez R.A. (2006) Nutritional characteristics of sorghum. *World Poultry Science Journal* 53 : 48-52.

Mudambi, S.R and Rao, S. (1986). Food Science. Wiley Eastern Limited, New Delhi.

Murty, D. S. and Subramanian, V. (1982). Sorghum roti. I. Traditional methods of consumption and standard procedures for evaluation. ICRISAT, 1982, Proc, Intl. Symp. On Sorghum grain quality. 28-31 October, 1981, Patancheru, A.P. India. Pp. 73-78.

Murty, D. S., Patil, H. D., Prasad Rao, K. E. and House, L. R. (1982). A note on screening the Indian sorghum collection for popping quality. *J. Food Sci. Technol.* 19: 79-80.

Novellie, L. (1977). Beverages from sorghum and millets in "Proc. of Symp. On Sorghum and Millets for Human Food". International Association for Cereal Chemistry. Symposium. Vienna. 1976. Tropical Products Institute, London. pp. 73-77.

Odunfa, S. A., Nordsrom, J. and Adeniran, S. A. (1994). Development of starter cultures for nutrient enrichment of ogia West African fermented cereal gruel. Report submitted to HBVC research grants program. USAID, Washington, U.S.A.

Panse, V. S. and Sukhatme, P. V. (1967). Statistical methods for Agricultural workers, ICAR, New Delhi, pp. 70-72.

Parthasarathy, Rao P., Basavaraj G., Ahmad W. and Bhagavatula, S. (2010). An analysis of availability and utilization of sorghum grain in India. *Journal of SAT Agricultural Research* 8 : 1-8.

Patil, U. S, Dalvi, U. S. and Chavan, J. K. (2003). Studies on the production of starch from mould infected discolored sorghum. *J. Food Sci. Technol.* 40: 115-117.

Patil, U. S, Dalvi, U. S. and Chavan, J. K.(2003). Studies on the production of starch Paulus, K. O. (1984). Modelling in Industrial Cooking. In: *Thermal Processing and Quality of Foods.* Zeuthen *et al.* (eds.), Elsevier Applied Science Publishers, London.

Peplinski, A. J and Pfeifer, V. F. (1970). Gelatinization of corn and sorghum grits by steam-cooking, *Cereal Science Today,* 15, 144, 149-51.

Potter, N.N. and Hotchkiss, J.H. (1996). Food Science. 5th Edition. CBS Publishers and Distributors, New Delhi.

Raghavandra Rao, S. N. and Desikachar, H. S. R. (1964). Pearling as a method of refining jower and wheat and its effect on their chemical composition. *J. Food Sci. Technol.* 1: 40-42.

Raghavandra Rao, S. N., Malleshi, N. G., Sreedharamurty, S., Viraktamath, C. S. and Desikachar, H. S. R. (1979). Characteristics of roti, dosa and vermicelli from maize, sorghum and bajra. *J. Food Sci. Technol.* 16: 21-24.

Rai, Mangla and Mauria S. (1999). Coarse cereal: Not for the poor alone. *The Hindu Survey of Indian Agriculture.* pp. 55-58.

Rai, Mangla. 2002. Amazing nutrient composition. *The Hindu Survey of Indian Agriculture.* pp. 59-62.

Ratnavathi, C. V., Bala, S. R. and Subramanian, V. and Rao, N. S. (2000). A study on the suitability of Unmalted Sorghum as a Brewing Adjunct. *Journal of the Institute of Brewing* 106: 383-387.

Ratnavathi, C. V., Gopalkrishna, D. and Hariparasema, K. (2010). Compositional grain quality of sorghum hybrids evaluated under advanced hybrid Trial. Annual Report–2010. Book-2 of AICSIP Co-ordinating Team Report agn-2010. pp. 55-59.

Reichert, R. D. and Young, C. G. (1976). Dehulling cereal grains and grain legumes for developing countries. I. Quantities comparison between attrition and abrasive type mills. *Cereal Chem.* 53: 829-839.

Robertson, S. K. and Perez-Maldonado, R. A. (2006). Nutritional characteristics of sorghum from QLD AGd NSW. *Aust. Poultry Sci Symp* 2006-18.

Salunkhe, D. K., Kadam, S. S. and Chavan, J. K. (1977). Nutritional quality of proteins in grain sorghum. *Qual. Plant. Pl. Foods Human Nutr.* 27: 187-205.

Salunkhe, D.K., S.S. Kadam and S. J. Jadhav, (1991). *Potato: Production, Processing and Products,* CRC Press, Soca Raton.

Seetharama N. and Rao B D (2004). Sustaining nutritional security. The Hindu Survey of Indian Agriculture. pp. 37-38.

Schoch, T. J. (1941), Physical aspects of starch behavior, *Cereal Chem.*, 18, 121-28.

Schrimshaw, N. S. (1974). Nutritional requirements for high protein foods, Proc. Int. Symp. Protein Foods and Concentrates, Central Food Technol. Res. Institute, Mysore, India, pp. 13-18.

Scott, J. H. (1951), *Flour Milling Processes,* Second ed., Chapman and Hall, London.

Seetharam A. (1997). Coarse cereals: Waiting for incentives. *The Hindu Survey of Indian Agriculture*. pp. 55-59.

Shepherd, A. D. (1974). How grain structure influences sorghum quality, presented to 5th E. African Cereals Res. Conf. Malawi.

Shibano, M., 1973, Construction and Function of Friction Type Rice Whitening Machine Technical note, JRMA, Japan.

Shreve, N. R. (1956). The Chemical Precess Industries, McGraw-Hili Book Co. Inc., Tokyo.

Singh, G. (1997). Data book on Mechanization and Agro-Processing Since Independence, CIAE, Bhopal-462038.

Smith, L. (1944), *Flour Milling Technology,* Third ed., Northern Publishing Co., Liverpool.

Srivastasva, M (1985). Essentials of Food and Nutrition Vol. II (Applied Aspects). The Bangalore Printing and Publishing Company Limited, Bangalore.

Srivastava, R.K and Singh, S. (1987). Utilization of Mushrooms for Preparation of Different Types of Preserved Products. Government Fruit Preservation and Canning Institute,.U.P., Lucknow.

Stivers, T. E., Jr. (1955). American corn milling systems for degermed products, *Assoc. Operators Millers Bull.*, 2168-79.

Subramanian, V. and Jambunathan, R. (1982). Properties of sorghum grain and their relationship to roti quality. In International symposium on sorghum grain quality. ICRISAT, Patancheru, India pp. 280- 288.

Subramanian, V. and Jambunathan, R. (1984). In Nutritional and Processing Quality of Sorghum (eds D. K. Salunkhe, J. K. Chavan and S. J. Jadhav) Oxford and IBH Publishing Company, New Delhi.

Sullins, R. D. and Rooney, L. W. (1974). Microscopic evaluation of the digestibility of sorghum lines that differ in endosperm characteristics. *Cereal Chem.* 51: 134–142.

Sulllins, R. D., Rooney, L. W. and Rosenow, D. T. (1975). The endosperm structure of high lysine sorghum. *Crop Sci.* 15: 599-600.

Vannalli, S., Kasturiba, B., Naik, R, K. and Yenagi, N. (2008). Nutritive value and quality characteristics of sorghum genotypes. *Karnataka J. Agriculture Sci* 20 : 586-588.

Viraktamath, C. S. and Desikachar, H. S. R. (1972). Processing of sorghum with special reference to its use as human food. In sorghum in seventies. *Proceedings of a Symposium*, (Eds. N.G.P. Rao and L.R. House) Oxford and IBM Publishing Co, New Delhi pp. 588.

Wang, Y. D. and Fields, M. L. (1987). Germination of corn and sorghum in the home to improve nutritive value. J. Food Sci. 43: 1113-1115.

Watson, S. A. (1970). Wet milling process and products. In Sorghum production and utilization. J. S. Wall and Ross, W. M. Eds. The AVI Publishing Co. Inc, Westport, Connecticut USA. pp. 602-626.

Watson, S. A., Sanders, F. W., Wakely, R. D. and Williams, C. B. (1955). Peripheral cells of the endosperms of grain sorghum and corn and their influence on starch purification, *Cereal Chem.*, 32, 165-82.

Winchester, M. D. and Makokha, A. O. (2011). Mineral contents in grain sorghum. African *Journal of Food Science* 5(7): 436 – 445.

Wolf, M. J., Buzan, C. L., Mac Masters, M. M. and Rist, C. E. (1952b). Structure of the mature Corn Kernel, II, Microscopic structure of pericarp, seed coat and hilar layer of dent corn, *Cereal Chem., 29,* 334-48.

Woods, A.E. and Aurand, L.W. (1977). Laboratory Manual in Food Chemistry. The AVI Publishing Company, Inc., Westport, Connecticut.

Chapter 13

Wheat: Coarse Millet (*Tritium aestivum*)

Introduction

Wheat is the principal food grain in many countries of the world. It is one of the most important cereals and is used as staple food in the form of flour. In India, a large proportion of wheat is used as the familiar *atta* and *maida* (white flour). The hard wheat's are also ground into *suji* (semolina). Whole wheat is ground into atta by the traditional stone grinder without prior separation of bran and germ from it.

Naming

There are many botanical classification systems used for wheat species, discussed in a separate article on Wheat taxonomy. The name of a wheat species from one information source may not be the name of a wheat species in another.

Within a species, wheat cultivars are further classified by wheat breeders and farmers in terms of:

- ☆ Growing season, such as winter wheat vs. spring wheat.
- ☆ Protein content. Bread wheat protein content ranges from 10 per cent in some soft wheat with high starch contents, to 15 per cent in hard wheat.
- ☆ The quality of the wheat protein gluten. This protein can determine the suitability of wheat to a particular dish. Strong and elastic gluten present in bread wheat enables dough to trap carbon dioxide during leavening, but elastic gluten interferes with the rolling of pasta into thin sheets. The gluten protein in durum wheat used for pasta is strong but not elastic.
- ☆ Grain colour (red, white or amber). Many wheat varieties are reddish-brown due to phenolic compounds present in the bran layer which are transformed to pigments by browning enzymes. White wheat has a

lower content of phenolics and browning enzymes and is generally less astringent in taste than red wheat. The yellowish colour of durum wheat and semolina flour made from it is due to a carotenoid pigment called lutein, which can be oxidized to a colourless form by enzymes present in the grain.

Major Cultivated Species of Wheat

Hexaploid Species

- **Common wheat** or **Bread wheat** (*T. aestivum*) – A hexaploid species that is the most widely cultivated in the world.
- **Spelt** (*T. spelta*) – Another hexaploid species cultivated in limited quantities. Spelt is sometimes considered a subspecies of the closely related species common wheat (*T. aestivum*), in which case its botanical name is considered to be *Triticum aestivum* subsp. *spelta*.

Tetraploid Species

- **Durum** (*T. durum*) – The only tetraploid form of wheat widely used today and the second most widely cultivated wheat.
- **Emmer** (*T. dicoccon*) – A tetraploid species, cultivated in ancient times but no longer in widespread use.
- **Khorasan** (Triticum turgidum ssp. turanicum also called Triticum turanicum) is a tetraploid wheat species. It is an ancient grain type; Khorasan refers to a historical region in modern-day Afghanistan and the northeast of Iran. This grain is twice the size of modern-day wheat and is known for its rich nutty flavour.

Diploid Species

- **Einkorn** (*T. monococcum*) – A diploid species with wild and cultivated variants. Domesticated at the same time as emmer wheat, but never reached the same importance.

Classes Used in the United States

- **Durum** – Very hard, translucent, light-coloured grain used to make semolina flour for pasta and bulghur; high in protein, specifically, gluten protein.
- **Hard Red Spring** – Hard, brownish, high-protein wheat used for bread and hard baked goods. Bread Flour and high-gluten flours are commonly made from hard red spring wheat. It is primarily traded at the Minneapolis Grain Exchange.
- **Hard Red Winter** – Hard, brownish, mellow high-protein wheat used for bread, hard baked goods and as an adjunct in other flours to increase protein in pastry flour for pie crusts. Some brands of unbleached all-purpose flours are commonly made from hard red winter wheat alone.

It is primarily traded on the Kansas City Board of Trade. One variety is known as "turkey red wheat" and was brought to Kansas by Mennonite immigrants from Russia.

- ☆ **Soft Red Winter** – Soft, low-protein wheat used for cakes, pie crusts, biscuits and muffins. Cake flour, pastry flour and some self-rising flours with baking powder and salt added, for example, are made from soft red winter wheat. It is primarily traded on the Chicago Board of Trade.
- ☆ **Hard White** – Hard, light-coloured, opaque, chalky, medium-protein wheat planted in dry, temperate areas. Used for bread and brewing.
- ☆ **Soft White** – Soft, light-coloured, very low protein wheat grown in temperate moist areas. Used for pie crusts and pastry. Pastry flour for example is sometimes made from soft white winter wheat.

Red wheat may need bleaching; therefore, white wheat usually command higher prices than red wheat on the commodities market.

As a Food

Raw wheat can be ground into flour or using hard durum wheat only, can be ground into semolina; germinated and dried creating malt; crushed or cut into cracked wheat; parboiled (or steamed), dried, crushed and de-branned into bulgur also known as groats. If the raw wheat is broken into parts at the mill, as is usually done, the outer husk or bran can be used several ways. Wheat is a major ingredient in such foods as bread, porridge, crackers, biscuits, Muesli, pancakes, pies, pastries, cakes, cookies, muffins, rolls, doughnuts, gravy, boza (a fermented beverage) and breakfast cereals (*e.g.*, Wheatena, Cream of Wheat, Shredded Wheat and Wheaties).

Nutrition

Hundred gram (100 g) (3.5 oz) of hard red winter wheat contain about 12.6 g (0.44 oz) of protein, 1.5 g (0.053 oz) of total fat, 71 g (2.5 oz) of carbohydrate (by difference), 12.2 g (0.43 oz) ofdietary fiber and 3.2 mg (0.00011 oz) of iron (17 per cent of the daily requirement); the same weight of hard red spring wheat contains about 15.4 g (0.54 oz) of protein, 1.9 g (0.067 oz) of total fat, 68 g (2.4 oz) of carbohydrate (by difference), 12.2 g (0.43 oz) of dietary fiber and 3.6 mg (0.00013 oz) of iron (20 per cent of the daily requirement).

Much of the carbohydrate fraction of wheat is starch. Wheat starch is an important commercial product of wheat, but second in economic value to wheat gluten. The principal parts of wheat flour are gluten and starch. These can be separated in a kind of home experiment, by mixing flour and water to form a small ball of dough and kneading it gently while rinsing it in a bowl of water. The starch falls out of the dough and sinks to the bottom of the bowl, leaving behind a ball of gluten.

In wheat, phenolic compounds are mainly found in the form of insoluble bound ferulic acid and are relevant to resistance to wheat fungal diseases. Alkylresorcinols are phenolic lipids present in high amounts in the bran layer (*e.g.* pericarp, testa and aleurone layers) of wheat and rye (0.1-0.3 per cent of dry weight).

Health Concerns

Several screening studies in Europe, South America, Australasia and the USA suggest that approximately 0.5–1 per cent of these populations may have undetected coeliac disease. Coeliac (also written as celiac) disease is a condition that is caused by an adverse immune system reaction to gliadin, a gluten protein found in wheat (and similar grains of thetribe Triticeae which includes other species such as barley and rye). Upon exposure to gliadin, the enzyme tissue transglutaminase modifies the protein and the immune system cross-reacts with the bowel tissue, causing an inflammatory reaction. That leads to flattening of the lining of the small intestine, which interferes with the absorption of nutrients. The only effective treatment is a lifelong gluten-free diet.

The estimate for celiac disease among people in the United States is between 0.5 and 1.0 percent of the population.

While gluten sensitivity is caused by a reaction to wheat proteins, it is not the same as a wheat allergy.

Recently non-celiac gluten sensitivity has been identified as a further gluten sensitivity condition that differs from celiac disease and wheat allergy.

Comparison of Wheat with other Major Staple Foods

The following table shows the nutrient content of wheat and other major staple foods in a raw form (Table 13.1). Raw forms of these staples, however, are not edible and cannot be digested. These must be sprouted, or prepared and cooked as appropriate for human consumption. In sprouted or cooked form, the relative nutritional and anti-nutritional contents of each of these grains are remarkably different from that of raw form of these grains reported in this table.

Commercial Use

Harvested wheat grain that enters trade is classified according to grain properties for the purposes of the commodity markets. Wheat buyers use these to decide which wheat to buy, as each class has special uses and producers use them to decide which classes of wheat will be most profitable to cultivate.

Wheat is widely cultivated as a cash crop because it produces a good yield per unit area, grows well in a temperate climate even with a moderately short growing season and yields a versatile, high-quality flour that is widely used in baking. Most bread are made with wheat flour, including many breads named for the other grains they contain like most rye and oat breads. The popularity of foods made from wheat flour creates a large demand for the grain, even in economies with significant food surpluses.

In recent years, low international wheat prices have often encouraged farmers in the USA to change to more profitable crops. In 1998, the price at harvest was $2.68 per bushel. USDA report revealed that in 1998, average operating costs were $1.43 per bushel and total costs were $3.97 per bushel. In that study, farm wheat yields averaged 41.7 bushels per acre (2.2435 metric ton/hectare) and typical total

Table 13.1: Nutrient Content of Major Staple Foods per 100g

STAPLE	Maize/ Corn[A]	Rice (White)[B]	Rice (Brown)[I]	Wheat [C]	Potato[D]	Cassava[E]	Soybean (Green)[F]	Sweet Potato[G]	Sorghum[H]	Yam[Y]	Plantain[Z]
Water (g)	10	12	10	13	79	60	68	77	9	70	65
Energy (kJ)	1528	1528	1549	1369	322	670	615	360	1419	494	511
Protein (g)	9.4	7.1	7.9	12.6	2.0	1.4	13.0	1.6	11.3	1.5	1.3
Fat (g)	4.74	0.66	2.92	1.54	0.09	0.28	6.8	0.05	3.3	0.17	0.37
Carbohydrates (g)	74	80	77	71	17	38	11	20	75	28	32
Fiber (g)	7.3	1.3	3.5	12.2	2.2	1.8	4.2	3	6.3	4.1	2.3
Sugar (g)	0.64	0.12	0.85	0.41	0.78	1.7	0	4.18	0	0.5	15
Calcium (mg)	7	28	23	29	12	16	197	30	28	17	3
Iron (mg)	2.71	0.8	1.47	3.19	0.78	0.27	3.55	0.61	4.4	0.54	0.6
Magnesium (mg)	127	25	143	126	23	21	65	25	0	21	37
Phosphorus (mg)	210	115	333	288	57	27	194	47	287	55	34
Potassium (mg)	287	115	223	363	421	271	620	337	350	816	499
Sodium (mg)	35	5	7	2	6	14	15	55	6	9	4
Zinc (mg)	2.21	1.09	2.02	2.65	0.29	0.34	0.99	0.3	0	0.24	0.14
Copper (mg)	0.31	0.22		0.43	0.11	0.10	0.13	0.15	–	0.18	0.08
Manganese (mg)	0.49	1.09	3.74	3.99	0.15	0.38	0.55	0.26	–	0.40	–
Selenium (μg)	15.5	15.1		70.7	0.3	0.7	1.5	0.6	0	0.7	1.5
Vitamin C (mg)	0	0	0	0	19.7	20.6	29	2.4	0	17.1	18.4
Thiamin (mg)	0.39	0.07	0.40	0.30	0.08	0.09	0.44	0.08	0.24	0.11	0.05
Riboflavin (mg)	0.20	0.05	0.09	0.12	0.03	0.05	0.18	0.06	0.14	0.03	0.05
Niacin (mg)	3.63	1.6	5.09	5.46	1.05	0.85	1.65	0.56	2.93	0.55	0.69
Pantothenic acid (mg)	0.42	1.01	1.49	0.95	0.30	0.11	0.15	0.80	–	0.31	0.26

Contd...

Table 13.1–*Contd...*

STAPLE	*Maize/ Corn*[A]	*Rice (White)*[B]	*Rice (Brown)*[I]	*Wheat*[C]	*Potato*[D]	*Cassava*[E]	*Soybean (Green)*[F]	*Sweet Potato*[G]	*Sorghum*[H]	*Yam*[Y]	*Plantain*[Z]
Vitamin B6 (mg)	0.62	0.16	0.51	0.3	0.30	0.09	0.07	0.21	–	0.29	0.30
Folate Total (μg)	19	8	20	38	16	27	165	11	0	23	22
Vitamin A (IU)	214	0	0	9	2	13	180	14187	0	138	1127
Vitamin E, alpha-tocopherol (mg)	0.49	0.11	0.59	1.01	0.01	0.19	0	0.26	0	0.39	0.14
Vitamin K1 (μg)	0.3	0.1	1.9	1.9	1.9	1.9	0	1.8	0	2.6	0.7
Beta-carotene (μg)	97	0		5	1	8	0	8509	0	83	457
Lutein+zeaxanthin (μg)	1355	0		220	8	0	0	0	0	0	30
Saturated fatty acids (g)	0.67	0.18	0.58	0.26	0.03	0.07	0.79	0.02	0.46	0.04	0.14
Monounsaturated fatty acids (g)	1.25	0.21	1.05	0.2	0.00	0.08	1.28	0.00	0.99	0.01	0.03
Polyunsaturated fatty acids (g)	2.16	0.18	1.04	0.63	0.04	0.05	3.20	0.01	1.37	0.08	0.07

A: Corn, yellow; **B**: Rice, white, long-grain, regular, raw, unenrichedl; **C**: Wheat, hard red winter; **D**: Potato, flesh and skin, raw; **E**: Cassava, raw; **F**: Soybeans, green, raw; **G**: Sweet potato, raw, unprepared; **H**: Sorghum, raw; **Y**: Yam, raw; **Z**: Plantains, raw; **I**: Rice, brown, long-grain, raw.

wheat production value was $31,900 per farm, with total farm production value (including other crops) of $173,681 per farm, plus $17,402 in government payments. There were significant profitability differences between low- and high-cost farms, mainly due to crop yield differences, location and farm size.

In 2007 there was a dramatic rise in the price of wheat due to freezes and flooding in the northern hemisphere and a drought in Australia. Wheat futures in September, 2007 for December and March delivery had risen above $9.00 a bushel, prices never seen before. There were complaints in Italy about the high price of pasta.

Flour Milling

The objective of modern flour milling is to obtain the maximum amount of white flour from the wheat endosperm without any bran or germ content (Figure 13.1). Conditioning of wheat by hydrothermal treatment prior to milling helps in the separation of barn and germ from the endosperm. If wheat is conditioned by hydrothermal treatment, bran and germ become rubber-like while the endosperm becomes soft. It also eliminates the difference in grinding characteristics between soft and hard wheat. When the conditioned wheat is sheared by the corrugations of first break roll during the milling operation, it splits open releasing small endosperm pieces and thus exposing the remaining endosperm which could be carefully scraped off the bran in successive break rolls. The yields of white flour and byproducts (called mm feed) from white flour milling are about 70 per cent and 30 per cent by weight, respectively. The mill feed is composed of 12 per cent bran, 3 per cent germ and 15 per cent shorts. Wheat consists of bran (12 per cent), germ (3 per cent) and endosperm (85 per cent). Modern flour milling consists of six steps: (1) receiving, drying and storage of wheat, (2) cleaning, (3) conditioning; (4) milling into flour and byproducts, (5) packaging and storage of finished products and (6) blending. Of them the most important operations namely, cleaning, conditioning and milling have been discussed here.

Cleaning

Wheat is thoroughly cleaned to remove all fine impurities and the dirt sticking to the surface of the grain. To remove loose fine impurities a set of cleaners is employed. Small pieces of sticks, stones, sand, etc., are removed by sieving and the light impurities like chaff, etc., are removed by aspirations. Then the wheat is allowed to pass over powerful magnetic separators to remove pieces of ferromagnetic materials. The seeds of other food grains, defective grains and weed are removed by disc evaporators. The next step in the cleaning operation is the removal or dirt sticking to the surface by scouring. Usually, wheat is moved by paddles against stationary emery-coated surface. Then the dirt and loose outer coating are aspirated off. The scratches and cracks formed in wheat during scouring help in increasing the rate of moisture absorption at the time of washing and conditioning. The final cleaning step is washing by water which allows the dirt and bits of metal to sink. The moisture content of wheat is increased by about one per cent during washing.

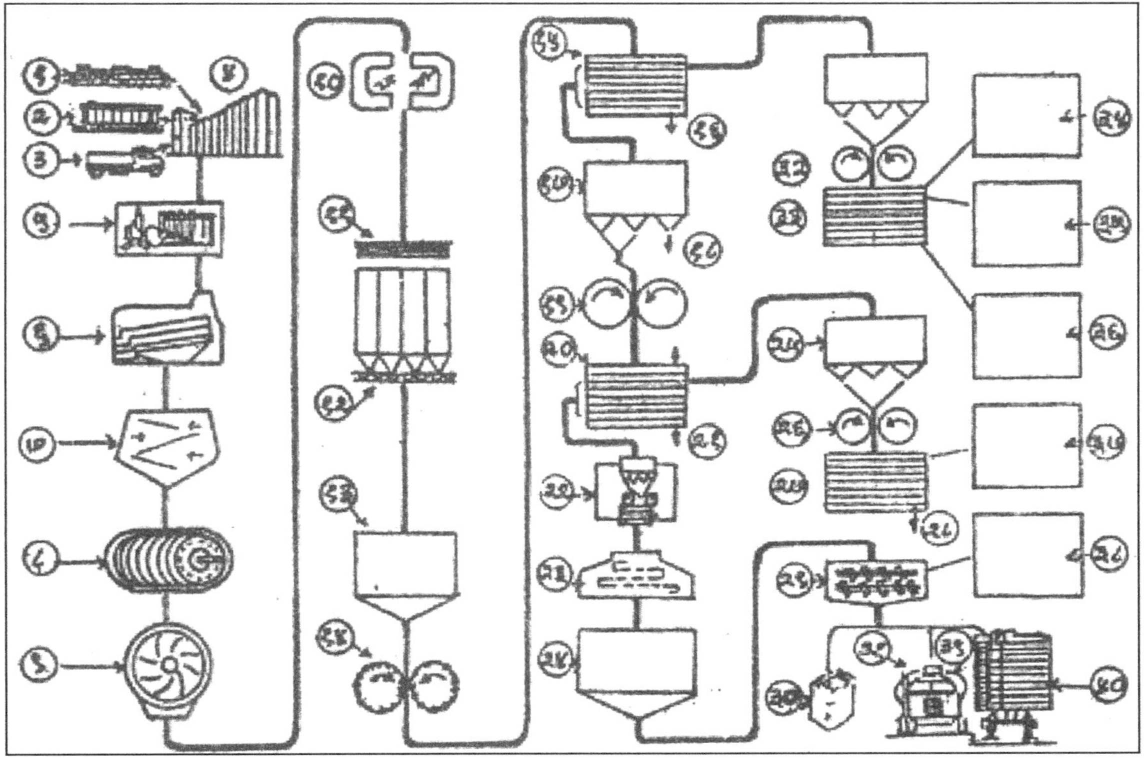

Figure 13.1: Flow Sheet of Wheat Processing.

1. Ship 2. Railway 3. Truck 4. Storage (godwon) 5. Examination and combining 6. Sieving 7. Air cleaning 8. Cleaning by disc 9. Scorers 10. Magnetic cleaning 11. Condition of wheat 12. Combing 13. Milling tank 14. First breaking by roller 15. Sieving 16. Maida 17. Purification 18. Brain + shorts 19. Reduction rolls 20. Sieving 21. Maida 22. Reduction rolls 23. Bleaching 24. Large storage 25. Purification 26. Germ rolls 27. Sieving 28. Maida 29. Addition of minerals and vitamins 30. Bagging 31. Maida transport 32. Reduction rolls 33. Sieving 34. Brain 35. Shorts 36. Maida 37. Germs 38. Patient Maida 39. Transport by truck 40. Transport by train.

Conditioning/Hydrothermal Treatment

The conditioning of wheat can be done either at room temperature, elevated temperature or at high temperature. But the temperature of wheat grain should not be raised above 47° C otherwise the gluten quality will be affected which deteriorates the baking quality of the flour. Generally the moisture contents of soft and hard Wheat's are increased to 15 to 17 per cent and 16 to 19 per 'Cent respectively by soaking and then the moisture of the grain is equilibrated by tempering for 18 to 72 hours in the tempering bin. In a modern system conditioning of wheat is performed in four stages. The conditioner mainly consists of three sections namely, preheating section, moistening section, and cooling section. In the first section wheat is preheated to the proper temperature, in the second section wheat is moistened to the desired moisture level and in the third section soaked wheat is cooled to the room temperature. Finally, the treated wheat is kept in a separate tempering bin for 18 to 72 hours.

Hydrothermal treatment of grain by direct steaming has been popular for the last few years. It has many advantages over heating by air because both moistening and heating are carried out simultaneously in a single operation. Moreover, the grain is heated within 20 to 30 seconds to about 47°C. But the grain temperature above 47° C may adversely affect the quality of the flour. The rapid rate of heating weakens the intermolecular bonds in various parts of the grain to a considerable extent resulting in easier separation of bran, more effective grinding of endosperm and stronger action on proteins and enzymes.

Grinding (Milling)

Milling of wheat is carried out by roller mills. The roller milling system is mainly divided into the break roll and reduction roll systems. In addition most of the flour mills keep a stand-by system known as scratch system. The scratch system is nothing but an extension of the break roll system. The break rolls and the reduction rolls are differentiated with the variation in their surface conditions. The surface of the reduction roll is smooth whereas the surface of break roll is corrugated. In the break rolls, the bran is cracked, the kernel is broken open. The endosperm adhering to bran is milled away successively in a few steps. Generally a series of four sets of break rolls are used. Each set of rolls takes stock from the preceding one. After each break, the mixture of free bran, free endosperm, free germ and endosperm still adhering to the bran is sifted and separated. The endosperm adhering to bran is passed through the next break roll while the middle size endosperms called middling's are sent to the reduction rolls for proper size reduction to flours. Therefore, the break rolls are mainly used for the production of middlings and the reduction rolls are used for grinding of free middlings into proper flour size. After each reduction of endosperm (middlings) the flour is sifted away from the bigger size middlings and the remaining middlings are passed to the next reduction rolls. The above operations are continued until the desired products are obtained. As many as 12 to 14 reduction rolls are used in most flour mills. But all reduction rolls are not used for all break products.

Storage of Finished Products

The flour and the mill feed (bran, germ and shorts) are bagged in waterproof bags, stitched and stored in cold dry condition in flat godowns. The flow sheet of all process is shown in Figure 13.1.

Components of a Wheat Mill

(1) Break Roll

Break roll consists of, twin pairs of corrugated steel rolls. One roll of a pair revolves faster than the other, differential speed being in the proportion of 2.5 to 1.

(2) Break Sifting System

This can be divided into two parts-plan sifters and purifiers. (a) *Plan sifter.* Plan sifter is a scalping system removing large bran pieces adhering with endosperm at the top. The next series, which are finer, remove the bran and germ. The next layer of still fine sieve removes the endosperm middling and the bottom rough flow. (b) *Purifier.* The middling containing finer bran particles are removed by purifier before they move to reduction roll.

(3) Reduction Rolls

The reduction roll comprises of two smooth rolls. The rolls in the reduction system are further divided into coarse rolls and fine rolls depending on the clearance between the rollers. It is possible to grind flour into very fine particles by gradual grinding. But under high grinding pressure the starch is ruptured and this should be avoided.

(4) Reduction Sifting System

The same plan sifting system is used here. After each reduction the product is separated by plan sifters where the finished flour is sifted by 120 mesh sieve (silk) and removed and oversized material is sent back to the reduction rolls for further processing.

(5) Scratch System

If the mill is functioning properly, *i.e.*, good release of endosperm is obtained on the break rolls, the scratch system can be bypassed, if not, the scratch system is employed to maintain proper release of endosperm from bran. The scratch system is an extension of the break system and thus used as stand-by system only.

Health Benefits

Wheat: The Whole Truth

The health benefits of wheat depend entirely on the form in which you eat it. These benefits will be few if you select wheat that has been processed into 60 per cent extraction, bleached white flour. 60 per cent extraction the standard for most wheat products in the United States, including breads, noodles and pastas, baked goods like rolls or biscuits and cookies means that 40 per cent of the original wheat grain was removed and only 60 per cent is left. Unfortunately, the 40 per cent that gets removed includes the bran and the germ of the wheat grain its most nutrient-rich parts. In the process of making 60 per cent extraction flour, over half of the vitamin B1, B2, B3, E, folic acid, calcium, phosphorus, zinc, copper, iron and fiber are lost.

Since 1941, laws in the United States have required "enrichment" of processed wheat flour with vitamins B1, B2, B3 and iron in response to the problems created by 60 per cent extraction. Since not nearly as much of these B vitamins and iron are replaced as are removed from 60 per cent extraction flour, "enriched" seems an odd word to describe this process.

If you select 100 per cent whole wheat products, however, the bran and the germ of the wheat will remain in your meals and the health benefits will be impressive! Our food ranking qualified whole wheat (in its original non-enriched form) as a very good source of dietary fiber and manganese and as a good source of magnesium.

The many benefits of whole wheat products are being recognized more and more by consumers. Even though many health-conscious individuals have been cutting back on their intake of total carbs and refined wheat products (by about 10 per cent between 1997-2007), the demand for whole wheat products has actually increased during that same time period. This trend fits in well with a Mediterranean diet approach to health, which looks to lower overall carbs but higher whole grains, including whole wheat.

Women Who Eat Whole Grains Weigh Less

A study published in the *American Journal of Clinical Nutrition* underscores the importance of choosing whole rather than refined wheat to maintain a healthy body weight. In this Harvard Medical School/Brigham and Women's Hospital study, which collected data on over 74,000 female nurses aged 38-63 years over a 12 year period, weight gain was inversely associated with the intake of high-fiber, whole-grain foods, such as whole wheat, but positively related to the intake of refined-grain foods, such as products made from refined wheat. Not only did women who consumed more whole grains consistently weigh less than those who ate less of these fiber-rich foods, but those consuming the most dietary fiber from whole grains were 49 per cent less likely to gain weight compared to those eating foods made from refined grains.

Whole Grains Reduce Risk of Metabolic Syndrome

First we were told, "Don't eat fat and you'll stay trim." After following this advice only to see obesity expand to never before seen proportions, we're told by the food gurus, "Eating fat is fine. Shun carbohydrates to stay slim."

In our opinion, neither piece of dietary advice is complete, accurate or likely to help us stay slim or healthy. Just as different kinds of fats have different effects in our bodies (*e.g.*, saturated and trans fats are linked to increased risk for cardiovascular disease while omega-3 fats decrease cardiovascular disease risk), some carbohydrates, such as whole grains, are healthful while others, such as refined grains and the foods made from them are not.

The latest research is clearly supporting this vital distinction. Refined grains and the foods made from them (*e.g.*, white breads, cookies, pastries, pasta and rice) are now being linked not only to weight gain but to increased risk of insulin resistance (the precursor of type 2 diabetes) and the metabolic syndrome (a strong predictor of both type 2 diabetes and cardiovascular disease), while eating more wholegrain foods is being shown to protect against all these ills. Common features of the metabolic syndrome include visceral obesity (the "apple shaped" body), low levels of protective HDL cholesterol, high triglycerides and high blood pressure.

In one of the most recent studies, which appeared in *Diabetes Care*, researchers who analysed data on over 2,800 participants in the Framingham Offspring Study, found that the prevalence of both insulin resistance and the metabolic syndrome was significantly lower among those eating the most cereal fiber from whole grains compared to those eating the least.

Prevalence of the metabolic syndrome was 38 per cent lower among those with the highest intake of fiber from whole grains. Conversely, study subjects whose diets had the highest glycemic index and glycemic load, both of which are typically low in whole foods and high in processed refined foods, were 141 per cent more likely to have the metabolic syndrome compared to those whose diets had the lowest glycemic index and glycemic load. In other words, compared to those whose diets were primarily composed of whole high fiber foods: whole grains, legumes, vegetables and fruits?

The researchers concluded, "Given that both a high cereal fiber content and lower glycemic index are attributes of wholegrain foods, recommendation to increase wholegrain intake may reduce the risk of developing the metabolic syndrome." Our perspective at the World's Healthiest Foods is that a way of eating that relies on the healthiest foods from all the food groups the whole foods that contain the healthiest fats, carbohydrates and proteins is the most effective, intelligent and most enjoyable way to not only lower your risk of developing the metabolic syndrome, but to stay slim, vital and attractive throughout a long and healthy life.

Whole Grains Substantially Lower Type 2 Diabetes Risk

Whole grains are a rich source of magnesium, a mineral that acts as a co-factor for more than 300 enzymes, including enzymes involved in the body's use of glucose and insulin secretion. The FDA permits foods that contain at least 51 per cent whole grains by weight (and are also low in fat, saturated fat and cholesterol) to display a health claim stating consumption is linked to lower risk of heart disease and certain cancers. Now, research suggests regular consumption of whole grains also reduces risk of type 2 diabetes. (van Dam RM, Hu FB, *Diabetes Care*). In this 8-year trial, involving 41,186 particpants of the Black Women's Health Study, research data confirmed inverse associations between magnesium, calcium and major food sources in relation to type 2 diabetes that had already been reported in predominantly white populations.

Risk of type 2 diabetes was 31 per cent lower in black women who frequently ate whole grains compared to those eating the least of these magnesium-rich foods. When the women's dietary intake of magnesium intake was considered by itself, a beneficial, but lesser-19 per cent -reduction in risk of type 2 diabetes was found, indicating that whole grains offer special benefits in promoting healthy blood sugar control. Daily consumption of low-fat dairy foods was also helpful, lowering risk of type 2 diabetes by 13 per cent.

Whole Wheat's Betaine Lessens Chronic Inflammation

People whose diets supplied the highest average intake of choline (found in egg yolk and soybeans) and its metabolite betaine (found naturally in beets, spinach and whole wheat), have levels of inflammatory markers at least 20 per cent lower than subjects with the lowest average intakes, report Greek researchers in the *American Journal of Clinical Nutrition* (Detopoulou P, Panagiotakos DB, *et al.*)

Compared to those whose diets contained < 250 mg/day of choline, subjects whose diets supplied >310 mg of choline daily had, on average:

- ☆ 22 per cent lower concentrations of C-reactive protein.
- ☆ 26 per cent lower concentrations of interleukin-6.
- ☆ 6 per cent lower concentrations of tumor necrosis factor alpha.

Compared to those consuming < 260 mg/day of betaine, subjects whose diets provided > 360 mg per day of betaine had, on average:

- ☆ 10 per cent lower concentrations of homocysteine.
- ☆ 19 per cent lower concentrations of C-reactive protein.
- ☆ 12 per cent lower concentrations of tumor necrosis factor alpha.

Each of these markers of chronic inflammation has been linked to a wide range of conditions including heart disease, osteoporosis, cognitive decline and Alzheimer's and type-2 diabetes. In an accompanying editorial in the *American Journal of Clinical Nutrition* entitled, "Is there a new component of the Mediterranean diet that reduces inflammation?," Steven Zeisel from the University of North Carolina at Chapel Hill noted that choline and betaine work together in the cellular process of methylation, which is not only responsible for the removal of homocysteine, but is involved in turning off the promoter regions of genes involved in inflammation. "Exposure to oxidative stress is a potent trigger for inflammation. Betaine is formed from choline within the mitochondria and this oxidation contributes to mitochondrial redox status," Zeisel continued. "If the association between choline and betaine and inflammation can be confirmed in studies of other populations, an interesting new dietary approach may be available for reducing chronic diseases associated with inflammation," he concluded.

Recommended daily intakes of choline were set in 1998 at 550 milligrams per day for men and 425 milligrams a day for women. No RDI has been set for betaine, which, since it is a metabolite of choline, is not considered an essential nutrient.

Practical Tip: Egg yolks are the richest source of choline, followed by soybeans. Spinach, beets and whole wheat products are primary sources of betaine.

Whole Grains Help Prevent Gallstones

Eating foods high in insoluble fiber, such as cereals and breads made from whole wheat can help women avoid gallstones, shows a study published in the *American Journal of Gastroenterology*. Studying the overall fiber intake and types of fiber consumed over a 16 year period by over 69,000 women in the Nurses Health Study, researchers found that those consuming the most fiber overall (both soluble and insoluble) had a 13 per cent lower risk of developing gallstones compared to women consuming the fewest fiber-rich foods. Those eating the most foods rich in insoluble fiber gained even more protection against gallstones: a 17 per cent lower risk compared to women eating the least. And the protection was dose-related; a 5-gram increase in insoluble fiber intake dropped risk dropped 10 per cent.

How do foods rich in insoluble fiber help prevent gallstones? Researchers think insoluble fiber not only speeds intestinal transit time (how quickly food moves through the intestines), but reduces the secretion of bile acids (excessive amounts contribute to gallstone formation), increases insulin sensitivity and lowers triglycerides (blood fats). Abundant in all whole grains, insoluble fiber is also found in nuts and the edible skin of fruits and vegetables including tomatoes, cucumbers, many squash, apples, berries and pears. In addition, beans provide insoluble as well as soluble fiber.

Whole Wheat Gets You Going

Wheat bran is a popular bulk laxative. A third of a cup per day is all that is needed. Research studies support this popular practice. A fiber-rich diet, primarily composed of whole wheat breads, cereals high in bran and supplemental "millers bran" was shown to alleviate the symptoms of diverticular disease (pain, nausea, flatulence, distension, constipation, etc.) in 89 per cent of patients enrolled in a study which examined the effects of fiber on bowel regularity. Diverticular disease, a condition often marked by inflammation and lower abdominal pains in which chronic constipation and excessive straining results in a sac or pouch in the wall of the colon, is typically treated with dietary roughage such as cereal fiber (*i.e.*, wheat bran), fruit and vegetable fiber and plenty of fluids.

Whole Wheat Promotes Women's Health and Gastrointestinal Health

The benefits of wheat's bran portion don't stop here; it has also been shown to function as an anti-cancer agent. Wheat bran is thought to accelerate the metabolism of estrogen that is a known promoter of breast cancer. In one study, pre-menopausal women, ages twenty to fifty, who ate three to four high fiber muffins per day made with wheat bran, decreased their blood estrogen levels by 17 per cent after two months. The women eating corn bran or oat bran did not show the same benefits.

Interestingly, whole grains such as wheat also contain *lignans*, which are phytonutrients that act as weak hormone-like substances. *Lignans* occupy the hormone receptors in the body, thus actively protecting the breast against high circulating levels of hormones such as estrogen. By accelerating the metabolism of estrogen and occupying estrogen receptors in the body, the components of wheat appear to have a dual function in protecting women against one of the leading

Wheat Food Products

Wheat Roti

Wheat Nan

Wheat Paratha

Wheat Dashami

Wheat Bakery Products

Wheat Bakery Products

causes of cancer death. The fact that only wheat bran and not corn or oat bran, is beneficial in preventing cancer-promoting changes in the colon, provides additional clues that wheat bran contains something special that makes it a true cancer fighter. Only the bran from wheat has been shown to reduce the concentration of bile acids and bacterial enzymes in the stool that are believed to promote colon cancer.

The protective dose for colon cancer may be more than 28 grams a day, since men who ate this amount had only one-third the rate of colon polyps (precancerous tumors) compared to those who ate only 17 grams/day. The amount of wheat bran needed for protection from other cancers is still unknown, but based on the health benefits of this food, it may be wise, if you are not sensitive to wheat or gluten, to include several servings of whole wheat grain foods such as bread, pasta and bran cereals every day in your diet.

Fiber from Whole Grains and Fruit Protective against Breast Cancer

When researchers looked at how much fiber 35,972 participants in the UK Women's Cohort Study ate, they found a diet rich in fiber from whole grains, such as whole wheat and fruit offered significant protection against breast cancer for pre-menopausal women. Pre-menopausal women eating the most fiber (>30 grams daily) more than halved their risk of developing breast cancer, enjoying a 52 per cent lower risk of breast cancer compared to women whose diets supplied the least fiber (<20 grams/day). Fiber supplied by whole grains offered the most protection. Pre-menopausal women eating the most whole grain fiber (at least 13 g/day) had a 41 per cent reduced risk of breast cancer, compared to those with the lowest whole grain fiber intake (4 g or less per day).

References

Anderson, R. A. (1962). A note on wet-milling of high amylose corn containing 75 per cent-amylose starch, *Cereal Chem.*, 39, 406-408.

Anderson, R. A. (1963). Wet-milling Properties of grains: Bench- Scale Study, *Cereal Sci. Today*, 8, No.6, 190-92, 195, 221.

Anon, 1964, Joint United States-Canadian tables of feed composition, *Natl. A cad. Sci.*, Nath. Res. Council Publication 1232.

Athalye (1992). Plastics in Packaging. Tata McGraw-Hill Publishing Company Limited, New Delhi.

Atkins, T. A. and Geddes, W. F. (1939). The relationship between protein content and strength of flours in gluten enriched flours, *Cereal Chem.*, 16, 223-31.

Banerjee, S. N. (1975). Report on the evaluation of the modern dal mill, Department of Food, Govt. of India.

Banwart, G.J. (1979). Basic Food Microbiology. The AVI Publishing Com- pany, Inc., Westport, Connecticut.

Belderok, Robert 'Bob'; Mesdag, Hans; Donner, Dingena A (2000), *Bread-Making Quality of Wheat*, Springer, p. 3, ISBN 0-7923-6383-3

Bender, A. E. (1973). Composition of infant foods, In Nutritional and Dietetic Foods, International Textbook Co. Ltd., London, U. K., p. 64.

Bhat, C.M., Sharma, R.N. and Sehgal, S. (1982). A Manual on Food Preservation at Home. Directorate of Publication, Haryana Agricultural University, Hissar.

Borasio, L. and Gariboldi, P. (1957). Illustrated glossary of rice processing machines, Food. and Agricultural Organization of the United Nations, Rome.

Brautlecht, CiA, (1953). *Starch, Its sources, Production and Uses,* Reinhold Publishing Corp., New York.

Brekke, 0 L. and Kwolek, W. F. (1969). Com dry milling: Cold tempering and degermination of com of various initial moisture contents, *Cereal Chem.*, 46, 545-49.

Brekke, O. L. (1967). Com dry milling: Pretempering low-moisture com, *Cereal Chem.*, 44, 521-31.

Brekke, O. L. (1968). Com dry-milling: Stress crack formation in tempering low-moisture com and effect on degerminator performance, *Cereal Chem.*, 45, 291-303.

Brekke, O. L. (1970). Dry milling artificially dried com: Roller milling of degerminator stock at various moistures, *Cereal Science Today,* 15, 37-42.

Brekke, O. L. and Weinecke, L. A. (1964). Com dry-milling, A comparative evaluation of commercial degermer samples, *Cereal Chern.*, 41, 321-28.

Brekke, O. L., Weinecke, L. A., Boyd, J. N. and Griffin, E. L., Jr. (1963). Com dry milling: Effects of first-temper moisture, Screen perforations and rotor speed on Beall cgcrminator throughput and products, *Cereal Chem., 40,* 423-29.

Bridgwater, W. and Beatrice Aldrich. (1966) *The Columbia-Viking Desk Encyclopedia.* Columbia University. p. 1959.

Burton, W.G. (1989). Specific gravity as a guide to the content of dry matter and of starch in potato tubers. In 'The Potato'. 3rd Ed. Longman Scientific and Technical. Essex, England.

Cauvain, Stanley P. and Cauvain P. Cauvain. (2003) *Bread Making*. CRC Press. p. 540.ISBN 1-85573-553-9.

Chakraverty, A. (2007). Post harvest technology of cereals, pulses and oilseeds. Oxford and IBH Publishing Co. Pvt. Ltd. New Delhi.

Chavan, J. K. and Kadam S. S. (1989). Nutritional improvement of cereals by fermentation. *CRC in Food Sci. and Nut*. 28: 349-399.

Colledge, Sue; University College, London. Institute of Archaeology (2007). *The origins and spread of domestic plants in southwest Asia and Europe*. Left Coast Press. pp. 40–.ISBN 978-1-59874-988-5. Retrieved 5 July 2011.

CSIR, (1980). The Wealth of India: Raw Materials, Councill of Scientific and Industrial Research, New Delhi.

Curtis; Rajaraman; MacPherson (2002). "Bread Wheat". Food and Agriculture Organization of the United Nations.

Department of Agriculture Appropriations for 1957: Hearings 84th Congress. 2d Session. United States House Committee on Appropriations. 1956. p. 242.

Desrosier, N.W. and Desrosier, J.N. (1987). The Technology of Food Preservation. CBS Publishers and Distributors, New Delhi.

Desrosier, N.W. and Tressler, D.K. (1977). Fundamentals of Food Freezing. The AVI Publishing Company, Inc., Westport, Connecticut. Distributors, New Delhi.

Diamond J (1997) *Guns, Germs and Steel, A short history of everybody for the last 13,000 years.* Viking UK Random House ISBN 0-09-930278-0

Direct quotation: Grundas ST: Chapter: Wheat: The Crop, in *Encyclopedia of Food Sciences and Nutrition* p6130, 2003; Elsevier Science Ltd.

Earle, F.,R., Curtis, J. J. and Hubbard, J. E. (1946). Composition of the component parts U1 the com kernel, *Cereal Chem.*, 23, 504-11.

Elias, D. G. and Scott, R. A. (1957). British flour milling technology, *Cereal Sci. TorJay*, 7, 180-84.

FAOStat. Retrieved 27 January 2015.

Feldman, Moshe and Kislev, Mordechai E., Israel Journal of Plant Sciences, Volume 55, Number 3 - 4/2007, pp. 207 - 221, *Domestication of emmer wheat and evolution of free-threshing tetraploid wheat* in "A Century of Wheat Research-From Wild Emmer Discovery to Genome Analysis", Published Online: 3 November 2008.

Fennema, O.R. Ed.(1976). Principles of Food Science, Part-l. Food Chemistry, Marcel Dekker Inc, New York.

Finney, K. F. and Barmore, M. A. (1948). Loaf volume and protein content of hard winter and spring wheats, *Cereal Chem.*, 25, 291-311.

Frazier, W.e. and Westhoff, D.C. (1983). Food Microbiology. Tata McGraw-Hill Publishing Company Limited, New Delhi.

Garibaldi, F. (1974). Rice Milling Equipment Operation and aintenance, Agricultural Services Bulletin No. 22, FAO, Rome.

Goel, A. K., Kumar, R. and Mann, S. S. (2007). Postharvest management and value addition. Daya Publishing House, Delhi.

Gopalan, C, Rama Sastri, B.V. and Bala Subramanian, S.C. (1982). Nutritive Value of Indian Foods. National Institute of Nutrition, R.CM.R., Hyderabad.

Grist, D. H. (1959). Rice (3rd ed), Longmans Green and Co. Ltd., London.

Hahn, R. R. (1969). Dry milling of grain sorghum. Cereal Sci. Today. 14: 234-237.

Harris, RS. and Karmas, E. (1975). Nutritional Evaluation of Food Processing. 2nd Edition. AVI Publishing Co. Westport, Connecticut.

Henry, W.A. and Morrison, F.B. (1923). *Feeds and Feeding: a handbook for the student and stockman.* The Henry-Morrison Co. Madison, Wisconsin, USA.

Heun, MR *et al.* (1997). "Site of Einkorn Wheat Domestication Identified by DNA Fingerprinting". *Science* 278: 1312–4. doi:10.1126/science.278.5341.1312.

Hogan, J. T. and Deobald, H. J. (1995), Measurement of the degree of milling of rice, *Rice* 1, 68 (10), 10.

Houston, D. F. (1973). Western Marketing and Nutrition Research Division, Agricultural Research Service, U. S. Department of Agriculture, Berkeley, California.

http://www.feedipedia.org/node/223

Ilyas, S. M. (2003). Importance of postharvest processing and value- addition for additional income and employment generation. Presented at National Symposium on Resource Management for Eco-friendly Crop Production, held at CSAUA and T, Kanpur, February, 26-28.

Ilyas, S. M., Wanjari, O. D. and Goyal, R. K. (2002). Prospects of income and employment generation through agro-processing and value-addition. Presented at Conference on Emerging Opportunities of Agro and Food Processing Industries in Jharkhand held at Indian Institute of Coal Management, Ranchi during October, 8-9.

Ilyas, S.M. and Coyal, R.K. (2002). Quality assurance in food processing-impact of WTO, In: Proc. National Seminar on Development of Food Processing Industries in U.P., held at Lucknow during January 16-17.

Jacob, T. (1987). Poisons in our Food. Publications Division, Ministry of Information and Broadcasting, Government of India, New Delhi.

Javaraman. J. (1981). Laboratory Manual in Biochemistry. Wiley Eastern Limited, New Delhi.

Jones, CR. (1940). The production of mechanically damaged starch in milling as a governing factor in the diastatic activity of flour, *Cereal Chem.*, 17, 133-69.

Karel, M., O.R. Fennema and D.S. Lund, Eds. (1975). Principles of Food Science Part II. Physical Principles of Food Preservation. Marcel Dekker, Inc. New York.

Kent-jones, D. W. and Amos, A. J. (1957). Modern Cereal Chemistry, Fifth Ed., Northern Publishing Co. Ltd., Liverpool.

Kerr, R. W. (1950). Chemistry and Industry of starch, 2nd Edition, Academic Press, New York.

Koga, Y. (1969). Drying, husking and milling in Japan IV and V., *Farming,* Japan.

Kordylas, J.M. (1991). Processing and Preservation of Tropical and Subtropical Foods. Macmillan Education Ltd., Houndmills, Basingstoke, Hampshire.

Krishnamurthy, K., Girish, G. K, Rarnasivan, T., Bose, S.K., Singh, Karan and Tomar, R. P. S. (1972). A new process of removal of husk of red gram using sirka. *Bulletin of Grain Technology*, 10: 18l.

Kuprits, Y., ed. (1967). Technology of grain processing and provender milling, Tekhnologiya perer atbotki zernai Kombikormovoe, Izd. 'Kolos' : Moscow (1 per cent 5). (Translation by Israel Program for Scientific Translation, Jerusalem, Israel).

Lev-Yadun, S; Gopher, A; Abbo, S (2000). "The cradle of agriculture". *Science* 288(5471): 1602–3. doi:10.1126/science.288.5471.1602. PMID 10858140.

Lockwood, J. F. (1952). *Flour Milling,* Third ed., Northern Publishing Co. Ltd., Liverpool.

Lockwood, J. F. (1960). *Flour Milling,* The Northern Publishing Co., Liverpool.

Macrae, R., R. K. Robinson and M. J. Sadler, Eds, (1993). Encyclopaedia of Food Science, Food Technology and Nutrition, Academic Press, London.

Matz, S. A. (1959). The Chemistry and Technology of Cereals as Food and Feed, Avi Publishing Co., Westport, Conn.

Michael Hogan C. 2013. *Wheat*. Encyclopedia of Earth. National Council of Science and the Environment. ed. Lakhdar Boukerrou.

Modi B. S. (1972). Factors affecting the performance of paddy separator, Unpublished M. Tech. Thesis, 1. 1. T. Kharagpur.

Mudambi, S.R and Rao, S. (1986). Food Science. Wiley Eastern Limited, New Delhi.

Neill, Richard. (2002) *Booze: The Drinks Bible for the 21st Century*. Octopus Publishing Group – Cassell Illustrated. p. 112. ISBN 1-84188-196-1.

Nestbitt, Mark., When and where did domesticated cereals first occur in southwest Asia? in R.T.J. Cappers and S. Bottema (Eds.) The Dawn of Farming in the Near East. Studies in Early Near Eastern Production, Subsistence and Environment 6, 2002 (1999). Berlin, ex oriente.

Nissen, M. (1967). The weight of potatoes in water. Further studies on the relation between the dry matter and starch content. *Eur. Potato* J. 10: 85-99.

Nutrient data laboratory. United States Department of Agriculture.

Odunfa, S. A., Nordsrom, J. and Adeniran, S. A. (1994). Development of starter cultures for nutrient enrichment of ogia West African fermented cereal gruel. Report submitted to HBVC research grants program. USAID, Washington, U.S.A.

Ozkan, H; Brandolini, A; Schäfer-Pregl, R; Salamini, F (October 2002). "AFLP analysis of a collection of tetraploid wheats indicates the origin of emmer and hard wheat domestication in southeast Turkey". *Molecular Biology and Evolution* 19 (10): 1797–801.doi:10.1093/oxfordjournals.molbev.a004002. PMID 12270906.

Palmer, John J. (2001) *How to Brew*. Defenestrative Pub Co. p. 233. ISBN 0-9710579-0-7.

Paulus, K. O. (1984). Modelling in Industrial Cooking. In: *Thermal Processing and Quality of Foods.* Zeuthen *et al.* (eds.), Elsevier Applied Science Publishers, London.

Potter, N. N. (1984). Food Science. 3rd Edition. The AVI Publishing Company, Inc., Westport, Connecticut.

Potter, N.N. (1973). Food Science. The AVI Publishing Company, Inc., Westport Connecticut.

Potty, VH. (1988). Horticultural Industry in India. UNIDO Consultation Meeting, Beijing, Nov. 22-24.

Powar, C.B. and Daginawala, H.F. (1986). General Microbiology Vol II Himalaya Publishing Company, Bombay.

Primo, E., Barber, S., Tortosa, E., Camacho, J., Ulldemolins, J., Jimenez, A and Vega, R, (1970). Chemical composition of thebyproducts obtained in the different steps of the rice milling process (In Spanish), Rev. Agroquim Teenol, Alimentos 10′: 244.

Quick, Graeme R. (1980). An In-depth Look at Farm Fuel Alternatives. Power Farming Magazine, Feb. 10-17 pp.

Raghavandra Rao, S. N. and Desikachar, H. S. R. (1964). Pearling as a method of refining jower and wheat and its effect on their chemical composition. *J. Food Sci. Technol.* 1: 40-42.

Rai, Mangla and Mauria S. (1999). Coarse cereal: Not for the poor alone. The Hindu Survey of Indian Agriculture. pp. 55-58.

Rai, Mangla. 2002. Amazing nutrient composition. The Hindu Survey of Indian Agriculture. pp. 59-62.

Reichert, R. D. and Young, C. G. (1976). Dehulling cereal grains and grain legumes for developing countries. I. Quantities comparison between attrition and abrasive type mills. Cereal Chem. 53: 829-839.

Salunkhe, D.K., S.S. Kadam and S. J. Jadhav, (1991). Potato: Production, Processing and Products, CRC Press, Soca Raton.

Satake Owner Manual, Type-03, 91973). Satake Enzz, Co., Ltd., Japan.

Saxena, R P., Singh, B. P. N.; Singh, A. K. and Singh, J. K. (1981). Effect of chemical treatment on husk removal of arhar (*Cajanus Cajan*) grain. Paper presented at the Annual Convention of the Indian Society of Agricultural Engineering (ISAE), New Delhi, India.

Schoch, T. J. (1941), Physical aspects of starch behavior, *Cereal Chem.*, 18, 121-28.

Schrimshaw, N. S. (1974). Nutritional requirements for high protein foods, Proc. Int. Symp. Protein Foods and Concentrates, Central Food Technol. Res. Institute, Mysore, India, pp. 13-18.

Scott, J. H. (1951), *Flour Milling Processes*, Second ed., Chapman and Hall, London.

Seetharam A. (1997). Coarse cereals: Waiting for incentives. The Hindu Survey of Indian Agriculture. pp. 55-59.

Seetharam A., Kadalli, G G and Halaswamy B H. (2001). Results of front line demonstrations and Technology for increasing Prod uction of finger millet and small millets in India. All India Coordinated Small Millets Improvement Project, University of Agrl. Sciences, GKVK, Bangalore 560 065.

Seetharama N. and Rao B D (2004). Sustaining nutritional security. *The Hindu Survey of Indian Agriculture*. pp. 37-38.

Shakuntala Manay, N. and Shadaksharaswamy, M. (1987). Foods: Facts and Principles. Wiley Eastern Limited, New Delhi.

Shepherd, A. D. (1974). How grain structure influences sorghum quality, presented to 5th E. African Cereals Res. Conf. Malawi.

Shewry, Peter R (2009), "Wheat", *Journal of Experimental Botany* 60 (6): 1537–1553, doi: 10. 1093/jxb/erp058

Shibano, M. (1973). Construction and Function of Abrasive Roll Type Rice Whitening Machine, Technical note, JRMA, Japan.

Shreve, N. R. (1956). The Chemical Precess Industries, McGraw-Hili Book Co. Inc., Tokyo.

Singh, G. (1997). Data book on Mechanization and Agro-Processing Since Independence, CIAE, Bhopal-462038.

Six Basic Classes of Wheat *Minnesota Association of Wheat Growers*

Smith, Albert E. (1995) *Handbook of Weed Management Systems*. Marcel Dekker. p. 411. ISBN 0-8247-9547-4.

Smith, L. (1944), *Flour Milling Technology*, Third ed., Northern Publishing Co., Liverpool.

Srivastasva, M (1985). Essentials of Food and Nutrition Vol. II (Applied Aspects). The Bangalore Printing and Publishing Company Limited, Bangalore.

Srivastava, R.K and Singh, S. (1987). Utilization of Mushrooms for Preparation of Different Types of Preserved Products. Government Fruit Preservation and Canning Institute,.U.P., Lucknow.

Srivastava, V.; Mishra, D.P.; Laxmi Chand, Gupta, RK. and Singh, B.P.N. (1988). Influence of soaking on various biochemical changes and dehusking efficiency in pigeon pea (*Cajanus caian* L.,) seeds. Journal of Food Science and Technology 25,: 267-271.

Stermer, R A, (1968). Environmental conditions and stress cracks in milled rice, *Cereal Chem.*, 45, 365.

Stivers, T. *E.*, Jr. (1955), American corn milling systems for degermed products, *Assoc. Operators Millers Bull.*, 2168-79.

Sugden, G. H. (1956). Various aspects of wheat conditioning, *Cereal Science Today*, I, 136-42.

Tanno, K Willcox; Willcox, G (2006). "How fast was wild wheat domesticated?". *Science* 311 (5769): 1886. doi:10.1126/science.1124635. PMID 16574859.

Tuite, J. and Foster, G. H. (1993). Effect of artificial drying on the hygroscopic properties of corn, *Cereal Chem., 40*, 630-37.

Verma, P.; Saxena, R. P.; Sarkar, B. C. and More, P. K. (1993). Enzymatic pre-treatment of pigeon pea (*Cajanus cajan*) grain and its interaction with milling. Journal of Food Science and Technology. 30: 368-370.

Wang, Y. D. and Fields, M. L. (1987). Germination of corn and sorghum in the home to improve nutritive value. *J. Food Sci.* 43: 1113-1115.

Watson, S. A. (1976a). Manufacture of corn and milo starches, In: *Starch: Chemistry and Technology*, Vol. II, R. L. Whistler and E. F. Paschall (Editors), Academic Press, New York.

Wolf, M. J., Buzan, C. L., Mac Masters, M. M. and Rist, C. E. (1952a). Structure of the mature Corn Kernel, Gross anatomy and structural relationship, *Cereal Chem.*, 29, 321-33.

Wolf, M. J., Buzan, C. L., Mac Masters, M. M. and Rist, C. E. (1952b). Structure of the mature Corn Kernel, II, Microscopic structure of pericarp, seed coat and hilar layer of dent corn, *Cereal Chem., 29*, 334-48.

Woodroof, J.G. and Luh, B.S. (1975). Commercial Fruit Processing. The AVI Publishing Company, Inc., Westport, Connecticut.

Woods, A.E. and Aurand, L.W. (1977). Laboratory Manual in Food Chemistry. The AVI Publishing Company, Inc., Westport, Connecticut.

World Wheat, Corn and Rice. Oklahoma State University, FAO Stat.

Chapter 14

Marketing of Millet Based Food Products

The market for health and wellness foods in India is estimated to be worth \$7.5-10 billion by 2015 according to a study undertaken by the Tata Strategic Management group; and the demand for fortified foods, whole grain, high fiber, low cholesterol natural and organic foods is expected will grow. However, the perception of consumers dubbing millets as poor man's crops has been a hindrance to their commercialization. Also, according to Nagarajan and Smale the minor millets namely, Finger millet (*Eleucine corocona*), foxtail millet (*Setaria italica*) and little millet (*Panicum milliaceum*) are considered 'minor' because of the lack of research investment they attract and their limited commercial importance in terms of area, production and consumption. They are insignificant even in national and international marketing system (pearl millets have none) in the regions where they are traditionally grown.

However, millets have high nu traceutical properties, besides, various nutritional studies suggested millets as better choices than fine cereals such as wheat and rice for the nutritional security of the rural and urban poor who have limited access to other sources of dietary components. In these regard, various attempt by various research institutes and organization for value addition in millets. However, they are hardly known by the consumers as they are not available on-shelf in the market. But for commercialization, the products need to be standardized oganoleptically with high functional properties added to them and acceptable to the consumers.

In the Kolli Hills of Tamil Nadu, despite a consumption preference among the farming communities for minor millets, they have been cultivated on a subsistence scale and marketing was considered the main constraint for scaling up the production. As a response, the MS. Swaminathan Research Foundation

(MSSRF) based in Chennai has led targeted conservation cum commercialization intervention programs over the last 7-9 years, to establish a robust local market for minor millets- within 100 km radius- by using promotion campaign, awareness and introduced new recipes in order to add value locally. The MSSRF established market linkages at the local and regional levels. To do so, self-help groups and enterprises were formed to institutionalize market operations. The various groups have also been brought under a single system of federation that was linked to the Tribal Cooperative Marketing Development Federation of India Ltd. (TRIFED). Marketing mechanisms and particular economic linkages were negotiated and resulted in a Memorandum of Understanding. Other initiatives in this regard include branding and innovative packaging. They have also encouraged lots of entrepreneurs to take up these technologies in their business.

In regard to sorghum, efforts of ICAR-IIMR led intervention in public private partnerships mode under NAIP (2008-2014) have led to commercialization of some millet products which have been mentioned in the previous chapter, though scaling up still remains a challenge; however, they have proved the point of possibility of commercializing convenient millet foods on par with rice and wheat so as to offer choices to urban health conscious consumers. Similarly there are efforts from major private player such as Britannia industries Ltd through their nutria-choice segment and ITC's multigrain biscuits which appear to play a major role in millets in the days to come. In recent times Britannia Industries have entered a research MoU with IIMR on sorghum biscuits and bakery products which is again a land mark. Apart from big players there are many small scale players who produce extruded snacks from sorghum and pearl millet in Mumbai, Rajasthan, Tamil Nadu and Andhra Pradesh. IIMR has entered more than 20 MoU's with private partners. However, advocacy and policy prescription would take the consumption to greater heights that would generate not only demand for millets but also raise standard of living of the poor dryland farmers growing millets.

Among the millets, finger millet based malt is more acceptable product both in traditional and non-traditional consuming areas. Some of these products have premium prices and are acceptable as up market products cater to high income groups. Gluten-free sorghum products are commercialized by an entrepreneur based in Kolkata catering to celiac patients. Processed small millets are commercialized in a big way by entrepreneurs from Nasik, Maharashtra and are being exported. Though substantial studies are not attempted on the bioavailability of minerals in millets the general awareness that is being created across the country and world over that millets are nutri-cereals and offer promise to health seeking population who desire to keep lifestyle diseases at bay.

Steps Towards Commercialization of Sorghum/Millets Products

Branding

A brand can be understood simply as a name termed for a product, service, commodity, etc. The most vital and important part of a brand is its name and logo that distinguishes one seller's product from those of others. In other word, a brand can easily be understood if you refer to a brand as a person.

Effective branding can result in higher sales of not only one product, but of other products associated with that brand. For example, if a customer loves sorghum biscuits and trusts the brand, he or she is more likely to try other sorghum products offered by the company - such as sorghum cookies and other sorghum ready-to-eat products. Brand is the personality that identifies a product, service or company (name, term, sign, symbol, or design, or combination of them) and how it relates to key constituencies: customers, staff, partners, investors etc.

eatrite™

Eat Jowar - Stay Healthy

The IIMR brand *Eatrite* was launch for commercialization of sorghum product technologies in the twin cities of Secunderabad and Hyderabad on a pilot scale with the tagline 'eat jowar-stay healthy' and 'health meet taste'. The name itself refers its health benefits and nutritional superiority over alternate products available on-shelf in the market. IIMR being a government agency received a wide momentum from across the organization and individuals and incidentally attracted lots of entrepreneurs from this initiative that many firms entered MoU with the institute to market or for technology licensing as they became confident of the demand for these health products.

Labeling

In labeling the most important part is to mark the product as per the standard classifications. For *e.g.* if the product is a vegetarian product, we have to mark it green, as is seen in *Eatrite* logo.

Barcode

These days the product can be uniquely marked with its details encrypted in a barcode. It really proves very helpful for maintaining stock inventory, faster billing at POSs/product search on internet using smart mobiles. Generally for encryption EAN-13 codes, code 126 formats etc are used.

Choice of Material

Depending upon the shelf life required product type and the finish/look we have to choose the material. If a product is fragile and might get crushed easily, then its better we pack it in a paperboard box packing or plastic containers etc. You can meet packaging experts to get a perfect understanding of your product packing needs.

Packaging

Packaging is a science, art and technology of enclosing or protecting products for distribution, storage, sale and use. It also refers to the process of design, evaluation and production of packages. Packaging depends upon the need and product. There are various packaging types like poly packaging, paperboard, foil, etc.

The product and the package are two different things. The product is the thing we want or need, but its packaging is what we will throwaway after taking or making used of the product. However, packaging adds value to the product with respect to customer's perception. A well design packaging can complement or even enhance the product it was designed to carry. Good design isn't just about the product - it's about good packaging as well.

Packaging Layout

Always the brand name should be at the centre of the layout with maximum prominence. The product identity should get equal prominence so as to describe the customer what is inside the pack. It is necessary to always mention how to use or make the product. Also if you can mention the benefits of the product, then people are like to understand your product thoroughly. Other necessary requirements are credentials, contact and mandatory packing information should be mentioned behind the pack like Quantity, Net Weight, MRP, Expiry Date, Batch No., Barcode etc.

Promotion and Commercialization under *Eatrite* Brand

"Eatrite" is a brand promoted by the Indian Institute of Millets Research, in India. This brand has been initiated under the National Agricultural Innovation Project (NAIP) which is quite uncommon for an institutional brand promotion is unique.

Eatrite name is chosen to reflect the characteristics of carbohydrates present in sorghum which are complex in nature that are released slowly, regulating hunger which helps naturally not to over eat but eating right quantities. Eating sorghum is also right because it has high fiber, minerals, vitamins and anti oxidants. Eating sorghum is right because it has manifold health benefits which are enlisted earlier in the report.

The range of products under this brand includes: Sorghum Rich Multigrain Flour, Sorghum Semolina, Sorghum Pasta, Sorghum Vermicelli, Sorghum Flakes and Sorghum Roasted Flakes and Sorghum Biscuits/Cookies. These products are tested for their nutrient profile and nutritional suitability by National Institute of Nutrition (NIN), Hyderabad and ascertained that they are recommended for those who are ailing from life style disorders such as diabetes and obesity and are beneficial to all age groups.

Initially we started with aggressive promotion such as organizing Road shows and health awareness campaigns in public parks, malls, etc., wherein the product technologies were displayed and samples are distributed. Later with the increase in awareness by the denizens in Hyderabad, mass media started approaching IIMR, which further spread its popularity. With the introduction of IIMR brand *EATERITE*, with the message 'Eat Jowar, Stay Healthy' it has gain a great momentum from consumers especially in urban areas where people have started looking for health foods. The impact of popularization and increase of awareness among the masses stimulated entrepreneurs who started approaching IIMR for either exclusive marketing or manufacturing and marketing the product technologies in their own brand.

IIMR has now signed Memorandum of Understanding with more than 20 entrepreneurs of which, MoU with multinational company Britannia industries Ltd is a most prominent achievement. Others ToR include adopting processing technologies at pilot scale for commercialization, technology licensing on multigrain atta and biscuits and taking up Eatrite products for commercialization. Many more are in pipeline from cities like Delhi, Bangalore and Chennai. Under this, 5 formats of business plans are commercialized for Jowar products which evolved under their relative merit assessed in terms of farmer's share in the consumer rupee.

Branding and labeling with nutritional profile, recipes, mandatory information and packing was unique in sorghum as there were no products available in the market except multigrain atta. This has huge impact on consumer perception on product diversification and flagging of nutritional superiority. The acceptability among health conscious urban consumers is great so much so the response of entrepreneurs is overwhelming to produce and market this future alternate nutri-cereal and made MoU with IIMR.

Salient Features of Commercialization and Scaling Up

'Eatrite' brand name was registered and launched by IIMR with unique selling proposition 'Eat Jowar Stay Healthy' for commercialization of sorghum product technologies developed under NAIP- Millets Value Chain.

Successfully commercialized *Eatrite* products on pilot scale in Hyderabad through Heritage Retail Stores across the city and through un-organized retail sector.

For sustainability, IIMR has Signed MoU with 17 prospective entrepreneurs from Hyderabad, New Delhi and Bangalore with variable emphasis on production, promotion and marketing sorghum product technologies.

A MoU is signed with *M/s* Britannia, an MNC (first time in the history of ICAR) by IIMR to work jointly on R and D on bakery and biscuit products.

The commercialization is solidly backed with the popularization measures adopting various communication strategies.

Horizontal expansion is envisaged in Bangalore, Chennai, Pune and Delhi. These convenient and healthy products are targeted primarily at urban population. The products are aggressively promoted through road-shows, nutritional campaigns, print and electronic media across the Hyderabad City on a pilot scale. Currently they are marketed by M/s Heritage Fresh retail chain stores and are explored for their market in unorganized market. The response has been very good and there is an increasing demand for their availability across the city and other major metros. Particularly, multigrain *atta,* pure *jowar* biscuits, flakes have good demand across all age groups.

Introduction of Eatrite Products in Retail Chain: A Pilot Testh in Hyderabad City

Eatrite products are now marketed through Heritage Fresh and ITC retail chain stores in Hyderabad. Simultaneously, outsourced event managers for popularization, of sorghum products (360 degree communication, brand designing logo, etc with BTL and ATL strategies implemented) in urban markets and New age Media. Massive awareness is created on sorghum as health and nutria food through Road Sh01 (100+) in public parks, malls and institutes *etc* in Hyderabad and in exhibitions· imparting awareness of sorghum to across 40000 consumers through fabricated *Jowar Rath* in Pune, Bangalore, Jabalpur, Chennai, Coimbatore, New Delhi etc. This whole commercialization started with the soft launching of IIMR brand *'Eatrite'* at ICAR Krishi Bhavan and other 2 NAFED Outlets in New Delhi.

Marketing of Millet Products

Under aegis of National Agricultural Innovation Project (NAIP), IIMR has developed many post-harvest technologies for efficient and enhanced use of sorghum for the benefit of the consumers and various stakeholders making relevant to the society. The ultimate goal is benefit not only providing convenient and healthy option for enhanced sorghum consumption but to revive the sorghum crop cultivation through interventions such as diversified food processing and nutritional tagging for stabilization of farm incomes of our primary stakeholder, sorghum farmer. Three products have been identified by IIMR as most suitable processing technologies fl any entrepreneurial activities. These are:

1. Jowar Based Multigrain Atta
2. Jowar Based Biscuits
3. Jowar Based Flakes

All of these technologies has been developed and standardized as a value addition in the Sorghum/Jowar to provide healthy foods at the affordable price to the customer.

Advantage of the Jowar (Sorghum) Based Products

- ☆ Complete family product- individual, family, young, aged, kids, health conscious and traditional.
- ☆ Healthy product- wholesome nutrition for all members of the family, no restriction for diabetic/old/ill members, energy powerhouse, perfect diet management.
- ☆ Perfectly substitutable- No variation in terms of preparation, appearance and taste of regular atta.
- ☆ Convenience- no extra effort required and time consuming.
- ☆ Addressing both health and taste.
- ☆ Available products either positioned "Health (cholesterol management and diabetic (Saffola), Gluten free, Diabetic) or "Taste" (Ashirvaad, Pillsbury, Annapurna etc.)
- ☆ Available biscuits are either positioned "Health (digestive, light, nutritious, diabetic) or "Taste" (Good day, Hide and Seek, Butter Bite, or loose (Nan-Khatai, fresh backed).

Table 14.1: State-wise Potential Area/Markets for Launching Jowar Products

State	*Areas*	*Supply Options*	*Market Entry*	*Effort Required*
Andhra Pradesh	Vijayawada, Vizag, Karnool	Central manufacturing	Easy, moderate marketing cost	Consumer acceptance study with launching strategy
Karnataka	Bangalore, Dharwad, Gulbarga, Bidar, Hubli	Central manufacturing	Easy, moderate marketing cost	Consumer acceptance study with launching strategy
Rajasthan	Jaipur, Ajmer, Jodhpur, Pali	Outsourcing/ Tie-up	Easy, moderate marketing cost	Opportunity analysis, market survey, pre-launch consumer study
Gujarat	Bharuch, Vadodara	Outsourcing/ Tie-up	Easy, moderate marketing cost	Opportunity analysis, market survey, pre-launch consumer study
Tamil Nadu	Chennai, Coimbatore	Central manufacturing	Easy, moderate marketing cost	Opportunity analysis, market survey, pre-launch consumer study

Industry Size Estimation of Sorghum

A. Multigrain Atta Industry

- ☆ The market size of multigrain atta is very small compared to the normal atta market.
- ☆ It is estimated to be around 15-25 thousand tons per annum. (on an average 1-1.5 per cent of total packed atta industry of India)
- ☆ Mostly organized players dominate the segment and national brands are available in the market.
- ☆ Few regional players are also there in the market.
- ☆ Multigrain atta is sold at premium. (Ashirvaad, Pillsbury, Safola).
- ☆ Growth of this particular segment is estimated to be 3- 4 per cent annually.

B. Biscuits Industry

- ☆ Biscuit industry size is estimated to be around 25-30 lakh tonnes per annum (Indian Biscuit's Manufacturing Association (lEMA).
- ☆ About 55-65 per cent of the total volume is catered to by the organized players. Glucose biscuits constitute 45-50 per cent, followed by Marie to the tune of 10-15 per cent in total biscuits consumption basket.
- ☆ Share of cookies is estimated at around 3-4 per cent of the total market size. * Growth of overall biscuit sector has been estimated at around 12-15 per cent per annum (Indian Biscuit's Manufacturing Association (IEMA).

C. Rice Flakes Industry

- ☆ Rice flake market size is estimated to be around 8-10 lakhs tons per annum. (Source: Foretell internal research)
- ☆ The unorganized players dominate the segment.
- ☆ No branded players exist in this segment.
- ☆ It is sold under private labels in most of the urban areas.
- ☆ The consumption growth is estimated to be 5-7 per cent annually.

SWOT Analysis of Jowar Food Products

The SWOT Analysis of the jowar food products is essential for establishing its uniqueness amongst the available products in the market in same category. Few clear representations from the table make jowar as one of the healthy food products because of its nutritional content. However, it has few weaknesses like its shelf life and lack of commercial business model but its opportunity gives an advantage and put its potential to emerge as one of healthy food products.

In food product segment like multi grain atta and biscuits, consumer are looking for healthy option. The awareness bringing the health consciousness as a most important factor while purchasing food item. The changing life style and with increase in number of diseases people are becoming health conscious and trying to have healthy foods as a part of their eating habit (Table 14.2).

Table 14.2: SWOT of Jowar Food Products

Strength	*Weakness*
Awareness of Jowar as healthy and nutritious food is favorable	Infant brand in the market
Appealing response for quality certification from NIN	Lack of awareness among the non-traditional respondents
Strong research and development facility	Lack of commercial business operation, sales/ distribution network and presence in the market
Only brand (IIMR's brand 'Eatrite') offering versatile product range based on Jowar in the market	Immature economy of scale
Easy adoptable in terms of appearance, preparation and taste when compared to traditional products	Lack of communication and advertisement about the product
Taste of the product is liked by the respondents	Shelf life of the product is short (1 month)
People are willing to consider the product if "Eatrite" launched in the market	The typical Jowar smell has received disliking response from non-traditional consumers
	Jowar has difficulties in preparation/cooking for urban consumers
	Inferior appearance attribute compared with traditional product
Opportunities	*Threat*
Growing consumer's eagerness to try new products with health benefit is high	Initial hesitation of the retailers to keep the products in non-traditional areas
Specialty consumers (diabetic, Obese) are excited on favorable result	Consistency about the quality is doubtful in consumer's mind, as supplier is not a branded Fast Moving Consumer Goods (FMCG) supplier
Able to substitute traditional product giving variety and choice	Sustainability in the short term is high due to marketing set-up cost
Health and Wellness food market size is around Rs. 10000 crores and expected to grow by CAGR 23 per cent (TSMG)	Increasing raw material price and uncertainly in the production of Jowar
Increasing disposable incomes, diminishing culinary skills and the rising need for convenience would be the driving factors of Jowar products in traditional areas.	Negative health related perceptions (non-edible varieties for some seasons, dental disease issues)
	The health product segment size is very small and many consumers fall in the 'Late Majority' category.

Product-wise SWOT Analysis

A. Jowar Multigrain Atta

The SWOT analysis of multigrain atta in table shows that people have developed a liking towards the multigrain atta based on jowar (Table 14.3). The health benefits of the atta *i.e.* its gluten free content making it suitable for diabetic patients. There are weakness also related to the product like smell and shelf life, however opportunity is very big as very few players are there in the market who are selling jowar based multigrain atta. Also, continuous search from the consumer for the better health food item also provide them and opportunity to tap the current multigrain atta market.

Table 14.3: SWOT Jowar Multigrain Atta

Strength	*Weakness*
The product is very much liked by most of the respondents who have tasted the product.	Jowar multigrain atta is only product available in the market rich in (containing) Jowar.
More than 90 per cent of the survey respondents expressed their willingness to purchase the product.	Existing players in the multigrain atta segment in the market are very few.
Health benefit of Jowar atta is the major acceptance criteria in traditional areas.	The multigrain atta consumers are not brand loyal as they seek more health benefits from such types of atta.
The respondents have felt no significant difference between Jowar atta and wheat atta.	Increasing health consciousness, disposable incomes, diminishing culinary skills and the rising need for convenience could be driving factor of Jowar products in urban traditional Jowar consuming areas.
Jowar atta is an ideal product for diet plans and weight management.	
Suitable for all family members and no restriction for diabetic, ill, kids and aged.	
Opportunities	*Threat*
Shelf-life- The shelf life of Jowar atta is 2 month. and dominated by the national brands.	Size of the multigrain atta industry in very small
Smell- Typical Jowar smell in the atta during preparation is not liked by a few consumers (especially the new consumers).	Growth of this segment also very less (3-4 per cent yearly)
Swelling- swelling attribute of chapatti was not liked by the Delhi consumers.	Higher price of Jowar seed would in turn increase the price of Jowar multigrain atta.
Stickiness and texture of Jowar atta is another issue to be addressed	Cost of marketing and distribution is very high for the new product.
No existing business experience, supply chain is still in formation stage.	Usage of Jowar as food product is restricted to the traditional consuming areas.
No professional manufacturing, marketing and sales department	
Economy of scale, learning curve, expertise in commercial operation and packaging are yet to be fine-tuned.	

B. Jowar Biscuits

The S.W.O.T of the jowar biscuits in table represent the fact that currently either no or very few players have jowar based biscuits in their product portfolio (Table 14.4). Also, with the increase in awareness about the content of normal biscuits, people are trying to have more balanced and nutritional biscuits in their normal day- to-day life. The opportunity for this segment is very high as it is still under very early stage.

The jowar based biscuit being new to market can capture the market size of the nutritious biscuits segment as its nutritional contents are high compare to other millets and cereal crops.

Table 14.4: SWOT of Jowar Biscuits

Strength	*Weakness*
Jowar Biscuit is very much liked by the most of the respondents.	Jowar biscuit is the only product available in the market rich in (containing) Jowar.
More 90 per cent of the survey respondents showed their willingness to purchase the product.	Product can compete with other cookies targeted at the health conscious consumer segment.
Health benefits associated with Jowar biscuit is the major acceptance criteria in the traditional areas.	The health biscuit consumers are not brand loyal as they seek more health benefits from such type of cookies.
Jowar biscuit is suitable for all family members and no restriction for diabetic, ill, kids and aged.	Increasing health consciousness disposable incomes, diminishing culinary skills and the rising need for convenience could be driving factor of Jowar products in urban centric traditional areas.
It is an ideal product for diet plans and weight management.	
Taste-wise Jowar biscuits are not very much different from the existing cookies.	
Opportunities	*Threat*
Shelf-life- The product shelf life is very short (2 months).	Share of "cookies" in the total biscuit market is very small. (5-7 per cent of total)
Smell- Typical Jowar smell in the biscuit is not liked by some consumers.	Healthy-Cookies are niche product and dominated by the organized players or brands.
Crispiness- Coconut biscuits were less crispy while normal biscuits were hard while chewing.	Growth of this segment (Cookies) also very less (3-4 per cent yearly (CAGR 2008-13) as compared to 12-15 per cent for all). (Source- Data monitor)
The Jowar biscuits gumminess (while chewing biscuits getting broken into particles and becoming ready to swallow) is less as compared to the normal cookies.	Higher price of Jowar seed would in turn increase the price of Jowar biscuit.
Fullness of taste after eating and mouth feel is not liked by some consumers.	Cost of marketing and distribution is very high for the new product.
No existing business experience, supply chain is still in formation stage.	Usage of Jowar as food product is restricted to the traditional consuming areas.
No professional manufacturing, marketing, sales department	
Lack of production and operational advantages like economy of scale, learning curve, expertise in commercial operation, packaging, etc.	

Financial Analysis Production Plan

Table depicts the production plan of the Multigrain atta, Biscuits and Flakes. It is considered that plant will run for 8 hours on a daily basis. The monthly working day is assumed to be 25.

The financial analysis has been done for the 4 year and the production of these products will be increased on yearly basis as depicted in the Table 14.5.

Table 14.5: Daily, Monthly and Annual Production Quantity of the Above Three Products

Particulars	*Daily Production in tons*	*Month Production in tons*	*Annual Production in tons*
Multi-Grain-Atta	1.2	30	360
Biscuits	0.36	9	108
Flakes	0.8	20	240

Table 14.6 packaging size has been identified keeping in mind the movement of the product from the shelf. Table indicates the pack size of the selected products.

Table 14.6: Packs Size for different Products

Particulars	*Pack Sizes*	*Total Annual SKU Quantity*
Multi-Grain Atta	1 kg	270000
	5 kg	18000
Biscuits	0.1 kg or 100 g	810000
	0.25 kg or 250 g	108000
Flakes	1 kg	180000
	5 kg	12000

Financial implication of putting up the plant for the production of these 3 products was worked out in Table 14.7. It is advisable that multi-grain atta and biscuits plant should be considered together at the machineries are common up to certain extent in the early processing of raw jowar and other material and preparation of input for manufacturing of both products. Flakes require altogether different machines and set-up. However, if an entrepreneur (early start-up) is planning for these technologies, he should go for all the three technologies together(This will not only help him in reducing or sharing the cost in terms of putting up manpower but also provides variety to the consumer leading to enhancement of the sales).

The fixed cost in the below calculation has been put separately for all the products, however while calculating the IRR (Internal rate of return) and NPV (Net present value), we have taken the fixed cost of the biscuit and flakes together as the machineries will be same in case as of biscuits in case of multigrain *atta*.

The plan has assumed that it should have internal storage in order to have greater control over the raw material to be used for the production. It is also advised that the production plant should be near to the area where sorghum and other inputs are easily available in order to have smooth supply of the input for the production.

Table 14.7: Financial for the First Year of Operations (all cost in Rs. Lakh)

Particulars	*Multi-Grain atta*	*Biscuits*	*Flakes*	*Total*
Building and Plant including plant machineries and other legal and setup cost	82.8	77.8	99.8	166.5
Variable cost				
Cost of raw Material	68.8	52	96	216.8
Operational Cost which includes electricity, depreciation of the machineries. Packaging of the products etc.	12.66	13.75	9.93	36.34
Marketing cost	18	18	18	54
Manpower cost	7.62	7.62	7.62	22.86
Channel Margin (@ 25 per cent), transportation and logistics (@5 per cent) and taxes (@ 4 per cent)	39.9	33.72	48.94	122.56
Total Variable cost	146.39	125.09	180.49	452.56
Total Sales. It is assumed that 10 percent of the total SKU will be sales return and 10 of smaller SKU will be used in promotional activities. Therefore actual quantity for sales will be 80 per cent of the smaller SKU90 per cent of the bigger SKU	148.5	131.22	195.3	475.02
Net profit before tax	2.1	6.34	14.8	23.24
Net Profit after tax (@30 per cent)	1.47	4.44	10.36	16.27

Based on the first year operations the projection for the next 3 years have been made in terms of total number of quantity to be produce on annual basis of each product.

Table 14.8 provides suggestive estimated quantities of the three products that need to be produced year on year basis in order to achieve the breakeven point and recover the initial investment in first 4 year of the operations.

Table 14.8: Annual Production Wuantity (in ton) for Next Three Years

Particular	*2'd Year*	*3'" Year*	*4th Year*
Multi-Grain Atta	420	420	420
Biscuits	120	120	120
Flakes	300	360	420

Based on the production plan for all four years, IRR and NPV have been calculated in table which comes out to be 9 per cent and 48.4 per cent respectively (Table 14.9).

It is also important to highlight that the quantities on which the first year calculation is based are close to the breakeven point of the variable cost. So, this means that quantities below this will lead to loss on the operational front. It was also identified during the financial calculation that single product or technologies based on sorghum will not be able to give you viable income. Therefore, it is suggested that combination of products should be consider before starting the fresh venture. However, if entrepreneur is already in this business and he is diversifying his product mix can go for the single product or technology.

Table 14.9: IRR and NPV of the Project

Targets			*Fixed Cost 166.5*	*Year*	*Revenue*	*Cost (Variable)*	*Revenue Cost-166.6*
Multi-Grain Atta	*Biscuit*	*Flakes*					
1200	360	800		1	475.0	451.6	23.3
1400	400	1000		2	563.0	513.0	50.0
1400	400	1200		3	612.0	548.2	63.8
1400	400	1400		4	660.8	583.0	77.8
						IRR	9 per cent
						NPV (Rs. lakhs)	48.40

Marketing Plan

It is suggested that for the first year, entrepreneur has to go for an intensive marketing strategies, in order to establish their brand.

The Marketing plan for product will have both strategies *i.e.* above territory line (ATL) and below territory line (BTL). A combination of mass campaign, road shows, paper insertion, cable TV advertisement and newspaper advertisement will be given in order to create awareness of the brand in front of consumer, leading to the fast movement of the product from the retailer shelf.

It is also suggested that these activities should be conducted on regular intervals *i.e.* month on month basis in order to grasp the attention of the consumers.

Apart from the above mentioned activities, entrepreneur should also have banners, posters, T-shirts; Caps etc. in order promote the brands in the market. Banners and posters should be put on the key locations which receives maximum footfall of the customers. Sales and marketing team should wear the T-shirts and caps to create a buzz in the market. T-shirts and caps should be also given during the mass campaign and road shows, so that brands reaches to the consumer household and create word of mouth publicity.

The bundling of the products can also be considered during the product promotion activities. In Table 14.10 few combinations have been suggested, which entrepreneur can consider for their bundling plan during the promotional activities.

The number of activities in terms of Below Territory Line (BTL) and Above Territory Line (ATL) should be take on month on month basis in the initial year are listed in Table 14.11. This is essential in the initial period or years in order to establish the brand in the market. Once people will start recognizing the brand, it will give the momentum to the sales of the products. Also, the tie-up chances with big retail chains to get their shelf space will increase.

Table 14.10: Suggestive Combination for Bundling of Three Products

Product Purchased	*Quantity*	*Free Gift (Product) with Purchased Product*
Multigrain Atta	5 kg	100 gm pack of biscuit free
Flakes	1 Kg pack	100 gm pack of biscuit free
Flakes	5 kg pack	250 gm pack of biscuit free
Multigrain atta + Flakes	5 Kg of Atta+ 5 kg of Flakes	1 kg of Multi grain Atta pack free

Table 14.11: Brief Activity Under BTL and ATL

Sl.No.	*Marketing Strategy*	*Activities per Month or on Quarter Basis*
1.	Below Tertiary line	
1.1	Road Shows on the most happening places and the road on which footfall of customers are high	3 months
1.2	Mass Campaign with the help of umbrella and Standees on the key locations of the market	2 months
1.3	Paper Insert	3 months
2.	Above Tertiary line	
2.1	News Paper Advertisement	1/quarter
2.2	Cable Television advertisement	1/month

Alternative Strategies for the Selling of the Products

The business plan also suggests an alternative marketing or sales plan. The entrepreneur or enterprise can also become a contract manufacture for the big players in the market. They can ensure the supply of the products for particular region with the timely supply commitment. This can also facilitate the development of the cluster based sorghum production center in the major sorghum producing state. This development will also lead to the increase in the production area and also provide a dryland based farmers to and opportunity to improve their livelihood.

It is proposed that if IIMR can take this initiative in tying the entrepreneurs or enterprise with the bigger players having greater experience and expertise in the marketing and selling of the products. The entrepreneurs can played a role in doing the backend manufacturing for those bigger players. The companies which will be buying the products can implement their quality standard for maintaining the quality of the products. The Table 14.12 provides and better idea with the differentiation between the roles of the different stakeholders involved in the partnership.

Table 14.12: Suggestive Roles of the different Players Involved in the Partnership

Particulars	*Entrepreneur or Enterprise*	*Indian Institute of Millets Research*	*Big Indian or Multinational Company*
Procurement and production	Manage and concentrate on the production of quality sorghum base products	Can help entrepreneur with backward integration with Farmers and provide technology for production of sorghum based products	Provide technical guidance on setting up the plants and help in implementation and usages of the quality standards for production
Marketing and sales	Focus on the quality production with attractive packing	Technical help in highlighting the key Uniqueness of the products	Through their established distribution network provide an necessary momentum for the sales

Risk Factors

Availability and Accessibility of the Jowar

The availability of jowar for the production of value added and nutritional product is a critical factor. The plant needs to have very strong backend integration in terms of procurement of the grain. It is advisable that plant should be established near the production area. Also, there should be integration with the farmers under the contract farming agreement in order to have smooth and continuous functioning of the plant. The procurement planning and mapping of the required quantity should be planned well ahead before the start of the season.

The price fluctuation in the sorghum is also one of the critical factors. If crop production goes down, price of sorghum grain hit the range of Rs. 4400/quintal. Therefore, the backward integration in terms of the procurement will lead to have better control over the input cost for the production.

Entrepreneur Skills and Capabilities

This is also important to highlight this factor as these products are premium products in the market and have very less market share in terms of regular atta or biscuits available in the market. The entrepreneur should have very strong business acumen and should be able to lead the business from the front. He should have the understanding of the market channels and should be able to crack the right channel. He should also have understanding about the various ways through he can procure the sorghum while keeping his operations unaffected due to price variation.

Social Impact

The products identified have huge social impact on the society. Sorghum based products are more nutritional in comparison to the wheat or rice based products.

Gluten free content makes it right food for the health conscious people. It is also recommended to the people having diabetes.

The children can have healthy food which helps in balanced growth and development of their body. This can also lead to the increase in area under the cultivation and can give the dryland farmers a better price for the product which potential is still not recognized by the market. The farm family engaged in the sorghum production in dry land area can earn good income as the product made from sorghum are fetching premium price from the market. This will also help in the development of sorghum based product industry and can provide employment to the rural youth by engaging them in the process of production, procurement, manufacturing of the products and its sales and marketing.

Based on analysis of data, the following emerges as some major conclusion:

- ☆ First time entrepreneur or enterprises, needs to get two or more than two jowar based food products to start their business. This will ensure that business model be more economical viable. However, if existing entrepreneur or enterprises wants to diversify their existing business they can go for any number of jowar based products as the processing of the raw material into the final product is same.
- ☆ Data analyzed indicates minimum quantity of each food products, which needs to be manufactured on monthly or yearly basis. These quantities will ensure the recovery of investment at the end of the 4th year of the operations.
- ☆ The identified key districts where the sorghum cultivation having potential may be explored for the setting up of the manufacturing plant. This will ease down the difficulty in raw material procurement for the manufacturing of the food products for the entrepreneurs and makes it more cost effective.
- ☆ It is clear that Jowar based food products are relatively new to the market. Hence, it is essential that entrepreneurs develop competent marketing skills for sales and marketing of these food products. These products also need focused marketing activities in terms of ATL or BTL. This is essential to develop the brand in the market place leading to customer preferences towards purchase of these food products.
- ☆ As an alternative strategy to establish sole entrepreneur mode tri-partite partnership between the entrepreneur, IIMR and a big player (who is already in business) is also suggested. The entrepreneur will have the role of manufacturing standard quality food products, IIMR can provide him with backward integration for the raw material procurement and forward integration by big players through development of supply chain for the food products. The entrepreneur can act as a contract manufacturer for the big companies. The company can be provide technical guidance to the entrepreneur for the manufacturing of the good quality food products and play the role of marketer in which they have develop their expertise.

Index

C

D

P

R

S

T

U

V

W

Z

Barnyard Millet

Browntop Millet

Finger Millet

Foxtail Millet

Kodo Millet

Figure 1.1a: Seeds of Various Types of Millets (p. 3)

Little Millet

Maize

Pearl Millet

Proso Millet

Rice

Figure 1.1b: Seeds of Various Types of Millets (p. 4)

Sorghum

Wheat

Figure 1.1c: Seeds of Various Types of Millets (p. 5)

(p. 68)

Barnyard Rice

Barnyard Chhiki

(p. 69)

Barnyard Idali

Barnyard Laddoo

Barnyard Biscuits

Barnyard Puri

(p. 69)

Brown Top Millet Pakoda

Brown Top Millet Cake

(p. 74)

Brown Top Millet Pizza

Brown Top Millet Chhiki

Brown Top Millet Cookies

Brown Top Millet Cakes

(p. 74)

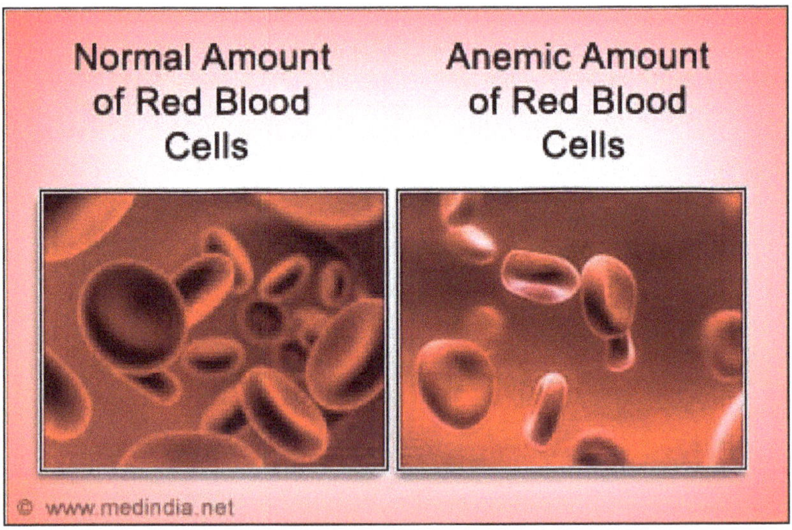

(p. 82)

Doughs

Masala Dough

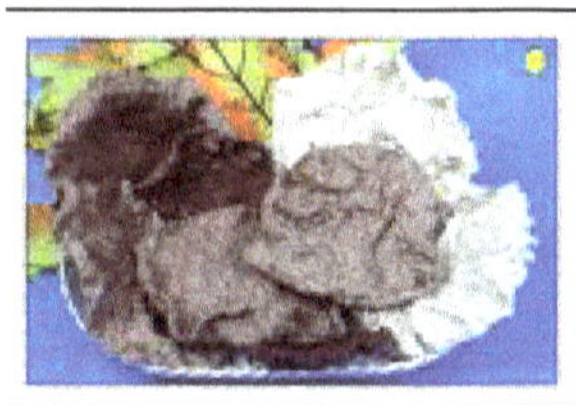

Papads

Extract

Whole Flour

Sweets

Flour

Laddoo

Soup

Pakoda

Masala Pakoda

Salty Soup

Laddoo with Sugar

Laddoo Jaggery

Seeds

Sweets

Meal

(p. 97)

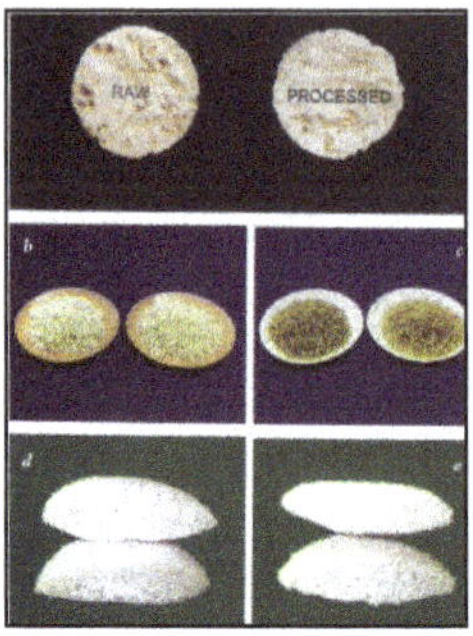

Idali

Parot and Soup

Rice

Sweets

Biscuits

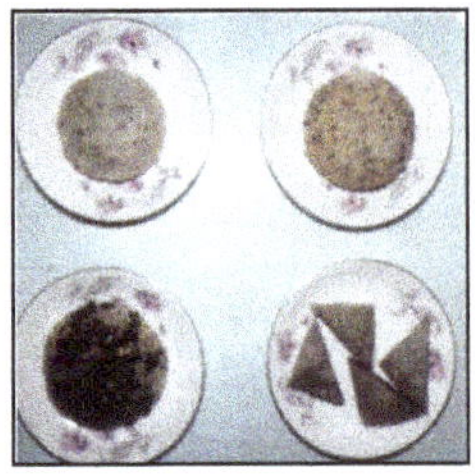

Cookies

Earheads

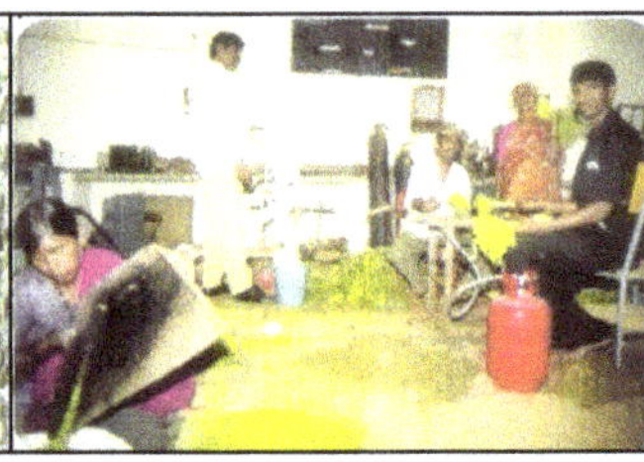

Processing Unit

Marketing Products

(p. 98)

(p. 101)

Foxtail Khichidi

Foxtail Upma

(p. 105)

(p. 106)

Foxtail Millet Thalpit (p. 107)

Foxtail Millet Pancake (p. 107)

Foxtail Millet Sweets

Foxtail Millet Upma

Foxtail Millet Laddoo

Foxtail Millet Masala Rice

(p. 108)

Kodo Millet ***Idali***

Kodo Millet ***Dosa***

Kodo Millet ***Dosa***

Kodo Millet ***Rabadi***

Kodo Rice

Kodo Millet ***Upama***

(p. 115)

Little Millet Bread

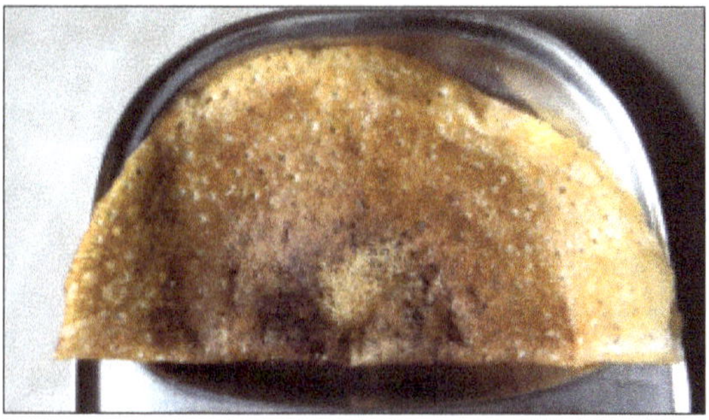

Little Millet Dosa

Little Millet ***Chakali***

Little Millet ***Roti***

(p. 121)

Corn Flakes (p. 129)

Corn Pops

Corn Cracker

Corn Crackers

Corn Starch

Corn Bakery Products

(p. 130)

Corn Oil

Corn Sweets

Bean and Cheese Burgers

Open Bean Burger

Broccoli Burger

Hara Tava Burger

Healthy Burger

Veggie Burger

Masala Toast

Chunky Cheese and Vegetable

Mexican

Rajma Burger

Spinach and Corn Open Burger

Mini Pav Bhaji Burgers

(p. 131)

Three Bean Burger

Veggie Burger

Masala Toast

Cheese and Spring Onion Sandwich

Bread Pakoda

Marinated Bell Pepper Open Burger

(p. 132)

Bajara Laddoo

Bajara Chapati

Bajara Suhali

Bajara Bakali

Bajara Khichari

Bajara Mathi

Chocolate Cake

Nan-khatai

Bajara Cookies

Kaju Biscuit

Bajara Eggless Cake

Melting Movement

(p. 161)

Bajara Sev

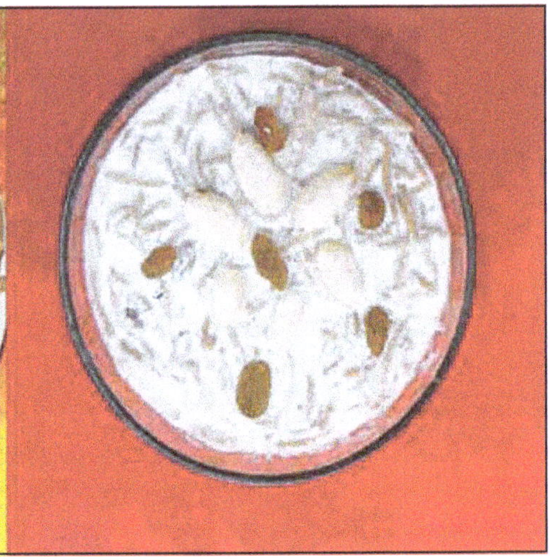

Bajara Sweet Vermicelli

Bajara Thalpit

Bajara Utaapav

Bajara Sweet Cakes

Bajara Khichadi

(p. 162)

Bajara Pan Cakes

Bajara Salty Cake

Bajara Idali

Bajara Masala Puri

Bajara Chapati

(p. 163)

Bajara Cakes

Bajara Muffins

Bajara Chocolate Biscuits

(p. 164)

Bajara Bread

Bajara Decorated Cake

Bajara Sweet Laddoo

Bajara Sweet Laddoo

Bajara Idli

(p. 165)

Bajara Processing Machines

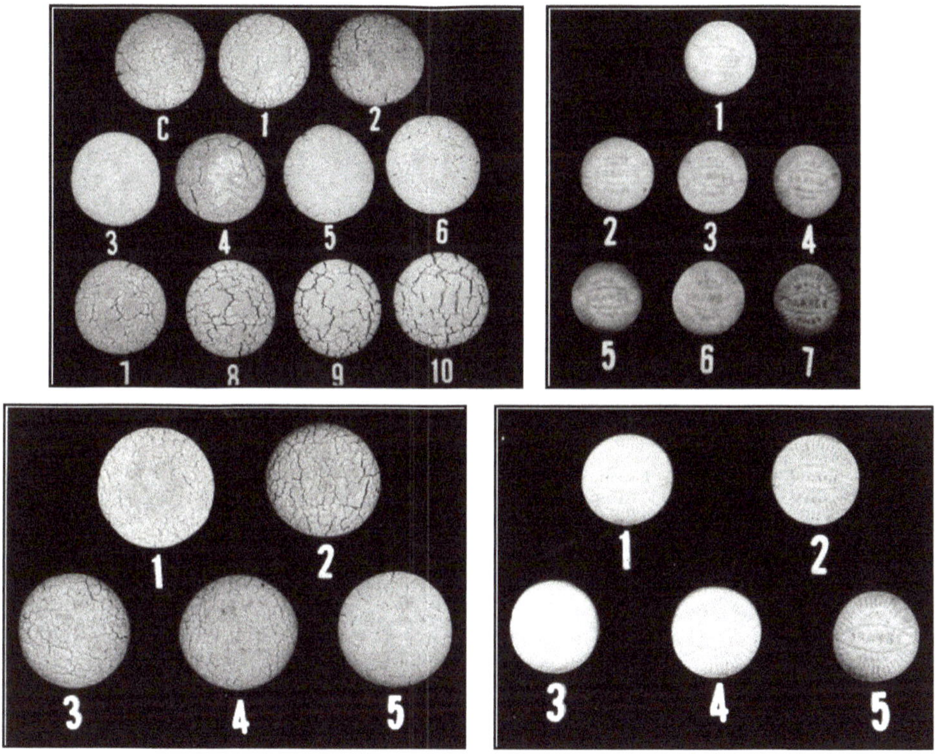

Bajara Various Types of Biscuits

Bajara Discolored Grains

(p. 166)

Bajara Beer

Bajara Masala Puri

Bajara Palak Puri

Bajara Roti
(p. 167)

Bajara Roti

Bajara Sweets

Polished Bajara Roti

Bajara Mehta Puri

Bajara Parotta

(p. 168)

Bajara Sweets

Bajara Chakali

Bajara Peanut Cookies

Bajara Beer

Bajara Masala Sev

Bajara Pan Cake

(p. 169)

Bajara Sweet Porgi

Bajara Khichadi

Bajara Masala Flakes

Bajara Flakes

Bajara Pops

(p. 170)

Foods and Alcoholic Beverages Prepared from Proso Millet

(a) ***Awaokoshi***, candied millet puffs, a specialty of Osaka, Japan (Source: Mugu-shisai, GFDL); (b) Millet with mushrooms (Source: Kathy Steger); (c) White millet soup (Source: Jean Hediger); and (d) Tongba, a millet-based alcoholic brew in Nepal and India (p.183)

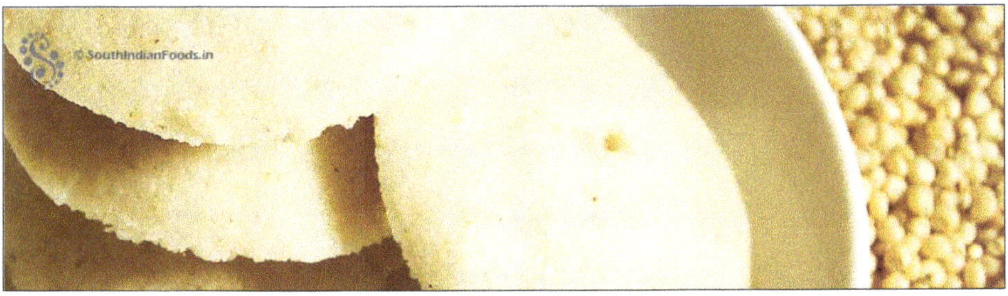

Proso ***idle*** (p. 183)

Proso Millet ***Pakoda*** (P. 184)

Rice Processed Products

Cooked Rice

Rice Cookies

Rice Crackers

Rice Extruded

Rice Sticks

Rice Pullava

(P. 196)

Rice Gel Cake

Rice Cake

Rice Oil

Rice Gel Kurdai

(P. 197)

Biryani Varieties

Egg Biryani

Chicken Biryani

Shrimp Biryani

Cooked Meat Biryani

Vegetable Biryani

Hyderabadi Biryani

Kaadai Biryani

Mushroom Soya Biryani

Pulao Varieties

Kashmiri Pulao

Zafrani Pulao

Jeera Pulao

Peas Pulao

Pulao

Mixed Vegetable Pulao

South Indian Variety Rice

Curry Leaf Rice

Curd Rice

Tomato Rice

Mint Coriander Rice

(P. 198)

Mango Rice

Lemon Rice

Fllu Sadam

Khuska Biryani

Coconut Rice

Vegetable Rice

Carrot Rice

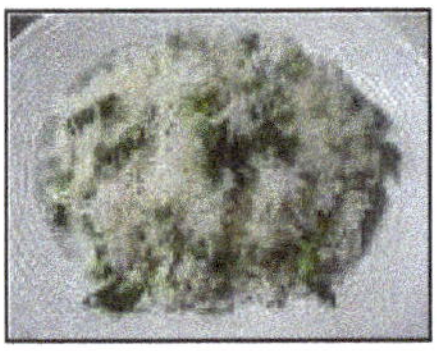

Asparagus Pulao

Capsicum Rice

Corn Rice

Gooseberry Rice

Spinach Rice

Methi Pulao

Okra Rice

Cauliflower Pulao

Potato Pulav

Vangi Bath

(P. 199)

Rice with Lentils (Kitchadi and Pongal Varieties)

Puli Pongal

Ven Pongal

Sambhar (Kadamba) Sadam

Bisi Bele Bath

Sweet Rice Dishes

Zarda Rice

Rice Pudding

Sweet Pongal

Phirni (Rice Kheer)

Nei Payasam

Akkaravadisal

Other Rice Dishes

Chicken Fried Rice

Turkey Fried Rice

Payaru Kanji

Tri Colour Rice Dhokla

(P. 200)

Figure 12.13: Sorghum ***Flakes*** **(P. 231)**

Figure 12.14: Sorghum ***Chiwada*** from Sorghum ***Flakes*** **(P. 232)**

Figure 12.15: Sorghum ***Poha*** from Sorghum ***Flakes*** **(P. 232)**

Figure 12.20: Sorghum ***Upma*** **(P. 240)**

Figure 12.21: Sorghum ***Idali*** **(P. 240)**

Figure 12.22: Sorghum Bread (p. 244)

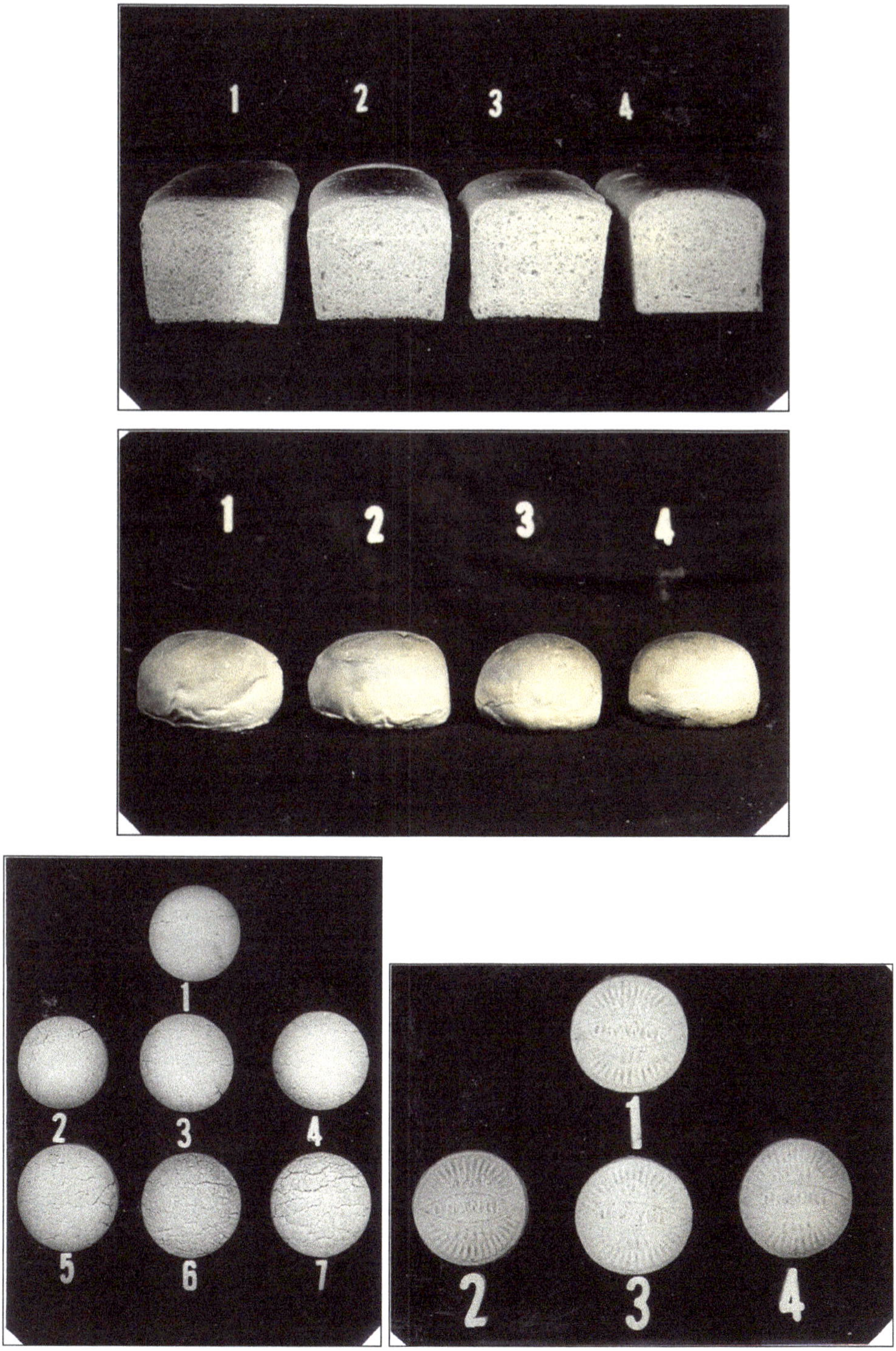

Figure 12.23: Bread, Buns, Cookies and Biscuits Prepared using Sorghum Flour (p. 245)

Figure 12.26: Sorghum ***Pops*** **(p. 252)**

Figure 12.27: Sorghum ***Pop*** Laddu (p. 257)

Figure 12.28: Sorghum Pop Grits Chikki (p. 258)

Sorghum Food Products

Sorghum Roti (p. 259)

Papads (p. 259)

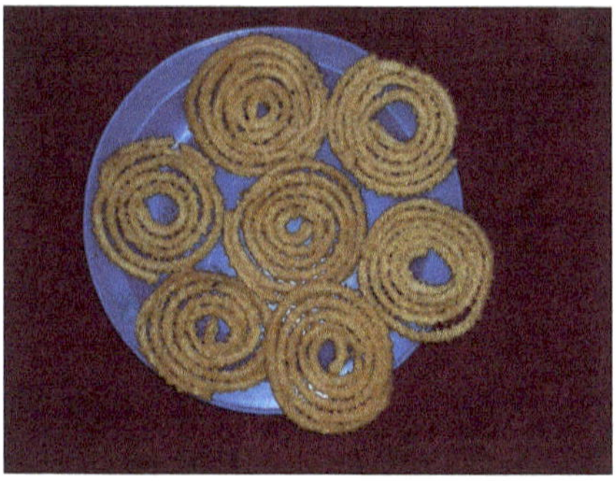

Sorghum Chakali (p. 259)

Noodles (p. 259)

Sweet Khir (p. 260)

Cookies (p. 260)

Upma (p. 260)

Pakoda (p. 261)

Sweet Shira (p. 261)

Idli (p. 261)

Dosa (p. 262)

Uttapa (p. 262)

Laddoo (p. 262)

Masala Chakali (p. 263)

Shev (p. 263)

Papadi (p. 263)

Plain Papadi

Shankar Pali

Cake

(p. 264)

Sorghum Starch (p. 265)

Sorghum Ethanol Industry and Storage (p. 265)

Sorghum Ethnic Cafe (p. 266)

Wheat Food Products

Wheat Roti

Wheat Nan

Wheat Paratha

Wheat Dashami

Wheat Bakery Products

(p. 289)

Wheat Bakery Products

(p. 290)

www.ingramcontent.com/pod-product-compliance
Ingram Content Group UK Ltd.
Pitfield, Milton Keynes, MK11 3LW, UK
UKHW022215051026
15624UKWH00001B/171

9 789351 309208